Mathematical Modeling for Intelligent Systems

Mathematical Modeling for Intelligent Systems: Theory, Methods, and Simulation aims to provide a reference for the applications of mathematical modeling using intelligent techniques in various unique industry problems in the era of Industry 4.0. Providing a thorough introduction to the field of soft-computing techniques, this book covers every major technique in artificial intelligence in a clear and practical style. It also highlights current research and applications, addresses issues encountered in the development of applied systems, and describes a wide range of intelligent systems techniques, including neural networks, fuzzy logic, evolutionary strategy, and genetic algorithms. This book demonstrates concepts through simulation examples and practical experimental results.

Key Features:

- Offers a well-balanced mathematical analysis of modeling physical systems
- Summarizes basic principles in differential geometry and convex analysis as needed
- Covers a wide range of industrial and social applications and bridges the gap between core theory and costly experiments through simulations and modeling
- Focuses on manifold ranging from stability of fluid flows, nanofluids, drug delivery, and security of image data to pandemic modeling, etc.

This book is primarily aimed at advanced undergraduates and postgraduate students studying computer science, mathematics, and statistics. Researchers and professionals will also find this book useful.

Mathematical Modeling for Intelligent Systems

Theory, Methods, and Simulation

Edited by
Mukesh Kumar Awasthi
Babasaheb Bhimrao Ambedkar University, India
Ravi Tomar
University of Petroleum & Energy Studies, India
Maanak Gupta
Tennessee Tech University, USA

CRC Press is an imprint of the
Taylor & Francis Group, an **informa** business
A CHAPMAN & HALL BOOK

First edition published 2023
by CRC Press
6000 Broken Sound Parkway NW, Suite 300, Boca Raton, FL 33487-2742

and by CRC Press
2 Park Square, Milton Park, Abingdon, Oxon, OX14 4RN

CRC Press is an imprint of Taylor & Francis Group, LLC

ISBN: 9781032272252 (hbk)
ISBN: 9781032272269 (pbk)
ISBN: 9781003291916 (ebk)

DOI: 10.1201/9781003291916

Typeset in Palatino
by codeMantra

Contents

Preface vii
Editors ix
Contributors xi

1 **Sensors, Embedded Systems, and IoT Components** 1
Ananya Sharma and Niharika Singh

2 **Security and Privacy in Mobile Cloud Computing** 17
A. Katal

3 **A Review on Live Memory Acquisition Approaches for Digital Forensics** 35
Sarishma Dangi and Diksha Bisht

4 **IoMT and Blockchain-Based Intelligent and Secured System for Smart Healthcare** 61
Hitesh Kumar Sharma, Ravi Tomar, and Preeti

5 **AI-Enabled Cloud-Based Intelligent System for Telemedicine** 75
Hitesh Kumar Sharma, Ravi Tomar, Preeti, and Prashant Ahlawat

6 **Fuzzy Heptagonal DEMATEL Technique and Its Application** 85
A. Felix, PP. Ajeesh, S. Karthik, and R. Dinesh Jackson Samuel

7 **A Comparative Study of Intrapersonal and Interpersonal Influencing Factors on the Academic Performance of Technical and Nontechnical Students** 101
Deepti Sharma, Rishi Asthana, and Vaishali Sharma

8 **PIN Solar Cell Characteristics: Fundamental Physics** 119
Aditya N. Roy Choudhury

9 **A New Approximation for Conformable Time-Fractional Nonlinear Delayed Differential Equations via Two Efficient Methods** 133
Brajesh Kumar Singh and Saloni Agrawal

10 **Numerical Treatment on the Convective Instability in a Jeffrey Fluid Soaked Permeable Layer with Through-Flow** 159
Dhananjay Yadav, Mukesh Kumar Awasthi, U. S. Mahabaleshwar, and Krishnendu Bhattacharyya

11 **Computational Modeling of Nonlinear Reaction-Diffusion Fisher–KPP Equation with Mixed Modal Discontinuous Galerkin Scheme** 171
Satyvir Singh

12 A Numerical Approach on Unsteady Mixed Convection Flow with Temperature-Dependent Variable Prandtl Number and Viscosity 185
Govindaraj N, Iyyappan G, A. K. Singh, S. Roy, and P. Shukla

13 Study of Incompressible Viscous Flow Due to a Stretchable Rotating Disk Through Finite Element Procedure 197
Anupam Bhandari

14 Instability of a Viscoelastic Cylindrical Jet: The VCVPF Theory 215
Mukesh Kumar Awasthi, Sudhir Kumar Pundir, Manu Devi, Dhananjay Yadav, Vivek Kumar, and A. K. Singh

15 Evaporative Kelvin–Helmholtz Instability of a Porous Swirling Annular Layer 231
Shivam Agarwal and Mukesh Kumar Awasthi

Index 245

Preface

The modeling of various engineering problems using mathematics is an important technique that is being used by many engineers and scientists around the globe. The real-life problems can also be modeled through mathematics, and by using analytical/numerical methods, one can solve these mathematical models. Bio-mathematical engineering can be used for modeling various diseases and the impact of different parameters involved. The security system can also be handled through mathematical techniques. Encryption and decryption play a vital role in the security system. The use of number theory and cryptography makes the study easier. The exact solution of the Navier–Stokes equation is not known to date and, therefore, the numerical methods are very helpful for the computation of various properties of fluid flows.

In recent years, scientists have been trying to solve practical problems using mathematical models and they are also getting success. As computing technologies have grown tremendously, the possibilities of solving complex problems have also increased. Most of the engineering systems are very complex, and it takes a lot of time and money to find their solutions. To design an engineering system and understand its processes, it is very important to understand and analyze all the parameters used in that system and mathematical models are very helpful in understanding and analyzing such parameters.

This book is intended as reference material for educators, researchers, and everyone who wishes to learn and have a grasp of the various advances in the fields of computer science, applied mathematics, computing, and simulation. This book will be a readily accessible source of information for the researchers in the area of applied computing as well as for the professionals who want to enhance their understanding of theoretical and application studies. This book is also beneficial for the researchers and professionals who are involved in the research on Intelligent Systems. This book is written for the students, professionals, researchers, and developers who are involved or have a keen interest in the domain of computational intelligence. One of the unique features of this book is it covers the various case studies and future directions in the domain of mathematical modeling.

This book is intended to be multidomain applications, which will help researchers and scholars to think beyond one specific field and work on cohort domain in mathematics and computer science. This book can be an idea for an audience working in the domain of modeling and simulations. It will present the reader with comprehensive insights about the various kinds of mathematical modeling and numerical computations for the problems arising in various branches of engineering starting from computer Science engineering, mechanical engineering, Electrical, and Electronic Engineering, Civil Engineering, etc.

This book contains 16 chapters, each of which presents commonly observed topics and problems which somewhere deep inside are solved through mathematical computation and usage of the latest computational intelligence. This book starts with discussing practical aspects and components of Intelligent systems and followed with the theoretical approaches to solve other day-to-day problems.

Chapter 1 is about Sensors, Embedded Systems, and IoT Components as the title suggests, the chapter presents a comprehensive understanding of the components of IoT and smart computing. Chapter 2 discuss the very critical aspect of Security and Privacy in Mobile Cloud Computing, followed by Chapter 3 about a Review of Live Memory Acquisition Approaches for Digital Forensics. The chapter discusses a practical approach around acquiring memory dump for different memory acquisition tools. Further, Chapter 4 is dedicated to discussing the need and applicability of the latest technologies like IoMT with Blockchain-Based Intelligent and Secured System for Smart Healthcare. Chapter 5 complements it by the implementation of an AI-Enabled Cloud-Based Intelligent System for Telemedicine.

The expansion of The DEMATEL method under the uncertain linguistic situation with heptagonal fuzzy numbers is given in Chapter 6, while the academic performance of technical and nontechnical persons is discussed in Chapter 7. Chapter 8 presents the fundamental aspects of the modeling of a PIN solar cell. Chapter 9 proposes the solution of conformable time-fractional nonlinear partial differential equations with proportional delay.

Chapters 10–15 are based on the fluid flows in various geometries. Chapter 10 is about the convective instability in a Jeffrey fluid-soaked permeable layer and Chapter 11 presents a numerical solution of the two-dimensional nonlinear Fisher–Kolmogorov–Petrovsky–Piscounov reaction–diffusion equation. The impact of transient fluid flow on a vertical plate in the laminar boundary layer with variable viscosity Prandtl number is given in Chapter 12. Chapter 13 complements the flow of viscous fluid due to a rotating and stretching disk. Chapters 14 and 15 discuss the interfacial instability in cylindrical geometries.

We wish the readers a successful study of the material presented, leading to new inspiration, a deepening understanding of the described concepts, and also fruitful applications to the contemporary challenges of science and engineering.

Dr. Mukesh Kumar Awasthi
Dr. Ravi Tomar
Dr. Maanak Gupta

Editors

Dr. Mukesh Kumar Awasthi is working as an Assistant Professor in the Department of Mathematics at Babasaheb Bhimrao Ambedkar University (A Central University), Lucknow, India. He completed his doctorate with a major in Mathematics from the Indian Institute of Technology Roorkee, Roorkee, India in 2012. He has qualified for various national-level competitive exams in the area of Mathematical Sciences. He has been a recipient of a research fellowship under the Council of Scientific and Industrial Research, India during his Ph.D. His research areas include Computational Fluid Dynamics, Heat transfer, Applied Mathematical Modeling, and Numerical analysis. He loves to teach applied mathematics courses to undergraduate, postgraduate, and Ph.D. students. He has been indifferent to academic and administrative positions during his academic profession. He has more than 100 publications (90 SCI indexed, 92 Scopus indexed) to his credit in the high impact factor journals of international repute. He has attended many symposia, workshops, and conferences in mathematics as well as fluid mechanics. He has got research awards consecutively four times from the University of Petroleum & Energy Studies, Dehradun, India. He has also received the start-up research fund for his project "Nonlinear study of the interface in multilayer fluid system" from UGC, New Delhi.

Dr. Ravi Tomar is currently working as Associate Professor in the School of Computer Science at the University of Petroleum & Energy Studies, Dehradun, India. Skilled in Programming, Computer Networking, Stream processing, Core Java, J2EE, RPA and CorDapp, his research interests include Wireless Sensor Networks, Image Processing, Data Mining and Warehousing, Computer Networks, big data technologies, and VANET. He has authored more than 51 papers in different research areas, filled four Indian patent, edited 5 books, and have authored 5 books. He has delivered Training to corporates nationally and internationally on Confluent Apache Kafka, Stream Processing, RPA, CorDapp, J2EE, and IoT to clients like KeyBank, Accenture, Union Bank of Philippines, Ernst & Young, and Deloitte. Dr. Tomar is officially recognized as Instructor for Confluent and CorDapp. He has conducted various International conferences in India, France, and Nepal. He has been awarded a young researcher in Computer Science and Engineering by RedInno, India in 2018, Academic Excellence and Research Excellence Award by UPES in 2021 and Young Scientist Award by UCOST, Dehradun.

Dr. Maanak Gupta is an Assistant Professor in the Department of Computer Science at Tennessee Tech University, United States. He received his Ph.D. in Computer Science from the University of Texas at San Antonio and has worked as a Postdoctoral Research Fellow at the Institute for Cyber Security. He also holds an M.S. degree in Information Systems from Northeastern University, Boston. His primary area of research includes security and privacy in cyberspace focused on studying foundational aspects of access control and their application in technologies including cyber-physical systems, cloud computing, IoT, and Big Data. Dr. Gupta has worked in developing novel security mechanisms, models, and architectures for next-generation smart cars, smart cities, intelligent transportation systems, and smart farming. He is also interested in machine learning–based malware analysis and AI-assisted cyber security solutions. His scholarly work is regularly published at top peer-reviewed security venues including ACM SIGSAC conferences and refereed journals. He was awarded the 2019 computer science outstanding doctoral dissertation research award from UT San Antonio. His research has been funded by the US National Science Foundation (NSF), NASA, the US Department of Defense (DoD), and private industry.

Contributors

Saloni Agrawal
Department of Mathematics
Babasaheb Bhimrao Ambedkar University
Lucknow, India

Shivam Agarwal
Department of Mathematics
Babasaheb Bhimrao Ambedkar University
Lucknow, India

Prashant Ahlawat
Department of Computer Science
GL Bajaj Institute
Greater Noida, India

PP. Ajeesh
Mathematics Division
School of Advanced Sciences
Vellore Institute of Technology
Vellore, India

Rishi Asthana
School of Engineering and Technology
BML Munjal University
Gurgaon, India

Mukesh Kumar Awasthi
Department of Mathematics
Babasaheb Bhimrao Ambedkar University
Lucknow, India

Anupam Bhandari
Department of Mathematics
School of Engineering
University of Petroleum & Energy Studies
Dehradun, India

Krishnendu Bhattacharyya
Department of Mathematics
Institute of Science
Banaras Hindu University
Varanasi, India

Diksha Bisht
Department of Computer Science and Engineering
Graphic Era Deemed to be University
Dehradun, India

Sarishma Dangi
Department of Computer Science and Engineering
Graphic Era Deemed to be University
Dehradun, India

Manu Devi
Department of Mathematics
Motherhood University
Roorkee, India

A. Felix
Mathematics Division
School of Advanced Sciences
Vellore Institute of Technology
Vellore, India

Govindaraj N
Department of Mathematics
Hindustan Institute of Technology and Science
Chennai, India

Iyyappan G
Department of Mathematics
Hindustan Institute of Technology and Science
Chennai, India

S. Karthik
Department of Mathematics
Vel Tech Rangarajan Dr. Sagunthala R&D Institute of Science and Technology
Chennai, India

Avita Katal
School of Computer Science
University of Petroleum & Energy Studies
Dehradun, India

Vivek Kumar
Department of Mathematics
Shri Guru Ram Rai (PG) College
Dehradun, India

U. S. Mahabaleshwar
Department of Mathematics
Davangere University
Davangere, India

Preeti
School of Computer Science and Engineering
University of Petroleum & Energy Studies
Dehradun, India

Sudhir Kumar Pundir
Department of Mathematics
S. D. (P.G.) College
Muzaffarnagar, India

Aditya N. Roy Choudhury
Department of Physics
Techno India University
Kolkata, India

S. Roy
Department of Mathematics
Indian Institute of Technology Madras
Chennai, India

R. Dinesh Jackson Samuel
Department of Mathematics
Oxford Brookes University
Oxford, UK

Ananya Sharma
Department of Computer Science
University of Petroleum & Energy Studies
Dehradun, India

Hitesh Kumar Sharma
School of Computer Science and Engineering, University of Petroleum & Energy Studies
Dehradun, India

Deepti Sharma
School of Management
BML Munjal University
Gurgaon, India

Vaishali Sharma
School of Management
BML Munjal University
Gurgaon, India

P. Shukla
Mathematics Division
School of Advanced Science
VIT University
Chennai, India

Niharika Singh
Department of Computer Science
University of Petroleum & Energy Studies
Dehradun, India

Brajesh Kumar Singh
Department of Mathematics
Babasaheb Bhimrao Ambedkar University
Lucknow, India

Satyvir Singh
School of Physical and Mathematical Sciences
Nanyang Technological University
Singapore

A. K. Singh
Mathematics Division
School of Advanced Science
VIT University
Chennai, India

Ravi Tomar
School of Computer Science and Engineering, University of Petroleum & Energy Studies
Dehradun, India

Dhananjay Yadav
Department of Mathematical & Physical Sciences
University of Nizwa
Nizwa, Oman

1

Sensors, Embedded Systems, and IoT Components

Ananya Sharma and Niharika Singh
University of Petroleum and Energy Studies

CONTENTS

1.1 Introduction 1
1.2 Introduction to Embedded Computing Systems 2
  1.2.1 Usage of Sensors and Actuators in Embedded System 3
  1.2.2 Embedded Computing System: Reliability Analysis 5
1.3 Internet of Things: Introduction, Applications, and Trends 5
1.4 Applications of IoT 6
  1.4.1 Data Privacy, Adaptability, Sensibility, and Security 6
1.5 Introduction to Sensors 7
  1.5.1 Environmental Monitoring 8
  1.5.2 Control Applications 8
  1.5.3 Trends in Sensor Technology 9
  1.5.4 Future Scope 10
1.6 Artificial Intelligence and Internet of Things 10
  1.6.1 Intro to Artificial Intelligence (AI) 11
  1.6.2 Convergence of AI and IoT 12
1.7 Conclusion 13
1.8 Summary 13
References 14

1.1 Introduction

The Internet of Things (IoT) defines a network of embedded computing devices that are linked together enabling communication and interaction within themselves and the external environment. According to internet usage worldwide statistics (https://www.statista.com/statistics/617136/digital-population-worldwide/) as of 2021, there were 4.95 billion internet users estimated across the world. The growth in the use of the internet in the past few years has been exponentially high, which has led to the development of IoT. The unifying thread running through these various IoT visions is that "things" are expected to become active participants in industries, data retrieval, and social processes (Hassan, 2018). IoT has a wide range of applications, including the digital transformation of industries, quality control, logistics, and supply chain optimization. Fitness wearables, smart home applications, health care, and smart cities have all been made possible by the IoT. A home automation system can be a great example as it portrays integration between devices and interoperability of all functions resulting in gathering information and conveying it to

DOI: 10.1201/9781003291916-1

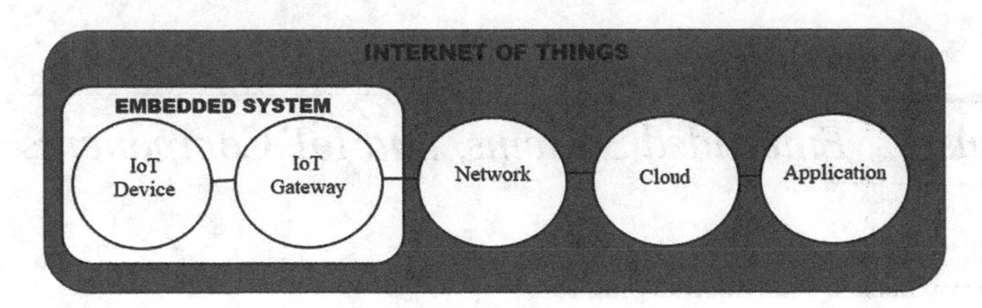

FIGURE 1.1
IoT infrastructure.

the user as per requirement. RFID (Radio-Frequency Identification) systems can be used to monitor objects in real-time; this enables mapping the real world into the virtual one (Atzori et al., 2010). Figure 1.1 illustrates the complete infrastructure of an IoT environment making it easier to understand how the IoT device and gateway are linked in an embedded system. The data is stored in the cloud over a network in order to reduce response time and fasten the data processing.

The sensors collect the data and send it to a control center where a decision is taken and an associated command is sent to the actuator in its response. According to the total data volume generation statistics, as of 2020, approximately 64.2 zettabytes (1 zettabyte = 10^{12} gigabytes) of data were created, consumed, and stored in that particular year, which is increasing day by day at exponential rates. The introduction of artificial intelligence (AI) and its subfields like machine learning (ML) and deep learning have paved the way for extracting useful insights from big data. The convergence of IoT and AI has enabled the viability of numerous real-life applications of IoT. AI proves to be helpful by avoiding expensive and unplanned downtime, increasing operational efficiency, enhancing risk management, and enabling new and improved products and services. According to the forecast by Lee et al. (2009), with a compound annual growth rate (CAGR) of 30.8%, India's AI spending would rise from US$300.7 million in 2019 to US$880.5 million in 2023. This shows that AI will help "all effective" IoT endeavors and data from deployments will have "limited value" without it. The applications of AI incorporated with IoT include IoT data mining platforms and tools, 5G-assisted IoT systems and applications, monitoring robots, and self-driving cars. The convergence of AI and IoT enables the systems to be predictive, prescriptive, and autonomous. AI can be applied to embedded systems. By making use of AI-embedded systems, developers can look to deliver efficient solutions.

1.2 Introduction to Embedded Computing Systems

The combination of software and hardware, which is specifically designed to perform a certain task, is known as an embedded system. These are controlled by the software and are responsible for meeting users' needs. The composition of an embedded system is shown in Figure 1.2. The software component consists of

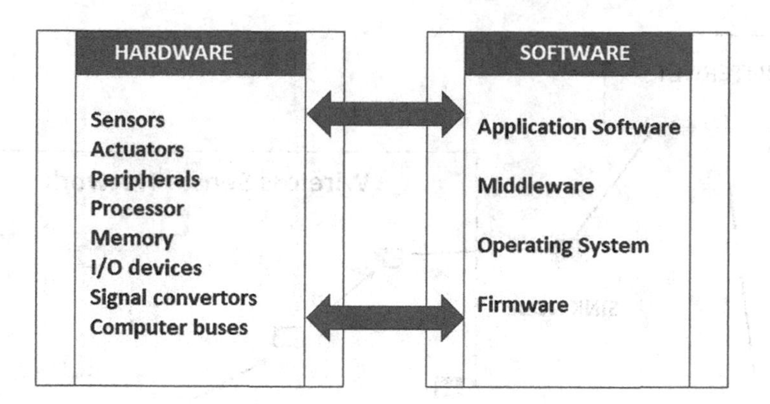

FIGURE 1.2
Composition of embedded systems.

- **Application software:** It performs the required functions and interacts with the user.
- **The mediator device:** It acts as a medium of communication between different layers of the software part.
- **An operating system:** It frames rules for the required function to be performed.
- **Firmware:** It is a program written for certain hardware.

The hardware comprises memory, I/O (input/output) devices, memory buses, sensors, actuators, and peripherals (cameras, printers, scanners, and keyboards). The applications of embedded systems include embedded Linux, Android, and Windows-embedded compact (Windows CE). It is also used in appliances such as washing machines, digital cameras, televisions, and refrigerators. The increasing global demands for communication devices and different electronics equipped with embedded systems are the fundamental cause driving the growth of the market of embedded systems. The embedded systems these days require less space and have low power consumption due to their small size.

1.2.1 Usage of Sensors and Actuators in Embedded System

Sensors and actuators are essential components of embedded systems. Sensors are devices that measure a physical characteristic of the environment and transform it into output signals. Essentially, sensors are needed to gain an understanding of the system. Sensor data can be used for a variety of purposes, including system operation and control, process monitoring, experimental modeling, product testing and qualification, product quality evaluation, fault prediction and diagnosis, warning generation, and surveillance. The different sensors that are being developed to pace up with developing devices include microminiature sensors, intelligent sensors with integrated preprocessing, and hierarchical sensory architectures. Sensor classification can range from extremely simple to complex and is designed on the basis of different requirements. The multimedia wireless sensor networks (WSNs) (Lee et al., 2009) are proposed to enable tracking and extracting data in the form of multimedia such as imaging, video, and audio. Figure 1.3 shows the overview of the battlefield, which is based on the principle of the multimedia network. It includes nodes that are deployed by using aircraft in order to capture the images of different scenarios such as the movement of the enemy and their equipment. These nodes are embedded with video cameras and microphones in order to capture the audio and

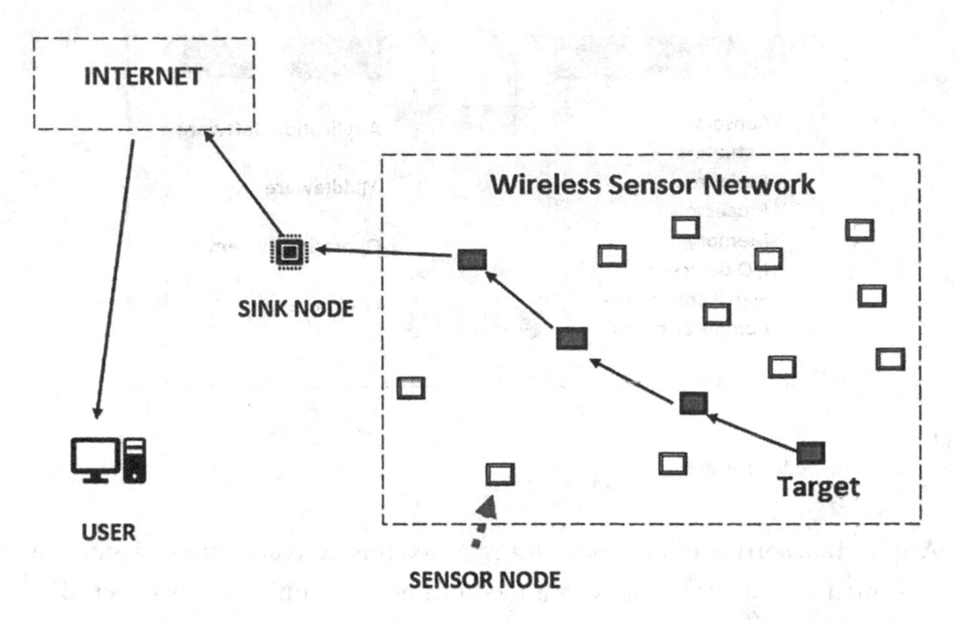

FIGURE 1.3
Multimedia network in a battlefield.

video of the enemies. The nodes are interconnected over a wireless connection for data compression, retrieval, and correlation. Sensors have their applications in real-life situations as well as in industries. They can be used in monitoring patients in medical applications; weighing scales; agriculture; and heating, ventilation, and air conditioning (HVAC) technologies.

Actuators, on the other hand, are devices that enable movement and are responsible for controlling the mechanism or a system. The actuators used in industries include electric motors, hydraulic motors, and pneumatic control waves. Sensors are being embedded at an exponential rate in almost every IoT device that is leading to a vast increase in their usage. Sensors and actuators are essential in an embedded system. The embedded system also contains a controller that generates control signals and manages the process that is being controlled by making use of sensors and actuators (Sarishma et al., 2021). Table 1.1 gives a comprehensive overview of the characteristics of sensors and actuators.

TABLE 1.1
Sensors and Actuators

	Sensors	Actuators
Inputs and outputs	They look at the input and initiate a particular action.	They track the outputs of devices.
Electrical signaling	They use signaling to analyze the environmental conditions to perform a particular task.	They measure the heat to determine the output.
Dependency	They are dependent on each other.	Actuators rely on sensors to do their job.
Conversion	It converts physical attributes to electrical signals.	It changes electrical signals to physical action.
Location	It is present at the input port.	It is present at the output port.
Application	They are used to measure temperature, vibration, and pressure.	They are used in operating dampers, valves, and couplings.

1.2.2 Embedded Computing System: Reliability Analysis

The use of embedded systems is growing exponentially in order to pace up the development of IoT devices, starting from everyday requirements like mobile phones, televisions, to complex healthcare applications. The systems that used 16-bit and 8-bit processors in the past are now being replaced with processors that complete the assigned tasks with less power consumption and response time. Security-sensitive embedded systems include aircraft flight control systems, surgical robotics, and patient monitoring systems used in hospitals, as well as nuclear power plant instrumentation and control systems. A challenge faced in embedded system development is the choice of hardware, as the cost of the product depends on the size of hardware, which is another challenge faced during development. Power consumption is another issue in the development of battery-powered applications. While in nonbattery-powered applications, excessive heat is released due to power consumption. A solution to this problem is to make the system run slowly, but this will lead us to missed deadlines as output will be delivered late due to increased running time. The fault-tolerant software, compiler, and instruction set design for embedded processors is explained in Azarpeyvand et al. (2010). The proposed method leads to time- and cost-effective fault-tolerant strategies since compiler-level fault-tolerant techniques are usable even after processor manufacture. Brute way of completing the task before a given deadline is to speed up the hardware so that the program runs faster, but this factor eventually makes the system expensive. Another main challenge is often software debugging because to really debug software, it needs to be run on the embedded hardware it's been designed for, and if the hardware design slips by a few months and the product launch date remains fixed, then the software debugging stage gets squeezed very often. While the hardware is capable of doing the task on paper, making it happen in software is generally the crucial problem. In Glaß et al. (2012), a flexible framework for cross-level Compositional Reliability Analysis (CRA) is proposed, which allows smooth integration of multiple reliability analysis approaches across different levels of abstraction. Reliability is always an important factor while selling a product, especially a safety-critical system like aircraft flight control, weapons, or nuclear systems. These are certain challenges that the developers must take care of while designing reliable embedded systems.

1.3 Internet of Things: Introduction, Applications, and Trends

"In a few decades, computers will be interwoven into nearly every industrial product," declared Karl Steinbuch, a German computer scientist, in 1964. These words came to reality when the concept of IoT was proposed in 1999 by the MIT Auto-ID Center (Liu & Lu, 2012). John Romke created a toaster in 1990 that functioned its power using the internet. This toaster was the very first IoT device. Later that year, some students at Carnegie Mellon University uploaded a Coca-Cola distribution machine to the internet in order to determine which of the machine's columns had the most iced coke. However, the whole technology was not given any name yet. When Kevin Ashton created the term "Internet of Things" in 1999, he gave a presentation about connecting RFID tags in supply chain management to the internet. The IoT is a world of interconnected things (smart things or smart objects) that are capable of sensing, actuating, and communicating with one another and with the environment, as well as sharing information and acting autonomously in response to real/physical world events, as well as triggering processes and creating services with or without direct human intervention (Hassan, 2018). However, a significant milestone that triggered the contemporary IoT craze occurred in 2008–2009, when more items were connected to the internet than there were individuals. By the year 2020, we had more than 50 billion devices connected to the IoT. IoT is expected to be a $19-trillion industry. It is also one of the six "Disruptive Civil Technologies" identified by the US National Intelligence Council as having the potential to alter US national power (Conference Report CR 2008). Most of the technologies built until now can be mentioned as the internet of people. So, whether it is a CRM application or an e-commerce application, largely it is believed that there is a person at the end of it. John Chambers, former CEO of CISCO assumed that there are going to be 500 billion things connected to the internet in the future which is nearly 100 times the world's population. The other benefit is that things can be present where people aren't. For all these reasons, IoT is dramatically different from the internet for people, and it will ultimately reshape all the fundamental industries that run the planet, power, water, agriculture, healthcare, construction, transportation, etc. These are a few ways by which the internet has changed our standard of living by making things to communicate with people by making use of IoT. It has the potential to provide this process a new dimension by allowing communication with and among smart things, opening the way for the vision of "anytime, anywhere, any medium, any object" communications (Atzori et al., 2010).

1.4 Applications of IoT

Today, IoT is one of the few technologies whose influence will extend across all businesses and directly into our homes. IoT deployment has already begun in areas such as logistics, energy monitoring, military, and industrial automation. Furthermore, IoT has an impact on all parts of the technical stack in terms of hardware and software innovation. But there are fields that are not able to make enough use of IoT standards for their development. Industries like manufacturing, transportation, logistics, mining, and agriculture are the physical industries (Section 1.4, Atzori et al., 2010). The mining industry also has a lot of interest and opportunities. Another application area is agriculture where advanced technology through feedback and coordination with sensor technologies such as electrochemical sensors is used to determine the level of nutrients in the soil. Its usage has shown a 23% reduction in the amount of fertilizer needed, 15% reduction in the amount of seed needed, and overall yield has increased to 27% by the automation of industrial IoT. The advantages of advanced tracking and precision agriculture are huge. Additional uses include location sensing and sharing, environmental sensing, remote control, ad hoc networking, and encrypted communication (Chen et al., 2014).

1.4.1 Data Privacy, Adaptability, Sensibility, and Security

The ever-growing number of devices can be treated as an incredible opportunity that brings immense risks. There are multiple challenges that IoT technology faces these days. The most important among them are data privacy, adaptability, sensibility, and security. Security stands for the protection of users' information from any type of malicious access such as preventing theft, corruption, and other types of damage while allowing the successful exchange of the data in the most efficient way, whereas privacy is the right of individuals to have control of when and how their personal information is collected and processed in the IoT environment. The Open Web Application Security Project (OWASP) (https://owasp.org/www-project-internet-of-things/) has identified the most common issues in IoT and demonstrated that many of these flaws are caused by a lack of understanding of well-known security applications such as encryption, authentication, access control, and role-based access control. All of these applications are explained in Bertino (2016). We can consider an example of location privacy. Most of the devices nowadays are equipped with GPS (global positioning system) receivers or other sensors that allow determining the position of the device. However, different applications require various levels of location accuracy, so the user can decide based on the collected information as to whether to disclose the coordinates or not. Personal health indicators like heart rate and no of steps taken, user whereabouts, payment credentials, calendar information, and contacts in the smartphone are all examples of data acquired by IoT services. In addition, the privacy impact assessment methodology of RFID serves as a prototype for a risk management–based approach that can be used in IoT scenarios. At the design stage, data privacy, security, and protection should be addressed. Meanwhile, IoT devices don't always have enough processing capacity to execute all of the necessary security layers and functionalities (Anuj Kumar et al., 2012). Usually, IoT devices have the functions to preserve users' privacy, access control and authentication, secure network protocols, cryptography, and edge computing, which means computation and data storage on the local device. If the data is transferred to the cloud, the techniques that can be applied to the datasets to avoid users profiling and linkage include data analysis and summarization, digital forgetting, and the concept of differential privacy. Some ways to protect data are to use trusted connectivity, install trusted software, check privacy settings, and study the privacy policy that details what the company is going to do with your data as one has the right to understand what happens to the information and whether the company shares any personal data with third parties. Still, there are loopholes in matters of data security because of which data adaptability and sensibility are considered difficult to handle. Security and privacy include challenges (Shukla, 2021) that need to be resolved for a secure digital future.

1.5 Introduction to Sensors

Sensors are devices that measure a physical characteristic of the environment like humidity, temperature, and pressure and transform it into output signals. Sensors that can see, hear, smell, and even taste are now available. Our life at home and at work would be much more difficult if we didn't have sensors. Sensors embedded in the road regulate the traffic lights at an intersection. When you arrive at the intersection, these sensors detect your presence. A sensor also causes the door to open automatically as you approach the grocery shop. A sensor is a device that measures physical input from its surroundings and provides an electrical output in reaction to that input in the realm of instrumentation and process control. Sensors detect level, temperature, flow, pressure, speed, and position, among other physical qualities. Sensors can be classified as either passive or active in terms of process control. An active sensor does not require an external source of electricity to function, but a passive sensor does. A thermocouple is an active sensor if it operates without the use of an external power supply. When a thermocouple is exposed to an increase in temperature, the voltage across it increases. A piezoelectric sensor is another type of active sensor. A passive sensor is a resistance temperature detector, or "RTD." It's a gadget, the resistance of which changes with change in temperature. An external supply or an excitation circuit is required to take advantage of this change in resistance by causing a voltage change. A strain gauge is an example of a passive sensor. Because the output of a sensor must be conditioned or amplified, almost every sensor used in process control will be connected to a transmitter.

1.5.1 Environmental Monitoring

Environmental monitoring is one of the most compelling industrial IoT applications driving operational transparency and efficiency. Across a variety of industries such as manufacturing, mining, oil and gas, and agriculture, monitoring and managing environmental conditions such as air quality, water quality, and atmospheric hazards is critical to prevent adverse conditions that may impact production process, product quality equipment, and worker's safety. However, the accessibility of environmental data requires the sensor network enabled by robust, scalable, and cost-effective connectivity. Low-power wide-area networks (LPWAN) are an excellent alternative for linking environmental sensor networks because they provide the wide coverage and indoor penetration required to cover large regions and reach previously unreachable locales. Compared to industrial alternatives like mesh networks, LPWAN solutions deliver much higher energy efficiency alongside easy installation and management. Three separate wireless sensors were created, built, and assessed in order to create IoT-based environmental monitoring systems (Mois et al., 2017). Another beneficial alternative is the ultra-low power design and simple star topology that has low device and network costs, fewer infrastructure requirements, and reduces setup and maintenance complexity. All contribute to a low total cost of ownership. Critical ambient conditions such as temperature, humidity, air quality, and pressure can be monitored and controlled to improve quality; measure workplace conditions such as air quality, heat, humidity, radiation, and noise to improve worker safety; ensure regulatory compliance and maintain quality assurance, and optimize energy consumption by connecting an LPWAN-enabled environmental sensor network. Many such algorithms and protocols are explained as well as analyzed in the work by Prabhu et al. (2018) along with their merits and demerits. Energy efficiency, overall system cost, sensor module response time, system accuracy, adequate signal-to-noise ratio, and radio frequency interference/electromagnetic interference (RFI/EMI) rejection during varying atmospheric conditions and in inhomogeneous environments are all requirements of a WSN for environmental monitoring (Prabhu et al., 2018). It addresses the complete procedure of the development of WSN for IoT devices from scratch considering all aspects of the required platform.

1.5.2 Control Applications

Sensors are characterized as low cost, low power, and highly reliable networks that collect data and make it available for processing (Atzori et al., 2010; Tomar & Tiwari, 2019). There are many types of sensors available for use in industrial applications. Some of the common sensors used are listed in Table 1.2.

Sensors play an important role by providing intelligence to the "things" so that they respond to all the tasks without any human interference. There is a sensor for almost every single requirement. Some of them include proximity, position, occupancy, motion,

TABLE 1.2

Commonly Used Sensors

Photoelectric sensors	They work by emitting a visible or infrared light beam from their light-emitting element. They can work differently depending on the type of sensor. Some may reflect their light back to themselves for detection, while others may emit their light to a separate element that watches for breaks in the signal.
Magnetic sensors	They work by detecting changes and disturbances in a magnetic field such as flux, strength velocity, and direction.
Hall effect sensors	These can be thought of as being in a similar category as magnetic sensors since they are devices that are attracted by external magnetic fields.
Presence sensors	They work by enabling or disabling devices based on whether presence is detected nearby. They are commonly used in buildings that turn the lights on or off based on motion in the room.
Flow sensors	These are devices that are used to measure the flow rate of something typically a fluid. These sensors are often part of a larger device such as a flow meter. Similar devices are also available for air and gas measurement but flow sensors are commonly used in various applications, including industrial processes, and HVAC.

velocity, optical, and chemical sensors (Sehrawat & Gill, 2019). The development of IoT-based applications relies heavily on these sensors.

1.5.3 Trends in Sensor Technology

In the developing world of the IoT, most of the things are sensors. Experts estimate that there will be as many as a trillion sensors connected to the IoT during this decade. The sensors are changing and evolving constantly and huge technology and market trends are driving them. Besides the fact that there are a whole bunch of sensors available probably joining things, social media groups on the IoT. Several trends can be seen in sensor development and they are primarily driven by the users and their requirements. Five major trends that can be observed include:

- **Miniaturization:** The demand for portable devices with various features, like improved security and cheap prices, has been increasing lately. Different kinds of designs and ideas are being incorporated across the sensor products to give them a smaller device with a lot more capability. Some applications of this trend include sports and fitness watches. The advantages are that these sensors can fit into small packages.
- **Digitization:** Almost all the sensors introduced in the present day are available in both analog and digital output. The common digital output protocols that can be seen in the digital interfaces globally are I^{2}C (interintegrated circuit) and SPI (serial peripheral interface). The advantages when we digitize the sensors is we get better performance in trends of resolution and accuracy and digital sensors draw less power. These are more compatible with IoT applications. Digitization can be seen across many industries. Examples include pneumatic control systems for factories that control air pressure and monitor various air systems that might be used for the machinery in a factory. There are sensors that can handle both low- and high-pressure systems. The ability to fit into different kinds of systems is made much easier through digital sensors. HVAC control systems that consider humidity, temperature, and airflow rate are all going digital. So, it makes it a lot easier to apply sensors in these kinds of applications.

- **Low power sensors:** The Low power sensor emphasize more on the running time of a sensor on a single charge. For e.g., TSYS02 and HTU21D are low power consuming temperature and humidity sensor(s), MS5837 is an low power altimeter sensor. These sensors consumes as low as 180 μA of power.
- **Multisensor modules:** The demand to put multisensor capabilities into single packages is also increasing. The advantages are space and power savings as several sensing elements are powered to a single power source. Some examples that include these combinations are HTU21D, which is a relative humidity and temperature sensor that takes information and measures both relative humidity and temperature, and MS8607, which goes into home appliances and automotive applications. One good example is that this sensor finds its way into an inlet manifold in a car and monitors relative humidity, air temperature, and barometric pressure, and all this information goes into the engine control unit in the vehicle. The fuel is adjusted to make sure of its proper usage and does not produce carbon monoxides and nitrous oxides. It monitors all three of those things, is a small package that fits into the intake manifold so that it doesn't restrict the air, and it is very low powered. HVAC systems are also examples of such kinds putting multiple sensing capabilities in the devices and creating better efficiencies.
- **Harsh environment sensors:** Many locations where sensors are needed are not so "friendly" to electronic devices and are termed harsh environments. One solution to this problem is the inclusion of sensors in packaging that can withstand some of these very harsh environments. A lot of sensors are finding their way into areas where there are very high EMI and RFI fields. So sensors are imbibed with some features to protect them. The MS5837-02BA pressure sensor is an example of this kind.

1.5.4 Future Scope

IoT and its devices are making progress rapidly with time. None of this would be possible without communicative, diligent data-collectors at the lowest field level called sensors. These sensors utilize the globally standardized I/O-link communication protocol that allows them to be configured via the system controller and flexibility adjusted to suit each new production job. The advantage of using small, disposable sensors is that the measurements can be done directly on the spot. Electrochemical and biochemical sensors are expected to become a commodity in the future because there are so many different possibilities with these kinds of sensors such as detecting viruses, safety, and explosive detection. In the work by Bogue (2007), the future use of microelectromechanical systems (MEMS) is discussed, as well as its capacity to create sensors at very low unit prices for big volume applications. The primary characteristics of any chemical sensor are specificity, accuracy, precision, resolution, and detection level. Its performance is thoroughly examined in the work by Prien (2007).

1.6 Artificial Intelligence and Internet of Things

IoT is the interconnectivity of devices, sensors, and systems. Some examples of IoT include sensors that can determine when something is either in proximity or as it moves through an environment and so forth. Smart homes and buildings are a perfect example of the convergence of artificial Intelligence (AI) and IoT where if someone has not been in a room for a certain amount of time, it actually shuts the lights off. It will set the air conditioner down or do whatever it needs to make sure that the building is operating efficiently. This is known as proactive or predictive maintenance and quality. AI simulates the ability of the computer to perform the assigned task by analyzing the data and giving desired outputs after processing the data. It can be observed in things like process automation and robotics. Nowadays, a lot of chatbots are embedded with ML. Things like smart assistants help with specific tasks that are also called decision optimization. But the intelligence of these devices increases eventually with time. The architectural alternatives for embedding AI in IoT are discussed in the work by Calo et al. (2017). The system developed controls the data flow rate and extracts the required data implementing the concept of edge computing.

1.6.1 Intro to Artificial Intelligence (AI)

AI is a technology that allows machines to sense, grasp, act, and learn how to accomplish a task in a way that is similar to human intelligence. Almost every company today uses AI in its goods. AI can be classified as either weak or strong. Weak AI is built and educated for a specific task, such as voice-activated assistants. It can either respond to your query or carry out a preprogrammed instruction, but it won't work unless you interact with it. Strong AI has generalized human cognitive abilities, which means it can solve problems and find answers without the need of human participation. A self-driving automobile is an example of powerful AI that employs a combination of computer vision, image recognition, and deep learning to operate a vehicle while keeping in a specific lane and avoiding unforeseen impediments such as pedestrians. Comparative analysis of weak and strong AI says weak AI becomes so strong while "strong AI" is almost as weak as it was decades ago (Liu, 2021). Healthcare, education, finance, law, and manufacturing are just a few of the fields where AI is being used to benefit both businesses and customers. AI in various modules is developed for the purpose of education and engineering, which include intelligent tutoring systems (Beck et al., 1996) and fuel shuffling in a nuclear powerpoint (Faught, 1986). Many technologies, such as automation, ML, machine vision, natural language processing, and robotics, incorporate AI. The use of AI creates legal, ethical, and security concerns. Hackers are gaining access to sensitive networks by employing advanced machine-learning technologies. Despite the dangers, the usage of AI tools is governed by a few restrictions. Experts, on the other hand, guarantee us that AI will only improve products and services. Data economy is one of the factors behind the emergence of AI. It refers to how much data has grown over the past few years and how much more it can grow in the coming years. The proliferation of data has spawned a new economy, and there is a perpetual war for data ownership among businesses seeking to profit from it. With the increase in data, the volume has given rise to big data that helps manage huge amounts of data. Data science helps analyze that data. As a result, data science is moving toward a new paradigm in which robots may be taught to learn from data and generate a range of useful insights, giving rise to AI. AI mimics human and animal intelligence. It involves intelligent agents, autonomous entities that perceive their environment, and take actions that maximize their chances of success at a given goal. AI is a method for computers to imitate human intellect through logic. It's a program with the ability to detect, reason, and act. By giving greater personalization to users and automating procedures, AI is transforming industries. ML is at the heart of AI. The use of algorithms to find meaning in random and unordered data is the first part of ML, and the second half is the use of learning algorithms to uncover the relationship between that knowledge and improve the

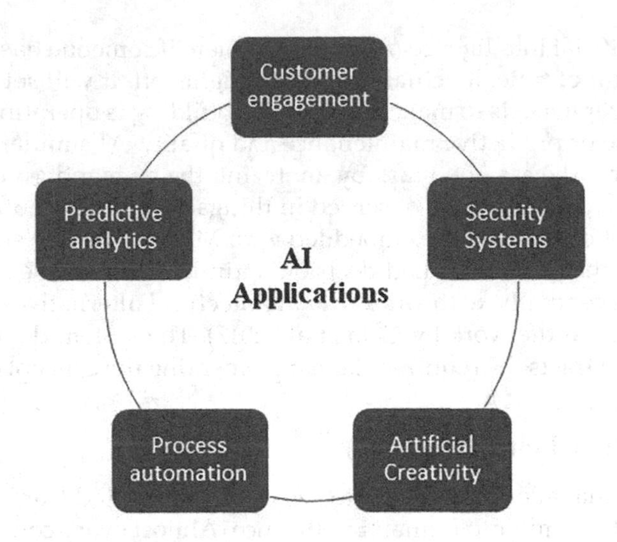

FIGURE 1.4
AI applications.

learning process. So the overall goal for ML is to improve the machines' performance on a specific task such as predicting the stock market. Emerging industries that might require AI in the coming years are shown in Figure 1.4.

1.6.2 Convergence of AI and IoT

The digital world is interconnected through the internet. But onboarding hardware into the network has presented myriad challenges over the years. With research and development, the past decade has seen about 50 billion devices being connected to the internet and the number is growing expansively. Institutions and researchers including municipalities, governmental bodies, and multinational corporations are harnessing IoT via collecting dynamic data points through various sources like cameras, multiple sensors, and personal devices. This is one of the factors behind the emergence of AI. AI has made it easier, in congruence with big data, to gather crucial insights from these data points. It involves intelligent agents, the autonomous entities that perceive their environment, and takes actions that maximize their chances of success at a given goal. AI is a method for computers to imitate human intellect through logic. It's a program with the ability to detect, reason, and act. It is transforming industries by allowing users to be more personalized and by automating procedures. The basic components of a neural network include an artificial neuron, and this neuron is connected to all other neurons and the interconnections are represented by a number called weight. The influence of an artificial neuron on other neurons depends directly on the value of weight. Three types of layers make up the usual architecture of a neural network:

- **Input layer:** The input layer receives data from the physical environment, such as a smartphone camera feed.
- **Hidden layer:** The real network's compute engine is the hidden layer. As a result, there could be a lot of hidden layers.
- **Output layer:** The outcome is produced by the output layer.

ML is the backbone of AI. The first half of ML entails utilizing algorithms to find meaning in random and disordered data, while the second half entails using learning algorithms to discover the relationship between that knowledge and improve the learning process. As a result, the overarching purpose of ML is to increase the computers' performance on a certain task, such as stock market prediction. More than 80% of commercial IoT initiatives will have an AI component by 2022 (Internet of things (IoT), 2019). A self-driving car is an example of advanced AI that uses a combination of computer vision, image recognition, and deep learning to drive a vehicle while staying in a defined lane and avoiding obstacles like pedestrians. Healthcare, education, finance, law, and manufacturing are just a few of the fields where AI is being used to benefit both businesses and customers. AI is used in a variety of technologies, including automation, ML, machine vision, natural language processing, and robotics. AI raises ethical, legal, and security concerns. If an autonomous vehicle is involved in an accident, for example, it is difficult to determine who is to blame. In addition, advanced ML technologies are being used by hackers to obtain access to sensitive networks. Despite the dangers, the usage of AI tools is governed by few restrictions. Experts assure us, however, that AI will simply improve products and services in the near future, not replace us, the people.

1.7 Conclusion

The IoT is paving the way for digital transformation at an exponential rate. Technical as well as physical industries are adopting IoT to digitize their products. The embedded systems give enhanced results because of their small size and low power consumption capability. The future scope of embedded software testing can be improved by the adoption of technologies that can process huge amounts of data with a reduced response time. The data collected from sensors such as liquid levels, distance measures, smoke detectors, soil moisture levels, temperature, humidity, and light data, as well as location's longitude and latitude, is used in burgeoning IoT applications. Apart from different control applications of sensors, they have also been helpful in environmental monitoring techniques. The data perceived from sensors becomes difficult to manage, so AI is incorporated in IoT. This combination not only automates the industries but also results in devices with optimized results. AI techniques can be used in order to embed human intelligence into devices and industrial products. AI models train the data for enhanced results. Overall, the importance and future scope of sensor technologies, embedded systems, and the need for IoT in the industrial and business sectors have been covered.

1.8 Summary

The IoT has been described here as the emerging resource for a future where all physical and technical devices will be connected through the internet. The importance of embedded systems has been covered with the supporting statements of their features. The lack of some characteristics in these systems leading to less reliable behavior of embedded systems has also been discussed. The sensors that collect huge amounts of data also play an important role in the study. The factors that make the technology of sensors more advanced and useful have been stated. The adoption of techniques discussed will result in desired results depending on the expectations of the consumer. Applications of IoT and its resulting outcomes have been studied to achieve a deeper understanding of the study. AI incorporated with IoT has given new hope in the development of a world of digitized things.

References

Atzori, L., Iera, A., & Morabito, G. (2010). The internet of things: A survey. *Computer Networks, 54*(15), 2787–2805. https://doi.org/10.1016/j.comnet.2010.05.010

Azarpeyvand, A., Salehi, M. E., Firouzi, F., Yazdanbakhsh, A., & Fakhraie, S. M. (2010). Instruction reliability analysis for embedded processors. In: *13th IEEE Symposium on Design and Diagnostics of Electronic Circuits and Systems* (pp. 20–23). https://doi.org/10.1109/DDECS.2010.5491824

Beck, J., Stern, M., & Haugsjaa, E. (1996). Applications of AI in education. *XRDS: Crossroads, the ACM Magazine for Students, 3*(1), 11–15. https://doi.org/10.1145/332148.332153

Bertino, E. (2016). Data security and privacy in the IoT. In: *EDBT*. https://doi.org/10.5441/002/edbt.2016.02

Bogue, R. (2007). MEMS sensors: Past, present and future. *Sensor Review, 27*(1), 7–13. https://doi.org/10.1108/02602280710729068

Calo, S. B., Touna, M., Verma, D. C., & Cullen, A. (2017). Edge computing architecture for applying AI to IoT. In: *IEEE International Conference on Big Data (Big Data), 2017* (pp. 3012–3016). https://doi.org/10.1109/BigData.2017.8258272

Chen, S., Xu, H., Liu, D., Hu, B., & Wang, H. (2014). A vision of IoT: Applications, challenges, and opportunities with China perspective. *IEEE Internet of Things Journal, 1*(4), 349–359. https://doi.org/10.1109/jiot.2014.2337336

Faught, W. (1986). Applications of AI in engineering. *Computer, 19*(7), 17–27. https://doi.org/10.1109/MC.1986.1663273

Glaß, M., Yu, H., Reimann, F., & Teich, J. (2012). Cross-level compositional reliability analysis for embedded systems. In F. Ortmeier & P. Daniel (Eds.) *Computer Safety, Reliability, and Security. Lecture Notes in Computer Science. SAFECOMP 2012* (vol. *7612*, pp. 111–124). Springer. https://doi.org/10.1007/978-3-642-33678-2_10

Hassan, Q. F. (2018). Introduction to the internet of things. In: *The Internet of Things A to Z: Technologies and Applications* (pp. 1–50). IEEE. https://doi.org/10.1002/9781119456735.ch1

https://www.idc.com/getdoc.jsp?containerId=prAP46899220

https://www.idc.com/tracker/showproductinfo.jsp?containerId=IDC_P33198

https://owasp.org/www-project-internet-of-things/

https://www.maximizemarketresearch.com/market-report/global-industrial-sensors-market/34465/

https://www.statista.com/statistics/273018/number-of-internet-users-worldwide/

https://www.statista.com/statistics/871513/worldwide-data-created/

https://www.wired.com/brandlab/ 2018/05/bringing-power-ai-internet-things/

Internet of things (IoT). (2019). *HI, road map AU*. https://eps.ieee.org/images/files/HIR_2019/HIR1_ch03_iot.pdf

Lee, I., Shaw, W., & Fan, X. (2009). Wireless multimedia sensor networks. In: Misra, S, Woungang, I. & Misra, S. (eds). *Computer Communications and Networks* (pp. 561–582). Springer. https://doi.org/10.1007/978-1-84882-218-4_22

Liu, B. (2021). 'Weak AI' is likely to never become "strong AI", so what is its greatest value for us? https://arxiv.org/pdf/2103.15294.pdf

Liu, T., & Lu, D. (2012). The application and development of IoT. In: *International Symposium on Information Technologies in Medicine and Education, 2012* (pp. 991–994). https://doi.org/10.1109/ITiME.2012.6291468

Mois, G., Folea, S., & Sanislav, T. (2017). Analysis of three IoT-based wireless sensors for environmental monitoring. *IEEE Transactions on Instrumentation and Measurement*, *66*(8), 2056–2064. https://doi.org/10.1109/TIM.2017.2677619

National Intelligence Council (1663). *Disruptive civil technologies—six: Technologies with potential impacts on the US interests out to 2025*. https://www.dni.gov/. Conference Report CR 2008, 07(April) p. 2008.

Prabhu, K., Dhineshkumar, P., Gobisankar, S., Gowrisankar, M., & Suryapranesh, M. B. (2018). *A comparative study of algorithms for measuring HVAC in IoT*. https://doi.org/10.15680/IJIRSET.2018.0702023

Prien, R. D. (2007). The future of chemical in situ sensors. *Marine Chemistry*, *107*(3), 422–432. https://doi.org/10.1016/j.marchem.2007.01.014

Sarishma, T. R., Kumar S., Awasthi M.K. (2021) To beacon or not?: Speed based probabilistic adaptive beaconing approach for vehicular ad-hoc networks. In: Paiva S., Lopes S.I., Zitouni R., Gupta N., Lopes S.F., Yonezawa T. (eds) *Science and Technologies for Smart Cities. SmartCity360° 2020. Lecture Notes of the Institute for Computer Sciences, Social Informatics and Telecommunications Engineering*, vol. *372*. Springer, Cham. https://doi.org/10.1007/978-3-030-76063-2_12

Sehrawat, D., & Gill, N. S. (2019). Smart sensors: Analysis of different types of IoT sensors. In: *3rd International Conference on Trends in Electronics and Informatics (ICOEI), 2019* (pp. 523–528). https://doi.org/10.1109/ICOEI.2019.8862778

Shukla, A. (2021). A study on Internet of things: Industrial application and difficulties. https://easychair.org/publications/preprint/5TVQ

Tomar R., & Tiwari R. (2019) Information delivery system for early forest fire detection using internet of things. In: Singh M., Gupta P., Tyagi V., Flusser J., Ören T., Kashyap R. (eds) *Advances in Computing and Data Sciences. ICACDS 2019. Communications in Computer and Information Science*, vol. 1045. Springer, Singapore. https://doi.org/10.1007/978-981-13-9939-8_42

2

Security and Privacy in Mobile Cloud Computing

A. Katal
University of Petroleum and Energy Studies

CONTENTS

2.1 Introduction 18
2.1.1 General Architecture 18
2.1.2 Features 20
2.2 Mobile Cloud Service Models 20
2.3 Green Mobile Cloud Computing 21
2.4 Introduction to Security and Privacy in Mobile Cloud Computing 22
2.5 Security and Privacy Challenges in Mobile Cloud Computing 23
2.5.1 Challenges Related to Data Security 23
2.5.2 Offloading and Partitioning Security Challenges 23
2.5.3 Virtualization 24
2.5.4 Challenges Related to Mobile Cloud Applications 24
2.5.5 Security Challenges Related to Mobile Devices 25
2.5.6 Security of Mobile Network 25
2.5.7 Problems Related to Mobile Terminal 25
2.5.7.1 Malware 25
2.5.7.2 Vulnerabilities in Software 26
2.5.8 Mobile Cloud 26
2.5.9 Privacy Challenges 27
2.6 Current Solutions for the Security and Privacy Issues 27
2.6.1 Data Security 27
2.6.2 Partitioning and Offloading 27
2.6.3 Virtualization 28
2.6.4 Mobile Cloud Applications 28
2.6.5 Mobile Devices 29
2.6.6 Mobile Network 29
2.6.7 Mobile Terminal 29
2.6.7.1 Antimalware 29
2.6.7.2 Software Vulnerabilities 30
2.6.8 Mobile Cloud 30
2.6.9 Privacy 30
2.7 Discussion and Conclusion 30
References 31

DOI: 10.1201/9781003291916-2

2.1 Introduction

Since 2007, MCC has been an important field of research for the technological and engineering sectors. Cloud computing is usually characterized as a collection of services made available through a Web cluster system. Such cluster structures are defined as a collection of reduced cost servers or Personal Computers (PCs) that arrange the machines' various resources as per a specific strategy as well as provide services for data processing that are secure, reliable, rapid, efficient, and consistent to customers. Meanwhile, smartphones have come to symbolize a wide range of mobile devices since they have become increasingly internet-connected as wireless network technology has progressed. Growing popularity and portability are two essential aspects of next-generation networking, which provides a variety of personalized communication networks via a number of network endpoints and access methods. The basic technology of Cloud Computing (CC) is the centralization of computers, services, and specialized applications as a utility to be sold to consumers in the same way that water, gas, or electricity are. As a result of the combination of ubiquity mobile networks and CC, a new computing paradigm, termed MCC, is created.

Instead of typical local PCs or servers, resources in MCC networks are virtualized and deployed to a group of multiple independent systems as development of cloud computing, and are made available to smartphones like cellphones, mobile terminals, and so on. Mobile Cloud Computing (MCC) has seen a faster expansion in terms of research as mobile phones have become a need for human life. The introduction of MCC represents a fundamental shift in computer science technologies as well as phone developers.

MCC is a relatively new subset of cloud computing in which mobile apps are created or driven by cloud computing technologies. It is constantly developing and is commonly used for data storage and processing in mobile devices. Mobile cloud technologies provide processing power and data management from smartphones and the cloud to a significantly greater population of cellular customers, delivering applications and mobile computing to nonsmartphone consumers (Noor et al., 2018). The number of mobile device users has increased dramatically in a very short period. Almost every person possesses at least one Smartphone Device (SMD), and as the number of SMDs has grown so has the use of the internet in SMDs. Because of the widespread usage of the internet, MCC is anticipated to play a significant role in the IT sector. A mobile cloud approach enables programmers to construct mobile user–specific apps which are not constrained by the phone's mobile operating system, computation, or storage abilities. MCC has attracted the interest of businessmen as a lucrative and beneficial solution provider that minimizes mobile app creation and maintenance costs while letting mobile consumers to easily acquire the on-demand technology.

While MCC will make and is making a significant contribution to our everyday lives, it will and has introduced a slew of new obstacles and concerns. In order to overcome the constraint of limited resources and processing capacity of mobile devices and to attain good access to CC equivalent to that of regular computers and laptops (Al_Janabi & Hussein, 2019), it is necessary to study MCC paradigm in detail.

2.1.1 General Architecture

Mobile phones are becoming increasingly commonplace in most people's daily lives. As a result, businesses have been compelled to develop programs that can be shared. Access is made easier by mobile phones. However, due to restricted resources, mobile devices

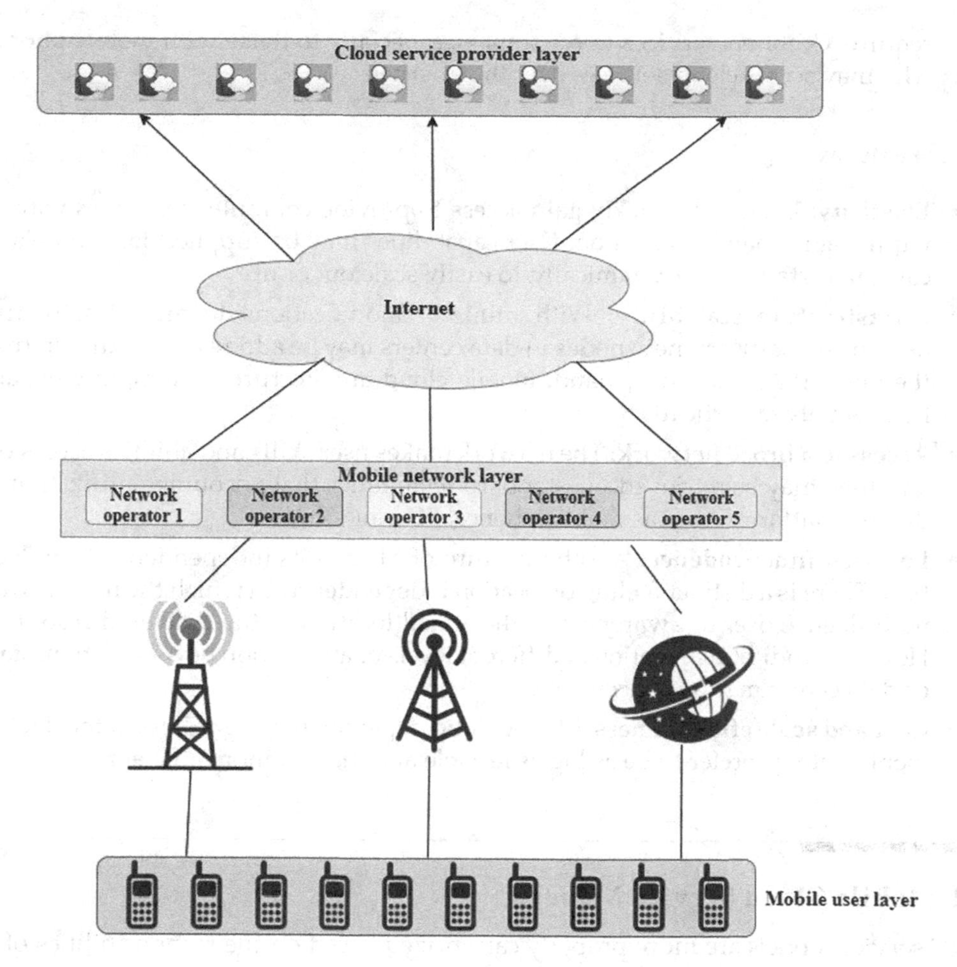

FIGURE 2.1
Architecture of mobile cloud computing.

provide certain design problems to mobile application developers. Cloud computing is being utilized to solve these issues. Today's lifestyle frequently necessitates staying in touch via mobile communication devices. Data transmission and reception are getting more convenient. MCC's architecture is depicted in Figure 2.1.

- **Mobile user layer:** This tier is made up of many different mobile cloud service consumers who use mobile devices to access cloud services.
- **Mobile network layer:** This layer is made up of several mobile network providers who handle mobile user queries and deliver information via base stations. These user requests are handled by the home agent's mobile network services, such as Authentication, Authorization, and Accounting (AAA). Mobile network carriers assist in identifying the subscribers' data held in their databases via their Home agent (HA).
- **Cloud service provider layer:** This layer is made up of numerous CC service providers who offer a variety of CC services. These CC services are elastic, meaning they may be scaled up or scaled down depending on what CC service consumers

require. CC offers services to consumers, especially to those with mobile phones, who may access cloud services over the Internet.

2.1.2 Features

- **Elasticity:** Users can quickly gain access to provide computer resources without requiring human intervention. User capabilities may be supplied fast and elastically, in certain cases dynamically, to easily scale out or up.
- **Infrastructure scalability:** With minimal modifications to an infrastructure design and software, new nodes in data centers may be added or withdrawn from the network. Based on demand, mobile cloud architecture may rapidly expand horizontally or vertically.
- **Access to a broad network:** The network makes user skills and abilities accessible, and they may be retrieved via standard techniques that encourage utilization by diverse platforms such as mobile phones, PCs, and PDAs.
- **Location independence:** Another feature of MCC is its independence from location. There is a distinct feeling of location independence, in which the user has had no influence over or awareness of the actual location of the delivered resources. However, indicating location at different levels of abstraction beyond nation, state, or data center may be conceivable.
- **Cost and scale effectiveness:** Mobile cloud deployments, regardless of the deployment strategy, prefer to be as big as feasible and make it more efficient.

2.2 Mobile Cloud Service Models

MCC service models are more properly categorized based on the responsibilities of computing objects within its project architect, whereas MCC service models may be categorized depending on the roles and relationships among cellular objects. Existing MCC services may be divided into three primary types based on this viewpoint: mobile as a service consumer (MaaSC), mobile as a service provider (MaaSP), and mobile as a service broker (MaaSB). Figure 2.2 depicts the MCC service models.

MaaSC evolved from the conventional client–server paradigm in the early stages by incorporating virtualization, fine-grained authentication, as well as other cloud-based technologies. Smartphones may outsource memory and communication activities to the cloud in order to enhance speed and program abilities. Under this design, the operation is one-way from the cloud to smartphones.

MaaSP differentiates with MaaSC in that the function of a smart phone is switched from that of a prospective customer with that of a provider. Smartphones with built-in sensors, for instance, can detect data from gadgets and their surroundings and then give sensing services to other smartphones through the cloud. Depending on their sensing and processing capabilities, smartphones provide a wide range of services.

MaaSB is a MaaSP extension that offers connectivity and packet forwarding capabilities to many other smartphones or edge devices. In certain cases, MaaSB is required since mobile devices often have poor sensing capabilities when compared to sensors devoted to particularly developed functionality and sensing sites. Mobile phones, for example, may be used to gather Nike Fuelband users' (Nike. Just Do It. Nike IN, n.d.) physical activity.

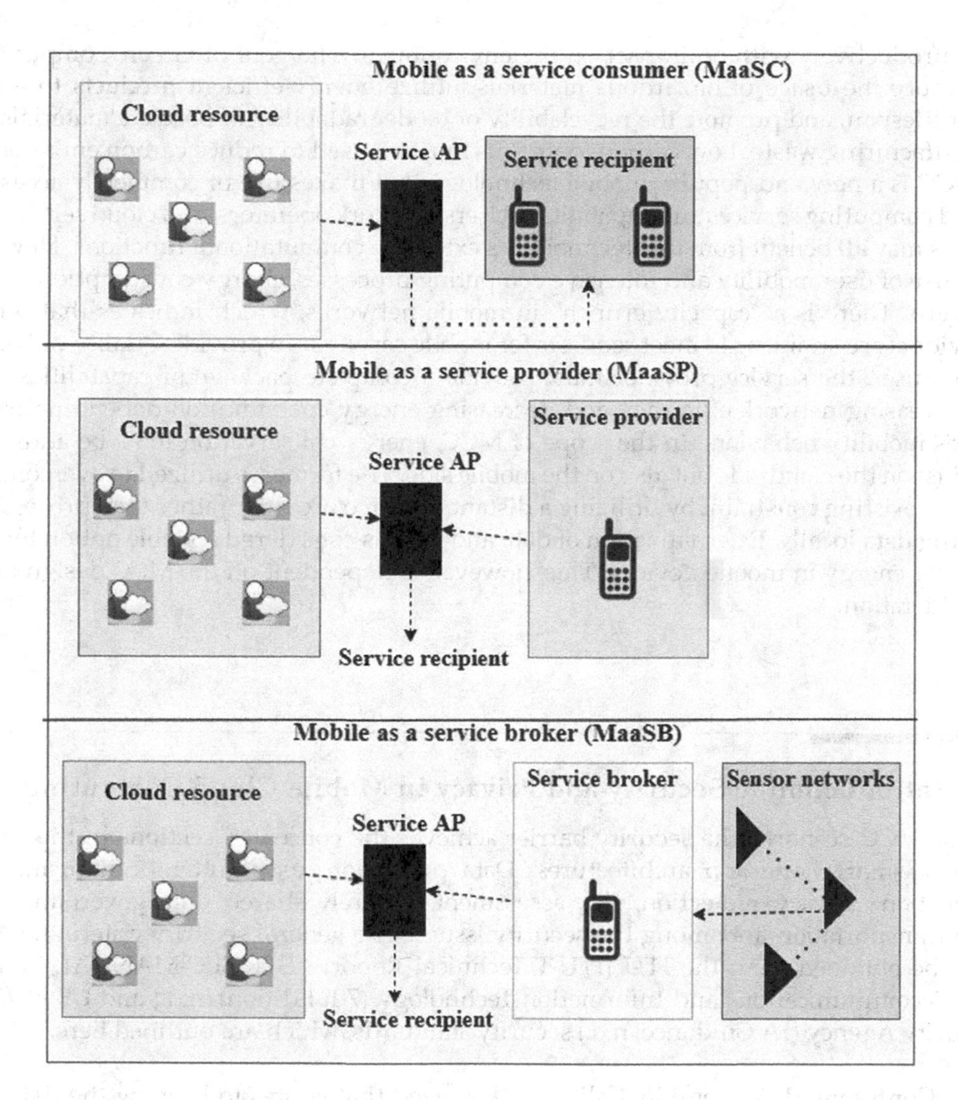

FIGURE 2.2
Service models of mobile cloud computing.

MaaSB extends the cloud boundaries to phones and wireless sensors. As a consequence, a smartphone may be configured to act as a gateway or proxy, providing network access via different communication protocols such as 3/4G, Bluetooth, and Wireless internet. Furthermore, the proxy mobile device may preserve the confidentiality and privacy of their interfaced sensors.

2.3 Green Mobile Cloud Computing

Green Computing (Green Chemistry) is responsible for creating, producing, utilizing, and disposing of computer systems as well as their equipment such as displays, printers, storage systems, and networking and communication systems to consume quickly and productively with no impact on the environment. The goal of Green Computing is to reduce the usage of hazardous materials, utilize energy-efficient products to extend their lifespan, and promote the recyclability or biodegradability of obsolete materials and manufacturing waste. Low carbon footprints must be used to reduce carbon emissions.

MCC is a new and popular mobile technology that makes use of completely accessible cloud computing services and capabilities. Users, network operators, and cloud service providers may all benefit from this technology's extensive computational functions. However, because of user mobility and intensive computing processes, energy consumption is a big concern. There is a "capacity crunch" in mobile networks, which indicates that service providers are straining to meet requests for mobile services. To provide Quality-of-Service to the user, the service provider must provide a complete package of capabilities, such as increasing network efficiency and decreasing energy consumption depending on the user's mobility behaviors. In the scope of MCC, energy conservation must be addressed not just on the cloud side but also on the mobile side. The former is utilized to overcome the latter's existing constraint by utilizing a distant resource provider rather than processing/storing data locally. Externalization of data and apps is considered a viable option for conserving energy in mobile devices. This, however, is dependent on the MCC design under consideration.

2.4 Introduction to Security and Privacy in Mobile Cloud Computing

In the MCC scenario, the security barrier achieves the concealed sections in the protection of smartphone app architectures. Data protection, user authentication, integrity protection, privacy protection, risk assessment, securely shared data governance, and safe transportation are among the security issues. The general security criteria for MCC may be obtained from the ITU [ITU-T Technical Report XSTR-SEC-MANUAL Security in Telecommunications and Information Technology (7th Edition), n.d.] and US National Security Agency (IA Guidance, n.d.) security standards, which are outlined here.

- **Confidentiality:** Confidentiality is a key need that relates to keeping the data of mobile users hidden in the cloud. It is a major barrier for mobile consumers to use different cloud services. Because data is sent and collected via public internet and stored or analyzed on public cloud infrastructure to deliver cloud services, there is a danger that it will be revealed to unauthorized individuals.
- **Integrity:** Data computations and storage are handled by the service provider in MCC. In this case, the integrity must assure the correctness and consistency of the data provided by users. In other terms, data integrity avoids undetected data change by unauthorized individuals or systems. Integrity infringement has an impact on commercial and other losses of mobile users.
- **Availability:** MCC ensures that all cloud solutions are always accessible at all times and in all circumstances as requested by phone users. Keeping services available includes preventing several forms of unavailability threats that lead to delays, modifications, or disruptions in network connectivity.
- **Authentication and access control:** The act or action of verifying the user's identity, participant's data, or program is known as authentication. Following the

successful authentication procedure, it is necessary to identify which resources are authorized to acquire and also what activities, like read, execute, change, or remove, are allowed. This is known as access control.

- **MCC privacy needs:** Safety goals such as secrecy, consistency, and identification encourage anonymity, and these goals explicitly or implicitly preserve the privacy of cloud service customers on smartphones.

2.5 Security and Privacy Challenges in Mobile Cloud Computing

2.5.1 Challenges Related to Data Security

The most serious data security issue happens as a result of phone devices' data being stored and analyzed in clouds located at network operator terminals. Information concerns include loss of data, data leakage, backup and recovery, data localization and data protection. The loss and compromise of data violates two security standards: consistency and privacy. Loss of data in this sense refers to user information that was damaged or lost through any physical force while process, transport, or storage. In a data breach incident, individuals' information is acquired, copied, or utilized by intruders. These two can be caused by malevolent insiders or by malicious apps from outside. Another issue to be concerned about is data recovery. However, because the users' data is kept on the premises of the service providers in cloud service models, firms need to ensure wherever the information is hosted or kept; thus, data localization is also a challenge. One customers' information should also be kept distinct from the other. When one user's data is mixed, combined, or confounded with the data of other users, it becomes considerably more susceptible. Whenever information is exported to cloud servers to improve storage space, phone devices lose physical control of the information at the same time. As a result, in the cloud storage scenario, one of the issues for mobile users is the accuracy of the data. Even though data centers are considerably more reliable and efficient than portable devices, they suffer a plethora of threats and vulnerabilities to data security.

2.5.2 Offloading and Partitioning Security Challenges

Offloading in MCC necessitates the use of wireless networks, over which there is no control, posing even additional danger. The integrity challenge will be breached if the offloaded content does not provide the desired outcome. Jamming assaults, availability attacks, and harmful content threats are among the other concerns. During the offloading process, wireless networks must be used to connect to the cloud. Furthermore, because content executions are offloaded to cloud or edge servers rather than mobile devices, there is a risk of compromising the privacy and consistency of exported data. The security problem occurs when mobile phones cannot easily check the authenticity of the outcomes after processing outsourced content if the result is erroneous or altered. The availability of cloud services can be impacted by jamming attacks that can occur between the information and the smartphone while segmentation, as well as between the smartphone and the cloud while offloading. Furthermore, the existence of harmful material between the partitioning and offloading stages might have an influence on the users' data confidentiality as well as breach mobile users' privacy.

2.5.3 Virtualization

Cloud service providers in MCC deliver cloud services to mobile consumers using virtualization methods. A replica of the portable smartphone's virtual machine (VM) is preinstalled in the cloud, as well as the smartphone's tasks are outsourced to the VM for execution. This VM is also known as a thin VM or a phone clone. VMs are preloaded in the cloud so that duties may be offloaded to the VM for execution, which is known as virtualization. The primary goal of virtualization is to provide several VMs that run on the same physical system. The implementation of this procedure in mobile apps may provide numerous problems, such as illegal user access and VM-to-VM attack. The primary goal of virtualization is to offer several VMs that run on the same physical system or mobile device while being separated from one another. A hypervisor, often known as a VM Monitor or Manager (VMM), is software that facilitates the creation, operation, and management of VMs and other virtual subsystems. Once implemented to MCC, even so, virtualization technologies create several security threats, such as security issues inside the VMs, unauthorized users, VM-to-VM strike, security mechanisms within the virtual environment, security issues inside the Hypervisors, and data security (Sgandurra & Lupu, 2016). Efficiency, coverage, complexity, security, and resilience are five essential properties of virtualization-based security techniques (Annane & Ghazali, 2019; Sahu & Pandey, 2018). The amount of rogue VMs that properly colocate with a targeted (victim VM) based on the number of VMs created by the hacker is approximately described as efficiency. Security relates to the secrecy of the VMs and their confidential material, which means no one must be able to delete or recover the characteristics unless they have the digital signature. Likewise, resilience relates to the way of tolerance to any sort of manipulation. Any new technique must achieve a balance between these five fundamental criteria. Many researchers have begun proposing frameworks, strategies, and techniques to address these types of issues in order to assure mobile user security.

2.5.4 Challenges Related to Mobile Cloud Applications

Cloud-based mobile app level can endanger the confidentiality and reliability of both information and programs via a number of means, such as the inclusion of malware (Pokharel et al., 2017; Prokhorenko et al., 2016). Malwares such as viruses, worms, Trojans, rootkits, and botnets (Qamar et al., 2019) are undesirable, programs, or protocols that are invasive or bothersome. These malwares are intended to run with malicious intent on mobile devices or to connect to apps without the consent of the user. As a consequence, mobile app performance can be adjusted. An intruder will often select a target system, insert malicious software into this one, and then relaunch it. The easiest solution to discover security threats will be to download and configure security software and antivirus programs on smartphones. Mobile gadgets, on the other hand, have computation and energy constraints; defending them from these attacks may be more challenging than securing traditional PCs. Several techniques for moving threat detection and security procedures to the cloud have been explored. Before mobile consumers may use a particular application, it must be subjected to some kind of danger assessment. Instead of running antivirus or penetration testing programs onsite, smartphones perform simple tasks such as transmitting execution logs to cloud safety servers. Giving out sensitive data, such as your current location and the user's vital information, raises privacy problems. Threats to private information might be reduced by choosing and assessing company requirements and requiring just specific services to be bought and migrated to the cloud.

2.5.5 Security Challenges Related to Mobile Devices

Physical hazards to mobile devices exist. If mobile devices are forgotten, lost, or stolen, data or apps may be lost, leaked, accessed, or accidentally disclosed to unauthorized users. Although password or pattern-based locking mechanisms are available, many smartphone users may not make use of them. Moreover, the identity module card contained within the smartphone could be stolen and accessed by unauthorized persons, and the number of users' phones lacks an antiattack feature. Intruders can use a variety of available attack techniques, such as sending a huge volume of unwanted activity and big texts to target smartphones in addition to making them underutilized or reducing their capabilities. The authors of Alakbarov (2021) address the constraints that arise in mobile device resources (energy usage, processing and memory capacities, etc.) as well as the delays that occur in communication channels. Racic et al. (2006) address this type of assault. They demonstrate that as a result of this attack. The power of a portable smartphone's batteries depletes up to 22 times faster than would under normal mode, rendering the gadget unusable in a short period. Malware developers and hackers are concentrating their efforts due to the popularity of smartphones and digital services. Malware allows attackers to get root access to mobile devices and manipulate them, which may then directly compromise the computational integrity of mobile platforms and apps.

2.5.6 Security of Mobile Network

Due to its mobility, a mobile network, based on a standard network, increases the network node and the access path of users. The networking module is expanded to smartphones and tablets and so on; smartphones can join the system in a number of ways, for example, phone users can access cell service, messaging, and other online services via 3G technology. As a result, additional security concerns, such as sensitive information leakage or malicious assault, would be introduced. For example, many types of public places (such as cafes, restaurants, and airports) can provide free Wi-Fi. In this situation, there is a risk of information leaking. Aside from this type of public Wi-Fi, even private Wi-Fi is vulnerable to security threats due to a flaw in Wi-Fi encryption process. Communication among smartphones and digital cloud providers is frequently accomplished using a variety of methods. It gives rise to security concerns.

2.5.7 Problems Related to Mobile Terminal

In general, mobile terminals have the following characteristics: an open OS; support for third-party applications; "personalization"; and wireless Internet connectivity at any time and from any location. Because of this, security concerns in mobile terminals are quite significant.

2.5.7.1 Malware

The mobile terminal's openness and adaptability constantly attract the attention of attackers. Some malware can be downloaded and carried without the user's knowledge, alongside helpful programs and systems. This allows the malware to get unauthorized access to personal information without the user's intervention. In order to combat malware, several security firms have created antivirus software for mobile devices. However, as harmful assaults get more complicated, antimalware solutions should perform a similar role to that of the desktop. They have limited capacity and resources, necessitating antimalware solutions that need considerable compute resources difficult to achieve.

2.5.7.2 Vulnerabilities in Software

- **Application software:** At the moment, the smartphone is the primary mobile terminal and the majority of smartphone customers are used to controlling their phones via smartphone management system, which manages the information inside the device through data syncing between the phone and the PC. This procedure is often carried out via FTP (File Transfer Protocol). The username and password of FTP are transmitted across the network and recorded in clear text in the configuration file. This leads to unauthorized access to the cellular telephone via FTP from PCs on the very same connection, resulting in personal data theft as well as unauthorized access, intentional deletion, and harmful manipulation. In reality, application software vulnerabilities are quite frequent. Since the software is somewhat lax, attackers can infiltrate the mobile phone via a flaw in the application program.
- **Operating system:** The OS is in charge of managing and controlling software and hardware resources. And because the software is so sophisticated, it may have coding flaws. In certain cases, attackers will utilize these flaws to destroy the mobile phone.
- **Others:** Furthermore, security concerns in mobile terminals are still caused by mobile users. Mobile users frequently lack security knowledge; second, mobile users themselves may make mistakes. As a result, companies must identify and prohibit unusual user activity.

2.5.8 Mobile Cloud

- **Platform reliability:** Because of the huge concentration of user information resources, the cloud platform is vulnerable to assault. The dangerous suspect's objective, on the one hand, is to acquire important information and services. Such attacks could come from malicious outsiders, a genuine CC user, or a part of the CC operator's team. The hostile attacker's goal, on the other hand, is to shut down the cloud service. For example, a DOS (denial of service) assault will damage the platform's availability and force the cloud's service to be shut down. Users risk data loss if they submit all of their data to Cloud Service Providers (CSPs) before subscribing for the more costly backup and recovery option. Such incidents have occurred on occasion at cloud providers in recent years. As a result, the cloud provider should incorporate modern security technology to assure the service, and customers must not be overly reliant on the CSP.
- **Privacy and data protection:** Privacy and security preservation are major problems in the mobile cloud. To start, inside the clouds, user information management and governance are segregated, which causes users' concerns over their own information resource to become a significant impediment to the widespread adoption of MCC. As a result, consumers' sensitive information is at a greater danger of being exposed (Feng et al., 2011).

2.5.9 Privacy Challenges

One of the key problems is privacy, which arises when mobile users' sensitive data or apps are processed and sent from mobile devices to heterogeneously dispersed cloud servers when using various cloud services. Such computers are located in diverse locations owned and managed completely by the providers. In this scenario, consumers cannot personally be monitoring the preservation of their data, so problems regarding data privacy issues are in the hands of service suppliers, and consumers are not held accountable for privacy violations. The use of cloud storage and processing in various places raises privacy concerns (Gupta et al., 2017a, b). Furthermore, it is critical for consumers to obtain information about the cloud hosting location because the legislation varies from one nation to another. There are several mobile applications accessible that may be dangerous owing to having different functionalities, gathering unintentionally users' personal information such as hobbies and whereabouts, and spreading unlawfully. Unwanted advertising emails, often known as trash emails, can infringe on consumers' privacy (Gupta et al., 2018). One of the key differences between mobile apps and PCs is context awareness, which is provided via sensors on mobile devices. Moreover, many systems involve and gather location information from users, which may then be used to directly market to clients depending on their whereabouts. As a result, location-based services pose privacy concerns since they must collect, store, and process user information.

2.6 Current Solutions for the Security and Privacy Issues

2.6.1 Data Security

In Alqahtani & Kouadri-Mostefaou (2014), the authors propose a system based on distributed multicloud storage, encryption keys, and compression algorithms techniques to ensure data protection in MCC. At the mobile phone end, data was originally divided into discrete segments based on user preference, and the sections are then secured and reduced. Finally, it keeps data on a decentralized multicloud platform. However, in order to increase security, the user can keep one section on its own storage. A nearly identical technique is given in Abdalla & Pathan (2014). The information is updated to a CSP end using a data security administrator that is split into two portions. The first one is a data fragmenter that splits information and sends it to many sites for preservation. The second is data merging, which combines data as the user downloads from the stored data. Wang et al. (2014) suggest a security strategy for protecting mobile users' data in the media cloud. The authors in (Lai et al., 2013) proposed an attribute-based data strategy. This technique is the combination of ciphertext attribute-based authentication (CP-ABE) and encryption algorithm. It is only suitable for CC and not for resource-constrained smartphones, as the current CP-ABE has severe performance constraints due to its large cipher size and processing power.

2.6.2 Partitioning and Offloading

Al-Mutawa and Mishra (2014) present a data-splitting technique for preventing cellular user data vulnerability while transferring to a remote and trusted or untrustworthy business. This concept is divided into three stages. First, based on the user's preferences, the data is categorized as nonsensitive or sensitive. The sensitive element is subsequently handled on the smartphone or by a trusted distant entity, whereas the nonsensitive part is transferred to another entity. TinMan is proposed by the authors of Xia et al. (2015) to maintain anonymity and to carry out security-oriented shifting from portable devices to the cloud. To limit vulnerability, critical data from mobile devices is isolated from mobile applications, which are subsequently downloaded and kept on a neighboring device. A reliable node is installed on a virtual computer in a reliable cloud. The TinMan is in charge of determining how efficiently and cleanly this unloading process is carried out. Saab et al. (2015) suggest a cloud-based smartphone app offloading that is both secure and energy-efficient. A profiling engine, a deception engine, and an offloading engine are all part of this system. The decision-making and profiling algorithms are used for continuous categorization of SMDs in order to enhance power consumption and safety. Following the deception, the entire or partial component is offloaded to the cloud using the offloading engine. Langdon et al. (2015) propose and test a method for offloading mobile applications while maintaining privacy. The privacy of the user information is maintained within the mobile device during unloading in this technique. The experiment results demonstrate that the proposed technique protects privacy while conserving energy throughout the offloading process.

2.6.3 Virtualization

Vaezpour et al. (2016) offer SWAP, a security-aware provisioning and migration approach for cellphone clones. Hao et al. (2015) explain SMOC, a secure mobile cloud platform. This technique helps phone devices to safely copy their OS and applications to cloud-based VMs. To maintain the protection of user information on smartphones, a hardware virtualization technique is used that enables data and application functionality to be separated from the mobile service device's operating system. Paladi et al. (2017) provide a security architecture for cloud infrastructure. Before deploying guest VMs, the trusted VM launching protocol is utilized.

2.6.4 Mobile Cloud Applications

Tan et al. (2015) offer a paradigm called STOVE (Strict, Observable, Verifiable Data, and Execution) for safely executing untrusted mobile applications. Once an unknown software starts and runs, it is not capable of causing harm to other programs or the OS, and it does not access any unauthorized programs or data. This model consists of three stages. In the first stage, the STOVE strictly restricts and completely isolates untrusted programs from other applications in the environment in which it operates. STOVE then uses formal verification methods to validate the tightly isolated code in the second stage. Finally, STOVE performs all data access in favor of the untrusted program so that all data access is visible. Tang et al. (2016) provide a token-based three-factor strong Application Programming Interface (API) security architecture. These are (a) fundamental identification, like API registration process with robust strong passwords; and (b) contemporary security techniques, such as messaging security requirements, online signatures, and internet cryptography; and (c) as a third prevention measure, the main protein inside API and its backward compatibility.

2.6.5 Mobile Devices

While customers retain and use confidential material, app developers should pay attention to the two issues: developers can implement a layer of security as an added step in the application level, and then they can ensure that customers' sensitive information also isn't kept in the identification module card. Moreover, cloud backup solutions are critical for customers because if a customer's mobile is stolen or their information is deleted, the user can restore his data from the cloud. At the moment, unique security mechanisms are integrated into mobile devices. There are numerous security programs for mobile devices available, each with its own set of functions such as detecting and preventing mobile viruses, controlling unwanted access, and protecting privacy.

2.6.6 Mobile Network

The two ways to protect mobile networks are data encryption and security protocol. In data encryption, only encrypted information is reasonably safe while transmission over the mobile network, regardless of how the mobile terminals access the network. The investigation of the security protocol is essential for all types of access methods. Rupprecht et al. (2018) aim to compile security information on mobile phone networks into a complete overview and to generate pressing open research topics. To accomplish this methodically, they create a framework that categorizes known attacks based on their purpose, suggested countermeasures, underlying reasons, and root causes. In addition, they evaluate the impact and efficacy of each assault and response. This technique is then applied to existing research on attacks and responses in all three network generations. This allows them to identify ten causes and four root causes of assaults (Tomar, 2020; Fore et al., 2016). Mapping the assaults to suggested countermeasures and 5G specification recommendations are helpful in identifying outstanding research topics and obstacles for the construction of next-generation mobile networks. The NEMESYS method to smart mobile network security is presented by the authors of Gelenbe et al. (2013). The NEMESYS project seeks to deliver novel security mechanisms for seamless network services in the intelligent phone environment and to enhance cellular data security through a better knowledge of the latest threats. It will gather data on the nature of cyber-attacks on smart devices and the network architecture in order to execute appropriate measures for protection.

2.6.7 Mobile Terminal

2.6.7.1 Antimalware

To safeguard the mobile terminal from attack, two steps must be taken: the first step is to investigate and eliminate virus. Virus monitoring might be relocated to the cloud to avoid the resource restrictions of mobile terminals. If virus is detected, lawful application form of the cloud can be allocated to a mobile terminal and run to remove it. CloudAV is a perfect example of antimalware software. CloudAV (CloudAV: N-Version Antivirus in the Network Cloud, n.d.) is a novel paradigm for malware detection on mobile devices that is built on offering antivirus as an in-cloud network service. CloudAV offers the following significant advantages: improved identification of dangerous software; the effect of antivirus loopholes is reduced; previously affected systems are detected retroactively; forensics abilities are enhanced; and modifiability and administration are improved. It also includes bridge hosting agents and a data network featuring ten antivirus programs and two behavioral recognition algorithms that improves identification substantially. The other is antimalware protection. Users must be cautious of their actions in order to avoid malicious software from being installed on their mobile devices.

2.6.7.2 Software Vulnerabilities

Consumers must give importance to smartphones OS version notifications and download patches or revised versions from of the OSs research and development firm as soon as possible. In the meantime, they must be cautious while installing third-party programs. On the other hand, in order to decrease software vulnerabilities, they need to implement a number of technological steps. Checking the validity and integrity of software, for example, is a crucial practice before using it.

2.6.8 Mobile Cloud

The MCC platform's dependability and availability are critical for both cloud providers and consumers. To begin with, CSPs should use contemporary security mechanisms such as VPN technology, authentication and authorization, cryptography, and other technical approaches to assure high supply from different attacks. Second, CSPs should provide comprehensive recovery and backup solutions in order to retrieve users' data in the event of a major assault. By using these methods, the cloud platform may improve service quality and boost customer confidence. Encryption technology is required for sensitive data to survive the transition from storage to transportation. However, because encryption reduces data usage, emphasis shifts to effectively interpreting and processing the ciphertext. The privacy homomorphism algorithm is the current focus of ciphertext processing research. Meanwhile, key management is a critical task for corporate users.

2.6.9 Privacy

Pasupuleti et al. (2016) suggest a method that protects the confidentiality of smartphone data that has been transferred to the cloud. A stochastic public-key cryptography method and a ranking keyword searching engine are employed in this situation. After creating an indexing for file collecting, the phone node secures both the information and the search while transmitting it to be hosted in the system. Bahrami and Singhal (2015) describe a lightweight authentication algorithm for smartphones to store information on clouds. This method is dependent on the effect of pseudo-random permutation. Since it is compact, the permutations process is done on a smartphone instead of the cloud to protect information confidentiality.

2.7 Discussion and Conclusion

Private cloud software solutions can be utilized to safeguard your privacy and security. Secured cloud-based services offer smartphones access control, identity management, on-demand cryptography, intrusion prevention, identification, and authorization in addition to confidentiality and protection in order to provide a secured channel of interaction between the cloud and the smartphone. Virtualization increases cloud resource efficiency but raises additional security concerns owing to the absence of complete isolation of VMs housed on a single server. VM safe monitoring, mirroring, and migration can assist to address some of the security concerns faced by virtualization. Smartphone users should be allowed to verify the level of security of managed services in order to establish a visible cloud environment. Cloud storage monitoring can be used to aid with the audits. The resource-constrained mobile devices pose a major threat to the rapid expansion of mobile computing. However, by integrating mobile computing with cloud computing, the expansion of mobile computing may be accelerated, resulting in the emergence of a new computing paradigm known as MCC. This chapter provides details about the introduction, architecture, and service models of MCC. It covers the details in depth about the security and privacy challenges in different domains of MCC and concludes with the solutions to the existing problems in MCC.

References

Abdalla, A. K. A., & Pathan, A. S. K. (2014). On protecting data storage in mobile cloud computing paradigm. *IETE Technical Review (Institution of Electronics and Telecommunication Engineers, India)*, *31*(1), 82–91. https://doi.org/10.1080/02564602.2014.891382

Alakbarov, R. G. (2021). Challenges of mobile devices' resources and in communication channels and their solutions. *International Journal of Computer Network and Information Security*, *13*(1), 39–46. https://doi.org/10.5815/IJCNIS.2021.01.04

Al_Janabi, S., & Hussein, N. Y. (2019). The reality and future of the secure mobile cloud computing (SMCC): Survey. *Lecture Notes in Networks and Systems*, *81*, 231–261. https://doi.org/10.1007/978-3-030-23672-4_18

Al-Mutawa, M., & Mishra, S. (2014). Data partitioning: An approach to preserving data privacy in computation offload in pervasive computing systems. In *Q2SWinet 2014- Proceedings of the 10th ACM Symposium on QoS and Security for Wireless and Mobile Networks*, pp. 51–60. https://doi.org/10.1145/2642687.2642696

Alqahtani, H. S., & Kouadri-Mostefaou, G. (2014). Multi-clouds mobile computing for the secure storage of data. In *2014 IEEE/ACM 7th International Conference on Utility and Cloud Computing*, pp. 495–496. https://doi.org/10.1109/UCC.2014.68

Annane, B., & Ghazali, O. (2019). Virtualization-based security techniques on mobile cloud computing: Research gaps and challenges. *International Journal of Interactive Mobile Technologies (IJIM)*, *13*(4), 20–32. https://doi.org/10.3991/IJIM.V13I04.10515

Bahrami, M., & Singhal, M. (2015). A light-weight permutation based method for data privacy in mobile cloud computing. In *2015 3rd IEEE International Conference on Mobile Cloud Computing, Services, and Engineering*, 189–196. https://doi.org/10.1109/MOBILECLOUD.2015.36

CloudAV: N-Version Antivirus in the Network Cloud (n.d.). https://www.usenix.org/legacy/event/sec08/tech/full_papers/oberheide/oberheide_html/index.html

Feng, D. G., Zhang, M., Zhang, Y., & Xu, Z. (2011). Study on cloud computing security. *Ruan Jian Xue Bao/Journal of Software*, *22*(1), 71–83. https://doi.org/10.3724/SP.J.1001.2011.03958

Fore, V., Khanna, A., Tomar, R., & Mishra, A. (2016). Intelligent supply chain management system. In *International Conference on Advances in Computing and Communication Engineering (ICACCE)*, pp. 296–302. https://doi.org/10.1109/ICACCE.2016.8073764

Gelenbe, E., Görbil, G., Tzovaras, D., Liebergeld, S., Garcia, D., Baltatu, M., Lyberopoulos, G., Gelenbe, E., Görbil, G, Görbil, G., Tzovaras, D., Liebergeld, S., Garcia, D., & Baltatu, M. (2013). NEMESYS: Enhanced network security for seamless service provisioning in the smart mobile ecosystem. In *Information Sciences and Systems 2013 Lecture Notes in Electrical Engineering*, vol. 264, pp. 369–378. https://doi.org/10.1007/978-3-319-01604-7_36

Gupta, M., Benson, J., Patwa, F., & Sandhu, R. (2017a). Dynamic groups and attribute-based access control for next-generation smart cars. In *Proceedings of the Ninth ACM Conference on Data and Application Security and Privacy*, pp. 61–72. 2019.

Gupta, M., Patwa, F., & Sandhu, R. (2017b). Object-tagged RBAC model for the Hadoop ecosystem. In *IFIP Annual Conference on Data and Applications Security and Privacy*, pp. 63–81. Springer, Cham.

Gupta, M., Patwa, F., & Sandhu, R. (2018). An attribute-based access control model for secure big data processing in Hadoop ecosystem. In *Proceedings of the Third ACM Workshop on Attribute-Based Access Control*, pp. 13–24.

Hao, Z., Tang, Y., Zhang, Y., Novak, E., Carter, N., & Li, Q. (2015). A secure SMOC mobile cloud computing platform. In *2015 IEEE Conference on Computer Communications (INFOCOM)*, pp. 2668–2676. https://doi.org/10.1109/INFOCOM.2015.7218658

IA Guidance (n.d.). https://apps.nsa.gov/iaarchive/library/ia-guidance/

ITU-T Technical Report XSTR-SEC-MANUAL Security in telecommunications and information technology (7th edition) (n.d.).

Lai, J., Deng, R. H., Guan, C., & Weng, J. (2013). Attribute-based encryption with verifiable outsourced decryption. *IEEE Transactions on Information Forensics and Security, 8*(8), 1343–1354. https://doi.org/10.1109/TIFS.2013.2271848

Langdon, R. J., Yousefi, P. D., Relton, C. L., & Suderman, M. J. (2015). Privacy-preserving offloading of mobile app to the public cloud. *Clinical Epigenetics*. https://doi.org/10.2/JQUERY.MIN.JS

Nike. Just Do It. Nike IN. (n.d.). https://www.nike.com/in/

Noor, T. H., Zeadally, S., Alfazi, A., & Sheng, Q. Z. (2018). Mobile cloud computing: Challenges and future research directions. *Journal of Network and Computer Applications, 115*, 70–85. https://doi.org/10.1016/J.JNCA.2018.04.018

Paladi, N., Gehrmann, C., & Michalas, A. (2017). Providing User Security Guarantees in Public Infrastructure Clouds. *IEEE Transactions on Cloud Computing, 5*(3), 405–419. https://doi.org/10.1109/TCC.2016.2525991

Pasupuleti, S. K., Ramalingam, S., & Buyya, R. (2016). An efficient and secure privacy-preserving approach for outsourced data of resource constrained mobile devices in cloud computing. *Journal of Network and Computer Applications, 64*, 12–22. https://doi.org/10.1016/J.JNCA.2015.11.023

Pokharel, S., Choo, K. K. R., & Liu, J. (2017). Mobile cloud security: An adversary model for lightweight browser security. *Computer Standards and Interfaces, 49*, 71–78. https://doi.org/10.1016/J.CSI.2016.09.002

Prokhorenko, V., Choo, K. K. R., & Ashman, H. (2016). Web application protection techniques: A taxonomy. *Journal of Network and Computer Applications, 60*, 95–112. https://doi.org/10.1016/J.JNCA.2015.11.017

Qamar, A., Karim, A., & Chang, V. (2019). Mobile malware attacks: Review, taxonomy & future directions. *Future Generation Computer Systems, 97*, 887–909. https://doi.org/10.1016/J.FUTURE.2019.03.007

Racic, R., Ma, D., & Chen, H. (2006). Exploiting MMS vulnerabilities to stealthily exhaust mobile phone's battery. In *2006 Securecomm and Workshops*. https://doi.org/10.1109/SECCOMW.2006.359550

Rupprecht, D., Dabrowski, A., Holz, T., Weippl, E., & Popper, C. (2018). On security research towards future mobile network generations. *IEEE Communications Surveys and Tutorials, 20*(3), 2518–2542. https://doi.org/10.1109/COMST.2018.2820728

Saab, S. A., Saab, F., Kayssi, A., Chehab, A., & Elhajj, I. H. (2015). Partial mobile application offloading to the cloud for energy-efficiency with security measures. *Sustainable Computing: Informatics and Systems, 8*, 38–46. https://doi.org/10.1016/J.SUSCOM.2015.09.002

Sahu, I., & Pandey, U. S. (2018). Mobile cloud computing: Issues and challenges. In *2018 International Conference on Advances in Computing, Communication Control and Networking (ICACCCN)*, pp. 247–250. https://doi.org/10.1109/ICACCCN.2018.8748376

Sgandurra, D., & Lupu, E. (2016). Evolution of attacks, threat models, and solutions for virtualized systems. *ACM Computing Surveys (CSUR), 48*(3), 1–38. https://doi.org/10.1145/2856126

Tan, J., Gandhi, R., & Narasimhan, P. (2015). STOVE: Strict, observable, verifiable data and execution models for untrusted applications. In *Proceedings of the International Conference on Cloud Computing Technology and Science, CloudCom*, pp. 644–649. https://doi.org/10.1109/CLOUDCOM.2014.116

Tang, L., Ouyang, L., & Tsai, W. T. (2016). Multi-factor web API security for securing Mobile Cloud. In *2015 12th International Conference on Fuzzy Systems and Knowledge Discovery (FSKD)*, pp. 2163–2168. https://doi.org/10.1109/FSKD.2015.7382287

Tomar R. (2020). Maintaining trust in VANETs using blockchain. *ACM SIGAda Ada LettersAda. 40*, 91–96. https://doi.org/10.1145/3431235.3431244

Vaezpour, S. Y., Zhang, R., Wu, K., Wang, J., & Shoja, G. C. (2016). A new approach to mitigating security risks of phone clone co-location over mobile clouds. *Journal of Network and Computer Applications, 62*, 171–184. https://doi.org/10.1016/J.JNCA.2016.01.005

Wang, H., Wu, S., Chen, M., & Wang, W. (2014). Security protection between users and the mobile media cloud. *IEEE Communications Magazine, 52*(3), 73–79. https://doi.org/10.1109/MCOM.2014.6766088

Xia, Y., Liu, Y., Tan, C., Ma, M., Guan, H., Zang, B., & Chen, H. (2015). Tinman: Eliminating confidential mobile data exposure with security oriented offloading. In *Proceedings of the 10th European Conference on Computer Systems, EuroSys 2015*. https://doi.org/10.1145/2741948.2741977

3

A Review on Live Memory Acquisition Approaches for Digital Forensics

Sarishma Dangi and Diksha Bisht
Graphic Era Deemed to be University

CONTENTS

3.1 Introduction....36
3.2 Related Work....37
3.3 Technical Background....38
3.3.1 Direct Memory Access....39
3.3.2 IEEE 1394....39
3.3.3 System Management Mode....40
3.3.4 x86 Architecture....41
3.3.5 Virtualization Support....41
3.4 Review of Memory Acquisition Techniques....42
3.4.1 Hardware-Based Memory Acquisition....42
3.4.1.1 Dedicated Hardware....42
3.4.1.2 Special Hardware Bus....42
3.4.1.3 Cold Booting....42
3.4.1.4 Limitations of Hardware-Based Techniques....43
3.4.2 Software-Based Memory Acquisition....43
3.4.2.1 User Level....43
3.4.2.2 Kernel Level....45
3.4.2.3 Virtualization Based....46
3.4.2.4 Limitations of Software-Based Techniques....46
3.4.3 Firmware-Based Memory Acquisition....46
3.5 Memory Dump Formats....46
3.5.1 Raw Memory Dump....47
3.5.2 Windows Crash Dump....47
3.5.2.1 Complete Memory Dump....47
3.5.2.2 Kernel Memory Dump....47
3.5.2.3 Small Memory Dump....48
3.5.3 Windows Hibernation File....48
3.5.4 Expert Witness Format....48
3.5.5 AFF4 Format....48
3.5.6 HPAK Format....49
3.6 Memory Acquisition Tools....49
3.6.1 Magnet RAM Capture....50
3.6.2 Belkasoft Live RAM Capturer....50

DOI: 10.1201/9781003291916-3

3.6.3 FTK Imager 50
3.6.4 DumpIt 50
3.7 Implementation 50
3.8 Performance Comparison 52
3.9 Conclusion 59
References 59

3.1 Introduction

Digital forensics is defined as the collection, preservation, examination, and presentation of digital evidence in a coherent, verifiable manner at a court of law using scientifically and technologically driven methods (Beebe, 2009). In the case of digital forensics, data present in digital systems is of great significance as it serves as strong baseline evidence in relation to cybercrimes being committed. Recently, Google suffered a major outage on December 14, 2020 that lasted for about 45 minutes, amounting to serious monetary and service loss (*2020_Google_services_outages @ En.Wikipedia.Org*, n.d.).

Since 2000s, malicious attackers have become sophisticated as well as complex in their approach to attacking and stealing. Various obfuscation based techniques (Harichandran et al., 2016), hiding via antiforensics based (Palutke et al., 2020) techniques, are now being actively used. Advanced malware can hide and is often encrypted, making it much more difficult for the incident response team to determine the true nature of the malware (Rathnayaka & Jamdagni, 2017). Memory analysis techniques can be used to acquire and examine certain malware behaviors in their native form while they are running in memory. Live memory forensics aids in acquiring and analyzing such activities within the digital systems.

Memory forensics is a specialized realm of acquiring, recovering, interpreting, and analyzing evidence present in the volatile memory of digital systems by using various approaches and techniques. Fast-paced evolution of cybercrime landscape has necessitated the presence of this area. Memory forensics is used to analyze physical or volatile memory; more commonly known as RAM (Random Access Memory) in order to collect the evidence by retrieving data from the confiscated device that was used at the commission of a crime. It is about acquiring the memory first and then identifying as well as analyzing it depending upon the case being investigated at hand. It apprehends the memory contents that is a prominent tool for malware analysis and digital forensic investigators.

Acquisition is a process of collecting evidence from the confiscated device through which the crime is committed by using various techniques. There are two types of methodologies that are followed for acquiring the image of digital evidence and these are Live Acquisition and Dead Acquisition.

The data stored on the hard disk drive and within the file system is not lost because the storage is nonvolatile in nature. However, for volatile memory such as RAM, the data is lost as soon as the power is cut off. Once the power has been cut off for a period of time, the state of the data becomes completely unknown. This traditional approach is more commonly known as "pulling the plug" where the focus of analysis is on nonvolatile storage. This is now more commonly referred to as Dead Memory Acquisition. It refers to the powered-off or dead system, which is turned off and no data processing is taking place. It is performed by pulling the plug methodology to avoid any malicious process from running and potentially deleting evidence. There are fewer chances of data modification. It has practical, legal constraints and lacks standardized issues.

In the early 2000s, most investigators simply used to capture the picture of the screen of what was open at the moment and would pull the plug out of the system. In cases where the system was live and encrypted, the investigator easily lost the mere opportunity to access the system state (Beebe, 2009). If pull the plug methodology is followed, it allows little or no insight into what operations the processor was executing at the time when the power was cut off. This severely limits the investigative process.

As the processor's computational ability and memory increased, the need to analyze it also increased since many applications started to execute only in-memory leaving no trace behind. Live Acquisition on the other hand refers to live systems that are up and running where information may be altered as data is continuously processed. It is performed by using specialized tools and techniques to extract volatile data from the computer before turning it off. It can easily retrieve volatile information. The acquisition of RAM can further be performed via two different ways, i.e., hardware- and software-based memory acquisition. Sometimes firmware-based memory acquisition is also possible (Stüttgen et al., 2015). The key contributions of this work are outlined as follows:

- To provide a critical and comprehensive review of the existing memory acquisition approaches.
- To present a detailed overview of memory acquisition tools prominent in the market and outline their limitations.
- To compare the performance of prominent memory acquisition tools on two different machines operating on Windows platform with 4 and 8GB RAM.
- To assess the performance of the machine when the tools are executing in terms of processor utilization, disk utilization, and time taken for acquisition.

The rest of this paper is categorized as follows: Section 3.2 outlines the related work in this area. Section 3.3 provides the technical background which is necessary to give a context for further discussion. Section 3.4 gives a comprehensive review of various hardware- and software-based acquisition approaches for memory acquisition. Section 3.5 discusses the various memory dump formats that are provided by using memory acquiring tools. Section 3.6 includes various prominent memory acquisition tools along with their detailed comparison. Section 3.7 provides the testbed setup and methodology required for implementing memory acquisition tools in the system. Lastly, we conclude our work.

3.2 Related Work

Memory forensics was highlighted and worked upon after the initial impetus provided by the DFRWS challenge in 2005 (Reust & Friedburg, 2006). A trend was observed in the research community toward acquisition, and analysis of volatile memory in particular post this challenge. Acquisition techniques faced some setbacks as discussed further, whereas analysis techniques continue to face resistance and are still worked on.

Figure 3.1 represents the research contribution in the area of volatile memory acquisition over the last decade. It is clear that the contributions started to significantly increase once the DFRWS challenge was launched. This impetus fed not only the progress of the research community but fed directly into the industry-based development. The result was a steep rise in the presence of write blockers and encryption alternatives which prevented reliable memory acquisition. This led to a downfall in research interest in the lieu of limited support from industry and government-backed organizations.

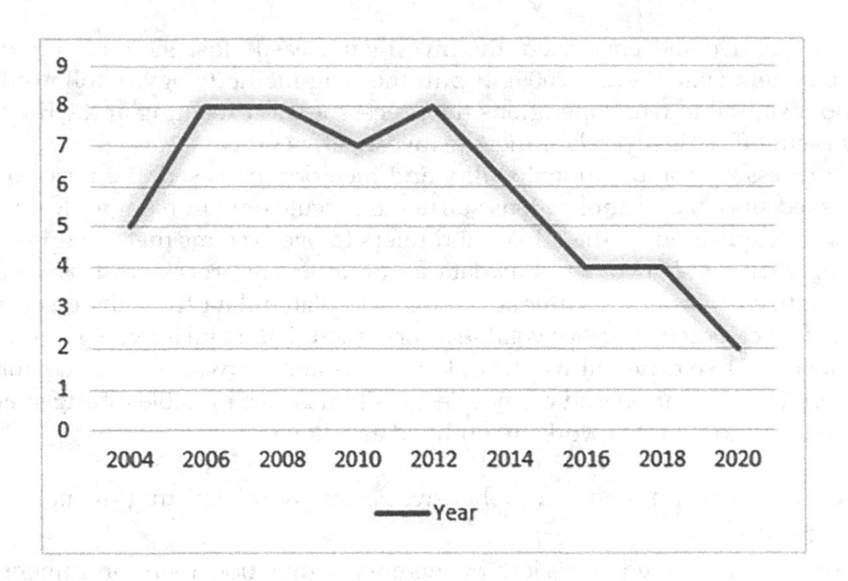

FIGURE 3.1
Year-wise contribution in research publications.

In this section, we analyze and review different research works that are done in the area of volatile memory acquisition for live systems. The methodologies, techniques, research works, and anything other than memory acquisition-based are out of the scope of this discussion. In particular, we have excluded analysis-based discussion from this work.

Brian D. Carrier and Grand (2004) presented a procedure for acquiring volatile memory using a hardware expansion card that can copy memory to an external storage device (Carrier & Grand, 2004). With the flow of passing time, Acquisition of Random-Access Memory (RAM) was described into two types, i.e., hardware acquisition and software acquisition as mentioned in the work of Vidas (2007). Vömel and Freiling (2011) in their work proposed various main memory acquisition approaches on the basis of two factors, atomicity and availability, and also provided analysis techniques for the Windows Operating system family. Classification of memory accessibility levels on the basis of pre- or postincident deployment and terminating vs nonterminating acquisition was provided in their work by Latzo et al. (2019). Through the years, many research works have been done in this field. Table 3.1 provides the review of these works with a focus on the purpose of work and the implementation, if any. Table 3.2 provides a review of the advantages and limitations of these works.

3.3 Technical Background

In this section, we provide the necessary technical background required for better understanding of rest of the work. Some generic concepts such as DMA, IEEE 1394, SMM, Virtualization support are described in this section. We have used Intel's x86-64 architecture to demonstrate these principles as it is widely deployed and well documented.

TABLE 3.1

Review of Existing Work on Memory Acquisition

Author's Name and Year	Purpose of the Work Proposed and Implementation (If Any)
Carrier and Grand (2004)	Procedure for obtaining volatile memory with a hardware expansion card, as well as general memory acquisition tool specifications. Tribble: The Proof-of-Concept device was designed to ensure that the protocol was successful
Vidas (2007)	The advantages and disadvantages of conventional incident management approaches versus an augmented model were discussed, as well as a basis for evaluating the captured memory. On systems with 512MB of RAM and an IBM ThinkPad R51, tests were performed.
Vömel & Freiling (2011)	An overview of the prevailing techniques and methods to collect and analyze a computer's memory. No specific implementation is provided.
Balogh & Mydlo (2013)	Proposed a new method for activating DMA, as well as a new approach for memory acquisition via DMA solution based on a custom NDIS protocol driver. The NDIS driver and Network Interface Card were used to implement a method that permitted the memory content to be dumped.
Stüttgen et al. (2015)	Variety of firmware-based manipulation techniques and methods are presented for identifying firmware-level threats. Two plug-ins for the Volatility and Rekall forensic frameworks were created for inspection of ACPI environment.
McDown et al. (2016)	Evaluated widely used seven shareware or freeware/open-source RAM acquisition forensic tools that are compatible to work with the latest 64-bit Windows Operating systems. Internet connection was not present while conducting the experiments on 2 and 4GB of physical memory.
Parekh & Jani (2018)	Analyzing the memory that stores relevant data, collection of evidences, extraction of crucial information using various memory forensic tools. No practical implementation was done. The tools are defined and compared theoretically.
Sokolov et al. (2018)	Proposed a modern instrumental approach for retrieving the contents of a volatile memory, as well as general implementation guidelines for retrieving data from memory. It just offers a theoretical framework for investigating security incidents, with no practical implementation.
Latzo et al. (2019)	A taxonomy of acquisition methods that generalizes the principle of ring-based privilege separation and is based on a well-defined partial order. It provides the techniques for memory acquisition and the process to achieve this.

3.3.1 Direct Memory Access

Traditionally, the processor was entirely used to initiate, execute, and terminate the memory access operations. This led to overloading as well as unavailability of the processor during the entire duration when memory transfer was being conducted, thus severely limiting the processor's ability to execute other workloads. As a solution to this issue, Direct Memory Access (DMA) was introduced. It's a method of transferring data between peripheral devices and the system's main memory without involving CPU. Since the CPU is only used to initiate the transfer and not to wait for it to finish, it is not used for the entire duration of read and write operations. The DMA controller sends an interrupt to the CPU to signal the end of the transfer. DMA requires direct access to the physical memory. Carrier and Grand (2004) proposed "Tribble," a Proof-of-Concept approach that creates a copy of physical memory using DMA operations (Vömel & Freiling, 2011).

3.3.2 IEEE 1394

IEEE 1394 (Witherden, 2010) also known as Firewire is a hardware bus potentially used for accessing physical memory. It is a serial expansion bus that allows one device on the bus to directly read/write from the physical memory to another. Most operating systems

TABLE 3.2

Advantage & Limitation of the Existing Work

Author's Name and Year	Advantage and Limitation of the Work
Carrier and Grand (2004)	Provides general process of volatile memory acquisition and more reliable hardware-based evidence than software-based solution. Prior to an incident, the hardware-based acquisition system must be installed.
Vidas (2007)	Techniques for RAM acquisition and Time Sliding Window are provided along with an overview of RAM analysis. Operating System level malicious activity detection can be improved; current tools only support specific runtime environments.
Vömel & Freiling (2011)	Classification of volatile memory acquisition is provided on the basis of atomicity and availability. The work is oriented toward traditional approaches without providing the path for future of acquisition
Balogh & Mydlo (2013)	Study and implementation of direct memory access that satisfies the requirements for rootkit detection and for forensic analysis. The installation of the modified driver can change the contents of the memory and could not be used if a sophisticated rootkit is already installed.
Stüttgen et al. (2015)	Rootkits and other malicious applications ability to exploit x86 systems at the firmware level is explored. Software-based memory forensic methods are unable to identify certain firmware rootkits.
McDown et al. (2016)	Memory acquisition software usage, user interface, processing time, loaded DLLs, and Windows registry keys are retrieved. The implementation is limited to 2 and 4GB of physical memory, while existing industry standards call for 16–32GB
Parekh & Jani (2018)	Comparison between different tools and techniques for memory acquisition is provided. Different tools have been compared but only theoretically and no practical implementation is done.
Sokolov et al. (2018)	Theoretical foundation for examining information security incidents, as well as a novel approach for accessing a hardware copy of the contents of a virtual machine. De-energizing the target computer results in the complete or partial destruction of information.
Latzo et al. (2019)	Classification on the basis of three factors: privilege level, pre- or postincident, and potential termination. On embedded systems, memory acquisition is still a hassle.

automatically configure 1394 devices upon insertion, and this can be done without requiring administrator privileges on the system. Even if the system is locked, it will function. The problem is that an experienced intruder with administrator access can easily change permissions.

3.3.3 System Management Mode

System Management Mode (SMM) is a mode in which system architectures contain specific operating modes that are entirely independent of all other aspects of the system's operation. Low-level management functions such as power management and legal interface emulation are handled by SMM. The unmaskable System Management Interrupt (SMI) initiates SMM. Following the trigger, the CPU is transferred to another address space known as SMRAM, where it stores its current context. A corresponding SMI handler is then executed as a result of this operation. SMI handlers are installed to SMRAM until and unless

D_ LAC bit in the System Management Control Register is unlocked (Latzo et al., 2019). Prior to a system's boot operation, BIOS performs this task. Once the installation is complete, the BIOS can lock the SMRAM, preventing any more modifications.

In SMM, only 4GB of RAM can be accessed because the address mode is equivalent to 16-bit real mode. The SMRAM can be physically located in system memory or a separate RAM. The saved context can be restored and resumed from SMRAM using the RSM instruction. The SMM's execution is fully transparent to the Operating System, which can detect its execution through time discrepancies. Interrupts are completely disabled in SMM.

3.3.4 x86 Architecture

The Intel IA-32 architecture, which supports 32-bit computing, is referred to as x86 architecture. It defines Intel's 32-bit processors instruction set and programming environment. Byte addressing is used in this architecture, and software running on this processor may have a linear address space of up to 4GB and a physical address space of up to 4GB. Virtual memory, paging, privilege levels, and segmentation are all supported by the IA-32's operational protected mode.

Fast memory registers are used by the CPU for temporary storage during processing in the IA-32 architecture. For logical and arithmetic operations, each processor core has eight 32-bit general-purpose registers. The EIP register, also referred to as the program counter, stores the linear address of the next instruction to be executed. This architecture also has five control registers that define the processor's configuration as well as the characteristics of the task being executed.

Segmentation and paging are two memory management mechanisms implemented in the IA-32 architecture. The 32-bit linear address space is divided into several variable-length segments using segmentation. Memory references are addressed using 16-bit segment selector that identifies memory-resident data structure known as a segment descriptor, and a 32-bit offset into the specified segment. The ability to virtualize the linear address space is known as paging. Every 32-bit linear address space is divided into fixed-length parts known as pages, which can then be mapped into physical memory in any order. This design also supports pages up to 4MB in size, with only a page directory needed for translation.

The Intel x86-64 architecture was introduced, which was identical to the IA-32 architecture but with 64-bit registers. It has a linear address space of up to 264 bytes and can accept 64-bit linear addresses. Only 48-bit linear addresses are used in the current version, which does not accommodate the full 64 bits. On these systems, virtual addresses are in canonical format, which means that 63:48 can be set to all 1s or all 0s depending on the status of bit 47. It also supports an additional level of paging structures called page map level 4 (PML4).

3.3.5 Virtualization Support

A virtual machine is a technology that simulates a computer environment, allowing an operating system and other programs to run inside of it. Since the operating system and applications do not often realize that they are within an emulated virtual environment when virtualization support is used, they can continue to run normally. If a virtual system is corrupted, suspending the virtual machine will preserve the memory and disk contents. Some virtual machines save the contents of the disk and memory in a raw file, while others save them in a proprietary format. The machine state is saved using trusted software in this case.

3.4 Review of Memory Acquisition Techniques

In this section, we provide the various techniques used for acquiring volatile memory. Primarily, these can be categorized into hardware-based, software-based, and firmware-based techniques.

3.4.1 Hardware-Based Memory Acquisition

Hardware-based memory acquisition techniques refer to the techniques that can directly access the system's main memory for acquiring the memory via hardware support. Due to the fact that it directly accesses the memory, it is regarded as more secure and reliable. It is capable of accessing memory without the use of operating system interfaces. These techniques do not use system memory while they are executing, thereby reducing the footprint and tampering. Currently, there are two hardware-based techniques, i.e., dedicated hardware and special hardware bus. Another means to acquire memory via hardware support is via cold booting. We cover it in this section since it does not fall under the software category.

3.4.1.1 Dedicated Hardware

Using a special hardware card was one of the first techniques for acquiring a forensic representation of a system's RAM storage. Carrier and Grand (2004) introduced the "Tribble" Proof-of-Concept, which uses DMA operations to construct a replica of physical memory. The PCI card is mounted as a dedicated hardware unit that can save volatile data to a storage medium connected to it. When an external switch is pressed, the card is triggered, and the imaging process begins. The contents of RAM are not altered by using this technique to create a copy.

This approach has a few disadvantages, such as the need to configure a PCI card prior to use and the fact that hardware cards can be reprogrammed to establish a different view of physical memory. Advanced attackers can prevent PCI cards from accessing the entire physical RAM.

3.4.1.2 Special Hardware Bus

As an alternative to PCI cards, another technique was proposed using the special hardware bus support available in the chipset. It includes reading of volatile memory via the IEEE 1394 (Firewire) bus (Witherden, 2010). Any hardware bus can potentially be used for physical memory access. In this technique, the interface is present by default in the majority of systems.

The use of this technique can cause random system crashes or malfunction. Also, this technique requires specific hardware requirements. This technique cause inconsistency after comparing images, hence it is not reliable to obtain a precise copy of a system's RAM.

3.4.1.3 Cold Booting

As volatile information is not immediately erased after powering off a machine, instead it gradually fades away. The information can be recovered for a nonnegligible amount of time as data stays present in the processor for some duration post power cut-off. This presence of data for a negligible period (nearly 10 seconds) is called data remanence (Vasisht &

Lee, 2008). It is achieved by artificially cooling down the RAM modules using liquid nitrogen to increase the data remanence from 10 to nearly 30 minutes. To access the retained memory, the target machine can be restarted, i.e., cold booted with a custom kernel.

3.4.1.4 Limitations of Hardware-Based Techniques

Hardware-based acquisition is relatively secure and reliable. However, it still faces some grave limitations. The first limitation of using a hardware-based acquisition system is that it must be set up before an incident occurs. It can seem costly and inefficient to mount a PCI card into several systems before an incident occurs. Second, since memory is continuously changing, verifying that the image is an exact copy and that data was not overwritten to target memory during acquisition would be difficult. By and large, hardware-based memory acquisition techniques seem infeasible for a majority of standard environments due to negligible attention being given to incident response with respect to cybercrimes.

3.4.2 Software-Based Memory Acquisition

Software-based memory acquisition techniques refer to the techniques that depend upon the functions and interfaces provided by the operating system and its associated privilege levels. These techniques are entirely dependent on the operating system kernel, which schedules processor access and manages all data flow to the storage locations. It is divided into three parts, i.e., User-level, Kernel-level, and Virtualization-based techniques, which are further categorized as follows.

3.4.2.1 User Level

User level is the least privileged level in most operating environments. Privilege separation is achieved by operating system by using different rings and the user operates in the least privileged ring. The user, however, retains complete control over the applications and targets that are operating on the same privilege level. Using the user-level privileges, memory dump can be acquired by using different methods such as Software via software emulators, via data dumper (dd) utility, via PMDump, via Processor Dumper utility (PD). All of these methods are outlined below (Figure 3.2).

User level is the least privileged level in most operating environments. Privilege separation is achieved by the operating system by using different rings and the user operates in the least privileged ring. The user, however, retains complete control over the applications and targets that are operating on the same privilege level. Using the user-level privileges, memory dump can be acquired by using different methods such as Software via software emulators, via dd utility, via PMDump, and via PD. All of these methods are outlined below.

3.4.2.1.1 Software Emulators

Software emulators can be used to acquire memory via user-level privileges. One of the emulators is Bochs that is an open-source x86-64 emulator (Hilgers et al., 2014). No special kernel driver is required for Bochs. An emulator attempts to replicate the original machine as closely as possible. It fetches, decodes, and executes instructions one by one, mapping them from source to target and vice-e-versa. The emulator, which is implemented as some kind of array, provides guest memory, which enables Bochs to generate a dump efficiently. It has a built-in debugger that can read memory from both virtual and physical addresses. Because of the significant performance overhead associated with instruction-wise interpretation, typical

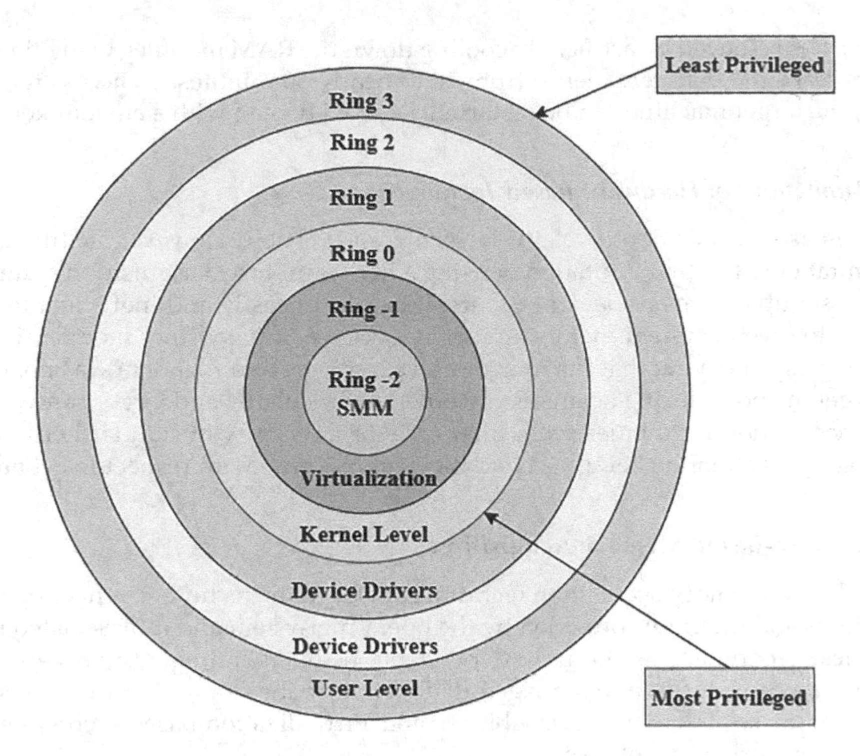

FIGURE 3.2
Privilege rings.

programs are not recommended for execution in an emulated environment. Emulators are also mostly used in software creation and debugging.

3.4.2.1.2 Data Dumper

Garner (2009) proposed Data Dumper, also known as dd, which is part of the Forensic Acquisition Utilities (FAU) suite. It employs the \\.*Device**PhysicalMemory* section object to construct a full memory dump of the target machine when run with administrative privileges. Access to this object at the user level has been constrained for security reasons. This approach has very little value as compared to modern computing platforms.

3.4.2.1.3 PMDump

PMDump, or Process Memory Dump, is a method that enables you to save the address space of a single process and dump the memory contents of that process into a file. It dumps process memory with its header information from */proc//maps* file. Data is dumped either to the file or through the network to a designated storage location. This technique is a closed source in nature that makes the development of additional parsing tools more difficult. It also has an effect on the target host's level of contamination.

3.4.2.1.4 Processor Dumper Utility

PD was proposed by Klein (2006); Vömel and Freiling (2011). It obtains the environment and state of a process by retrieving an executable's data and code mappings. By default, all collected data is redirected to standard performance. This approach makes it more difficult to create additional parsing methods because it uses a proprietary data format. It has an impact on the level of contamination on the target host, and it also requires the specification of a process ID, which involves the use of a process listing utility in the first place.

3.4.2.2 Kernel Level

Kernel in the system is responsible for maintaining volatile memory and its associated operations along with managing and segregating privileges of different levels. Kernel-based interfaces and utilities can be used for memory acquisition purposes. To monitor the debuggee and access its memory, software debuggers use kernel functionality. It has many different alternatives which can be used to take a memory dump such as: kernel drivers, hibernation files, software crash dumps, and software debuggers.

3.4.2.2.1 Kernel Drivers

Kernel drivers are commonly used to implement memory acquisition tools that run at the kernel level. Kernel drivers are more reliable and easier to deploy. Even after a potential incident has occurred, investigators can initiate the acquisition process. *Pmem* is one of the physical memory dumping techniques used with kernel-level access. Physical memory can be acquired by either requesting kernel support or manually mapping physical frames on its own. It can acquire the physical memory of both user processes and the kernel.

3.4.2.2.2 Hibernation Files

Hibernation is a built-in feature that is being adopted by most of the recent operating system kernels. The hibernation file is known as *hiberfil.sys* which is located in the root directory. It contains runtime-specific information for running processes in the volatile memory. The system state is frozen and a snapshot of the current working set is retained when the operating system is about to reach an energy-saving sleep mode and is suspended to disk. The *hiberfil.sys* file allows investigators to capture a consistent image of volatile data.

3.4.2.2.2.1 Software Crash Dumps When a system crashes, modern operating systems have the ability to save crash-related data in the form of a crash dump. The dump may be confined to the kernel's address space, an application's address space, or the entire physical address space. The system state is frozen in the event of a critical failure, and the main memory and CPU details are saved in the system's root directory. The user can use the built-in *CrashOnCtrlScroll* feature or a third-party utility like *NotMyFault.exe* to force a crash. The contents of memory will be preserved to a file if the machine is programmed for a complete crash dump.

3.4.2.2.3 Software Debuggers

A kernel's debugging support is used by most software debuggers. Debugging system calls is included in most modern operating systems. The debugger can fully access the debuggee's memory. The debuggee is not terminated by a debugger, but it may be interrupted for a random period. Since windows is proprietary in nature, a dedicated debugger is provided by Windows itself i.e., *Windbg*. *Windbg* can be used to understand the memory dump and map it as process and user level.

3.4.2.2.4 Operating System Injection

Schatz (2007) developed Body-Snatcher, a Proof-of-Concept application that injects an independent operating system into a target machine's kernel (Schatz, 2007). It is accomplished by freezing the host computer's state and relying on acquisition operating system functions to establish a stable snapshot of the volatile data. Because of its low-level approach and high complexity, the prototype is highly platform-specific.

3.4.2.3 Virtualization Based

Virtualization is a process of emulating complete, isolated, and reliable system environments known as virtual machines on top of a host computer. Virtual machine monitor (VMM) is a special software layer that is responsible for sharing, managing, and restricting access to the available hardware resources. A virtual machine has the ability to be suspended, which means it can interrupt its execution process. In the case of VMware-based machine, all volatile data is saved to a*.vmem* file located in the virtual machine's working directory (Shaw et al., 2014). The use of virtualization makes memory acquisition procedure comparatively easier.

3.4.2.4 Limitations of Software-Based Techniques

The software-based acquisition procedures rely on the underlying operating environment which may be untrustworthy in nature. Since it manages the scheduling of processor access and also maintains control over all data flow to storage locations, there is no way to stop using the base kernel for a software solution. Another disadvantage of using a software solution is that it will still require process and kernel memory to function. This poses the challenge of overwriting data in the memory that may serve as potential evidence in itself.

3.4.3 Firmware-Based Memory Acquisition

Firmware-based memory acquisition combines a commercial PCI network card with the SMM to remotely acquire the contents of memory, CPU registers, and peripheral state. SMM is primarily designed for firmware or basic input–output system (BIOS). A driver is required for a commercial PCI network card to acquire the system's physical memory. So, drivers for PCI network cards are placed in the SMM because the code in the SMM is trusted and locked. As a result, the malware does not pose a modification threat to the driver.

After acquiring the system's physical memory, the system converts it to virtual memory, which is used by the operating system, which determines the memory's semantics. The CPU registers are obtained using a firmware (BIOS)-based method that reads the registers using SMM code from the BIOS. It retains the CPU registers used by the OS in protected mode because the hardware saves them automatically before switching to SMM.

Another technique is used only when one wants to check a few variables that do not require the complete memory dump. The system is stopped, memory locations or particular registers are checked, and the system is then resumed. SMM is used to bring the machine to a halt and retrieve the contents of the memory (Wang et al., 2011). One disadvantage is that the debugger cannot freeze the target OS in SMM mode for an extended period. Another issue is that performing this technique efficiently on machines with a large amount of memory is difficult.

3.5 Memory Dump Formats

Memory dump formats are the file formats in which the memory dump acquired from the system is saved when the system crashes or memory dump of the system is required for analysis. It is the bridging factor between acquisition and analysis of the main memory. They can have binary, octal, or hexadecimal data in them. It is of great significance as the investigators analyze the memory dump file to identify the cause of the crash or abnormal termination of the system while using different analysis techniques. Memory dump can be used to figure out what was going on in the machine when it crashed and one can find the appropriate solutions to fix the problem. Memory dumps are useful for diagnosing device bugs and examining memory contents in the event of a program failure.

3.5.1 Raw Memory Dump

It is the most frequently used and widely supported memory dump format by modern analysis tools. Memory acquisition tools like Magnet RAM Capture and DumpIT use this file format to acquire the memory. There are no headers, metadata, or magic values in these raw memory dumps. Any memory ranges that were deliberately missed or that could not be read by the acquisition tool are padded in the raw format, which aids in preserving spatial integrity. RAW memory dump can be converted into Windows crash dump using *raw2dmp* plugin.

3.5.2 Windows Crash Dump

When Windows crashes, it frequently generates a memory dump file. This file contains significant details that will help you determine why your system stopped working. The DMP_HEADER or DMP_HEADER64 structure appears first in these crash dumps. In order to create a memory dump in Windows, the device needs a paging file on the boot volume that is at least 2 MB. A memory dump will not operate even if there is a big paging file on another disk but none on the boot volume. When the system is running Microsoft Windows 2000 or later, every time the system abruptly stops, a new file is generated. dumpchk.exe is a Microsoft application that checks memory dump files for data. Memory dumps are available in three different formats in Windows crash dump, as follows:

3.5.2.1 Complete Memory Dump

The largest memory dump file is the complete memory dump, which comprises everything that was in physical memory. When the machine abruptly stops working, it preserves all of the contents of system memory. If the Complete memory dump option is chosen, the boot volume must have a paging file large enough to contain all of the physical RAM plus 1MB. It doesn't have anything to do with the platform firmware's use of physical memory. The previous file is overwritten only if any other problem occurs during acquisition or if any other complete memory dump (or kernel memory dump) file exists.

3.5.2.2 Kernel Memory Dump

Kernel memory dump records all the contents that were used by the kernel at the time of capturing the memory. Since it only includes the content used by the kernel, this memory dump is considerably smaller than the full memory dump. When your machine suddenly stops functioning, it speeds up the process of recording information in a log. To accommodate kernel memory, the page file must be large enough. Kernel memory for 32-bit systems is usually between 150MB and 2GB. Unallocated memory and memory allocated to user-mode programs are not included in these memory dumps. It omits only those parts of memory that aren't likely to have played a significant role in the issue. The previous file is overwritten only if the *"Overwrite any existing file"* option is selected, or if any other issue arises, or if another kernel memory dump file (or a complete memory dump file) exists.

3.5.2.3 Small Memory Dump

In Windows crash dump, this is the smallest memory dump file that can be generated. It saves the tiniest amount of details that can be used to figure out why the machine suddenly shuts down. This choice specifies that Windows 2000 and later build a new paging file any time the device stops unexpectedly and needs a paging file of at least 2MB on the boot volume. These files' record is kept in a separate folder. When space is restricted, this type of dump file can be useful. If some other problem arises or a second small memory dump file is made, the previous file is protected (unlike other formats). The following information is normally included in a small memory dump file:

- The error or stop message and its parameters.
- The processor context (PRCB) for the processes that caused Windows to stop working normally.
- The process information and kernel context (EPROCESS) for the processes that have been stopped.
- The process information and kernel context (ETHREAD) for the thread that stopped.
- The kernel-mode call stack for the thread hat halted the execution of the operation.
- A list of loaded drivers.

3.5.3 Windows Hibernation File

Hibernation is the method of turning a machine off while maintaining it in its current state. The contents of RAM are saved to the hard disk or other nonvolatile storage, and when the process is resumed, the machine is in the same state as before hibernation. When the machine enters hibernation mode, a hibernated file is generated in the root system folder, containing the contents of a full memory dump. The hibernation file on Windows systems is called *hiberfil.sys* and is located in the root directory.

3.5.4 Expert Witness Format

Encase Forensics uses the Expert Witness Format, or EWF, to acquire a memory for EnCase-based software. EnCase has become one of the standardized file formats as a result of its popularity in forensic investigation. For analyzing these memory files, there are only a few resources available. Investigators should be experienced with the *EWFAddressSpace, Mounting with EnCase, and Mounting with FTK Imager* methods for converting EWF memory dumps.

3.5.5 AFF4 Format

The Advanced Forensics File Format 4 (AFF4) is an open-source format for digital evidence and data storage. Multiple image sources, arbitrary data, and storage virtualization are all integrated into the standard forensic file format. It supports features such as writing and reading zip file style volumes, writing and reading AFF4 image streams, writing and reading RDF metadata, logical file acquisition, etc. AFF4 format is believed to have a huge impact on the future of memory acquisition to create a unified storage format for forensic evidence present worldwide.

3.5.6 HPAK Format

HBGary software cooperation developed the HPAK file format. HPAK allows the physical memory and page file(s) of a target device to be embedded in the same output file. Since this is a proprietary format, the only tool that can create an HPAK file is FastDump. This file format requires the HPAK command-line option; otherwise, the memory dump is created in the default file format, raw, which excludes the page file (s).

3.6 Memory Acquisition Tools

In this section, we outline the memory acquisition alternatives which are available in the market in a comparative manner. The comparison is provided in Table 3.3. We also briefly describe the four leading acquisition tools which are further used for performance comparison in our work.

TABLE 3.3

Comparison of Memory Acquisition Tools

Tools/Features	Magnet RAM Capture	Belkasoft Live RAM Capture	FTK Imager	DumpIT
Release year	2014	2014	2013	2007
Latest version	v1.2.0	v1.0	v4.5.0.3	v1.3.2
Copyright	Magnet Forensics Inc.	Belkasoft Evidence Center	Access Data Group Inc.	Moonsols
Developer	Magnet Forensics	Belkasoft Evidence Center	Access Data	Moonsols
Acquisition time (4GB)	75 seconds	60 seconds	79 seconds	65 seconds
Acquisition time (8GB)	54 seconds	31 seconds	36 seconds	34 seconds
Average memory usage (4GB)	5.7MB	7.6MB	6.8MB	11.5MB
Average memory usage (8GB)	6.8MB	7.8MB	7.0MB	12.3MB
GUI/CLI	GUI	GUI	GUI	CLI
Source	go.magnetforensics.com	belkasoft.com	accessdata.com	qpdownload.com
Portable software	Single portable executable folder	Portable executable folder	Requires installation	Portable executable folder
Supported image format	.raw	.mem	.mem	.raw
Cost effective	Yes	Yes	Yes	Yes
OS supported	Windows Vista, XP, 2003, 2008, 2012, 7, 8, 10	Windows Vista, XP, 2003, 2008, 7, 8, 10	Windows Vista, XP, 2003, 2008, 2012, 7, 8, 10	Windows Vista, XP, 7, 8, 10

3.6.1 Magnet RAM Capture

Magnet RAM Capture is a free imaging tool that enables investigators to capture the physical memory of a target system, allowing them to recover and analyze valuable objects that are often only present in memory. Since Magnet RAM Capture has a small memory footprint, investigators can use it while reducing the amount of data that is overwritten in memory. It exports captured memory data in Raw (DMP) format, which can be easily imported into prominent analysis tools like Internet Evidence Finder. Processes and programs running on the device, network links, proof of malware infiltration, registry hives, username and passwords, decrypted files and keys, and evidence of operation not usually stored on the local hard disk are all examples of evidence found in RAM (*ram-capture-twitter @ go.magnetforensics.com*, n.d.).

3.6.2 Belkasoft Live RAM Capturer

Belkasoft Live RAM Capturer is a small free forensic tool that enables you to retrieve all of the contents of a computer's volatile memory, even if it's protected by an active antidebugging or antidumping scheme. Separate 32-bit and 64-bit builds are available to keep the tool's footprint as small as possible. Live RAM Analysis can be used to analyze memory dumps captured with Belkasoft Live RAM Capturer in Belkasoft Evidence Center (*ram-capturer @ belkasoft.com*, n.d.).

3.6.3 FTK Imager

FTK Imager is a data preview and imaging tool for forensically sound data acquisition by producing copies of data without tampering the original evidence. Following the creation of an image, the Forensic Toolkit (FTK) is used to conduct a comprehensive forensic analysis and generate a report of the findings (*ftkimager3 @ marketing.accessdata.com*, n.d.). FTK Imager makes it simple to create forensic images, display files and directories, preview their contents, mount an image for read-only viewing, recover deleted files, generate hash reports, and more.

3.6.4 DumpIt

DumpIT is a free memory forensics tool created by Matthieu Suiche from MoonSols. This tool is used to create a physical memory dump of Windows system. It is compatible with both 64-bit and 32-bit versions of Windows operating systems. It is a very easy tool that can be used by any user with administrative privileges. Only a confirmation question is prompted before the raw memory dump is created in the current directory. Since it is a very simple-to-use tool, it does not possess any analysis capabilities. Tools such as Mandiant Redline can be used for analysis purpose.

3.7 Implementation

Acquiring the memory dump of a system requires specific tools which can be used to extract and dump the volatile memory onto the physical memory of the system. In this section, we provide the testbed setup details and further compare four leading memory acquisition tools. The four memory acquisition tools for Windows, as described in the previous section, were implemented on two different physical machines with different architecture and RAM.

1. **System 1 Configuration:** The system with Intel(R) Core (TM) i3–6006U CPU @ 2.00GHz x64-based processor with 4GB of RAM running 64-bit Microsoft Windows 10 operating system.
2. **System 2 Configuration:** The system with Intel(R) Core (TM) i5–1035G1 CPU @ 1.00GHz x64-based processor with 8GB of RAM running 64-bit Microsoft Windows 10 operating system.

When these tools were executed in the system, it was assured that the same processes were functioning to have a common standard for analysis of the tool's performance. At the time of acquiring memory, the system was connected to the Internet. We provide the methodology for executing these tools, in order to assist others as follows:

- *Methodology for Magnet RAM Capture*
    1. Visit the site and enter your personal details and click on download.
    2. Download the tool from the link sent in the entered email account.
    3. A single portable executable file will be downloaded in the system.
    4. Users will need to specify two choices prior to start acquisition i.e., path to save the acquired memory data, and whether the segmentation is needed or not.
    5. Run the executable file named "MRCv120" in the system.
    6. Select the segment size if segmentation is to be done. By default, the segmentation is set as off.
    7. Select the location to save the captured memory dump file and click on "Start" button.
    8. It will begin acquiring the memory dump of the system.
- *Methodology for Belkasoft Live RAM Capture*
    1. Visit the site and enter the personal email account details.
    2. Download the tool from the link sent in the provided email account.
    3. A portable executable zip folder will be downloaded in the system.
    4. Extract the files from the zip folder.
    5. Run the executable file named "RamCapture64" in the system.
    6. Choose the folder path to save the output and click on "Capture" button.
    7. It will begin acquiring the memory dump of the system.
- *Methodology for FTK Imager*
    1. Visit the site and enter the personal email account details.
    2. Download the tool from the link sent to the Email account.
    3. An executable file named "AccessData FTK Imager 4.5.0 (x64)" will be downloaded in the system.

4. Run the executable file and install the AccessData FTK Imager.
5. Open the tool and in the "File" menu click on "Capture Memory" option.
6. Choose the destination path and then click on "Capture memory" button.
7. It will begin acquiring the memory dump of the system.

- *Methodology for DumpIT*

1. Visit the site and download the DumpIT tool latest version.
2. A portable executable zip folder will be downloaded into the system.
3. Extract the files from the zip folder.
4. Run the executable file named as DumpIt in the system.
5. It will ask for the confirmation then enter either yes or no according to your choice.
6. If yes, then it will begin acquiring the memory dump of the system.

3.8 Performance Comparison

Performance comparison of the abovementioned tools was conducted and results were analyzed on the basis of following:

1. Processor utilization (average during the entire process of acquisition).
2. Processor utilization (with respect to time, i.e., beginning, middle, and end).
3. Disk utilization (average during the entire process of acquisition).
4. Disk utilization (with respect to time, i.e., beginning, middle, and end).
5. Time taken (average during the entire process of acquisition).
6. Time taken (with respect to different runs of tool).

The abovementioned tools are thoroughly compared in Table 3.3 with respect to certain features. These tools were run on two Windows-based systems with varying configurations. The average processor utilization done by the tools over the acquisition procedure is graphically represented in Figures 3.3 and 3.4. Figures 3.5 and 3.6 represent processor utilization by the tools with respect to time. The average disk utilization done by the tools over the acquisition procedure are graphically represented in Figures 3.7 and 3.8. Figures 3.9 and 3.10 represent disk utilization by the tools with respect to time. The average time taken by the tools over the acquisition procedure is graphically represented in Figures 3.11 and 3.12. Another detailed comparison with respect to acquisition time taken by the tools in different runs is represented in Figures 3.13 and 3.14.

The time was calculated manually using a stopwatch. Belkasoft Live RAM Capture was the fastest for the acquisition of both 4 and 8GB RAM, with DumpIT just a couple of seconds behind. While the tools are being executed to acquire the memory, the tools themselves will also utilize the processor and disk space. When it comes to 4GB RAM, FTKImager

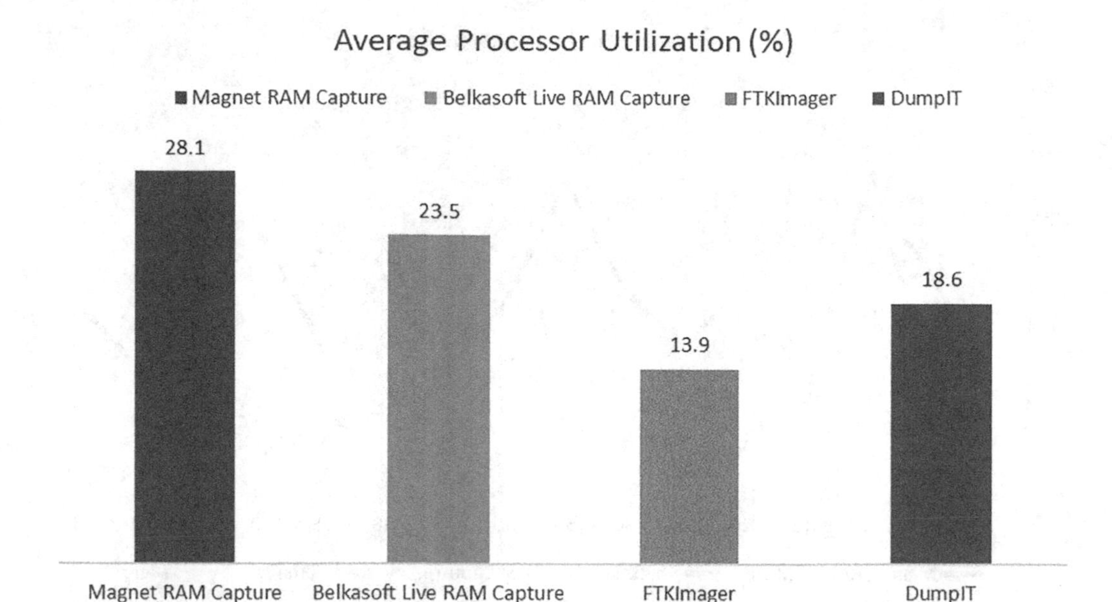

FIGURE 3.3
Average processor utilization during the entire acquisition process (4GB).

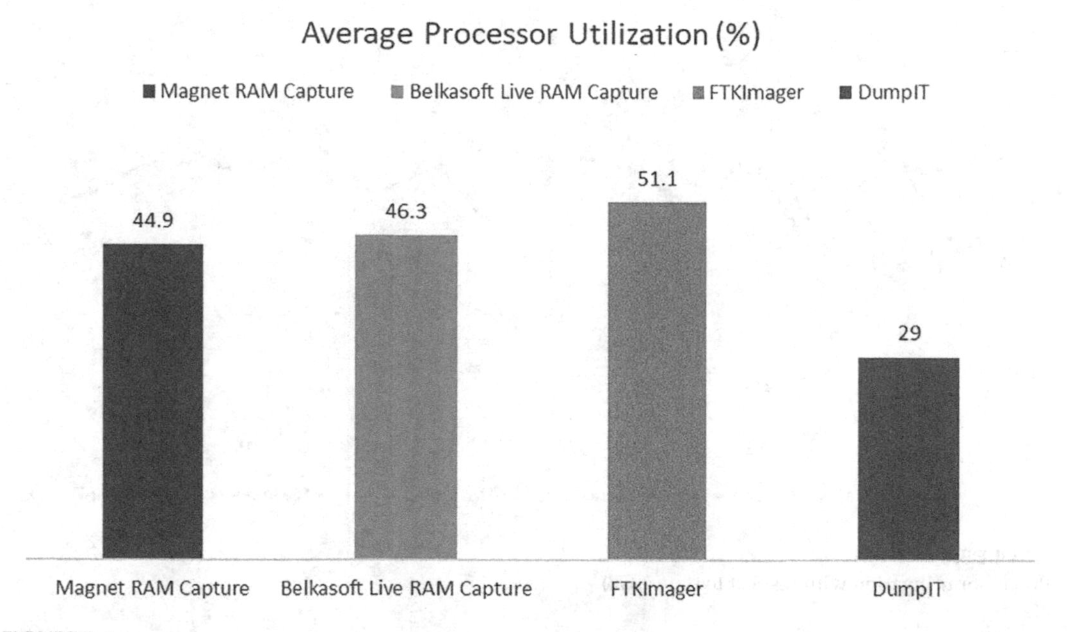

FIGURE 3.4
Average processor utilization during entire acquisition process (8GB).

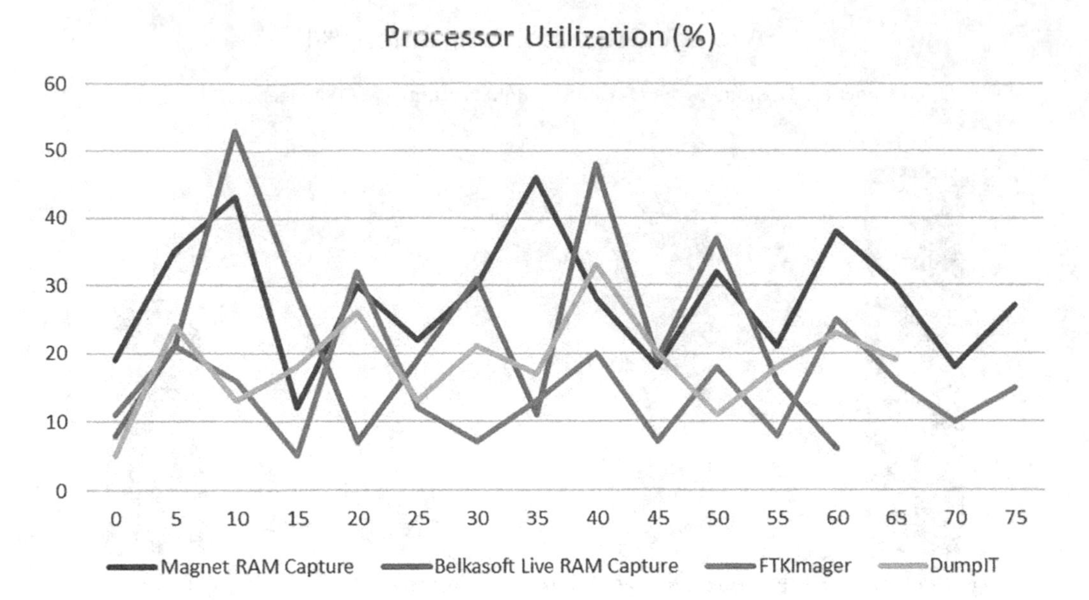

FIGURE 3.5
Processor utilization with respect to time (4GB).

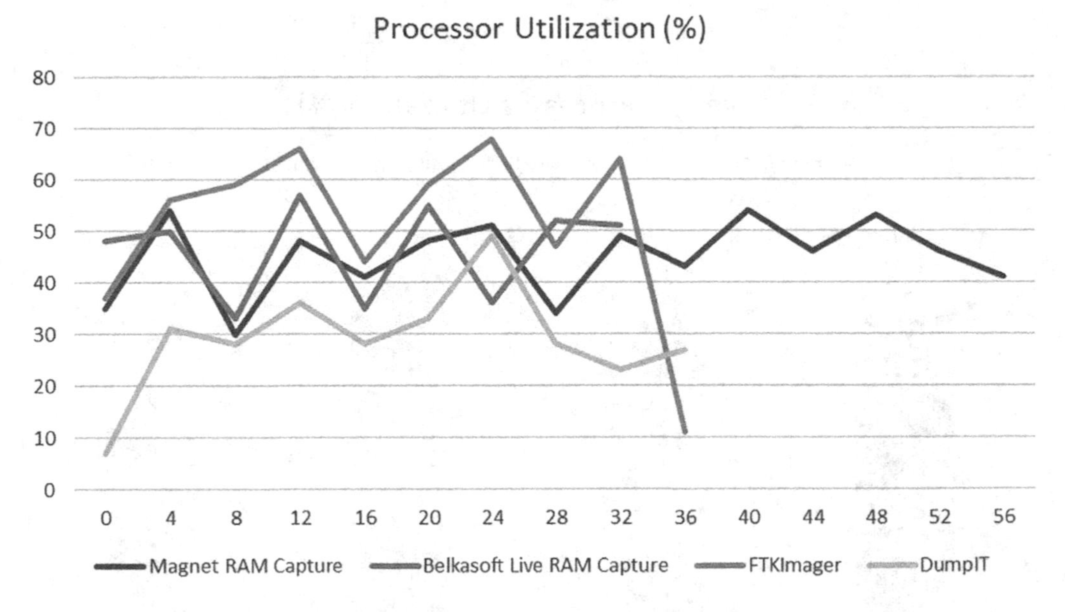

FIGURE 3.6
Processor utilization with respect to time (8GB).

utilizes the least percentage of the processor, whereas Magnet RAM Capture utilizes the highest percentage of the processor. As soon as the RAM size increases, the results were completely changed as now FTKImager utilizes the highest percentage of processor, with DumpIT being the best because it utilizes the least percentage of processor. When

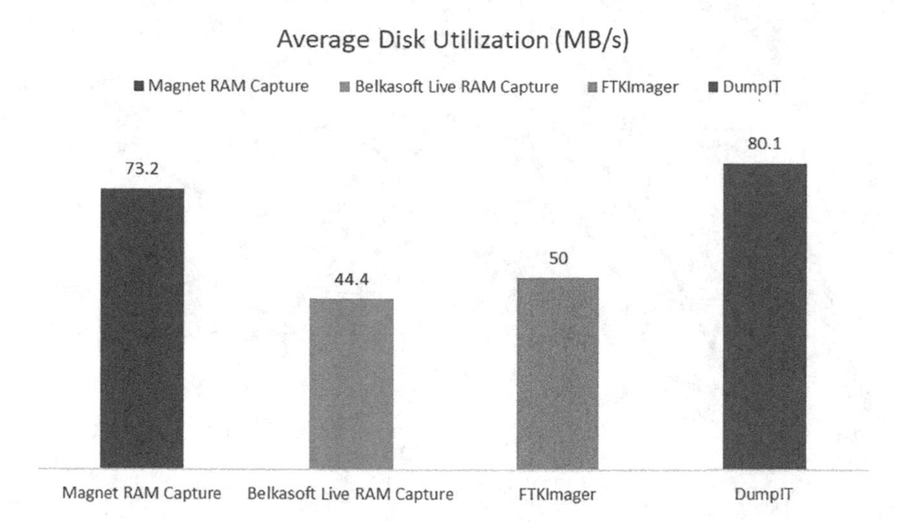

FIGURE 3.7
Average disk utilization during the entire acquisition process (4GB).

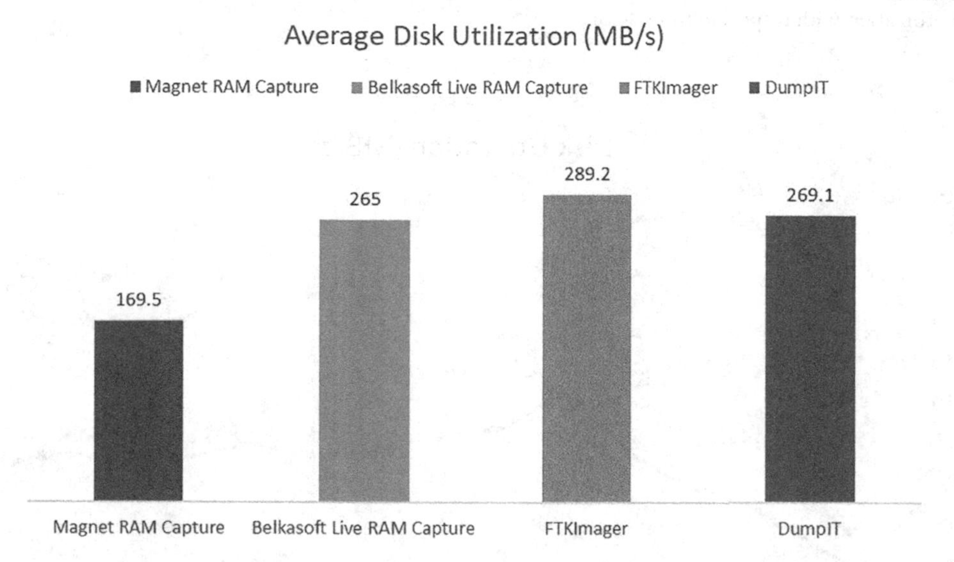

FIGURE 3.8
Average disk utilization during the entire acquisition process (8GB).

compared according to disk utilization in 4GB RAM, Belkasoft Live RAM Capture utilizes the minimal space while DumpIT utilizes the maximum space in the disk. Whereas in 8GB RAM, Magnet RAM Capture utilizes minimal space while the rest of the tools have only the slightest differences in their outcomes in utilizing the space in the disk. The variation in result of processor and disk utilization are not constant as it might increase, decrease, or remain constant. The result depends upon the various processes going on while executing the acquisition tools. The result also depends upon the configuration of windows as varying configurations will result in varying outcomes.

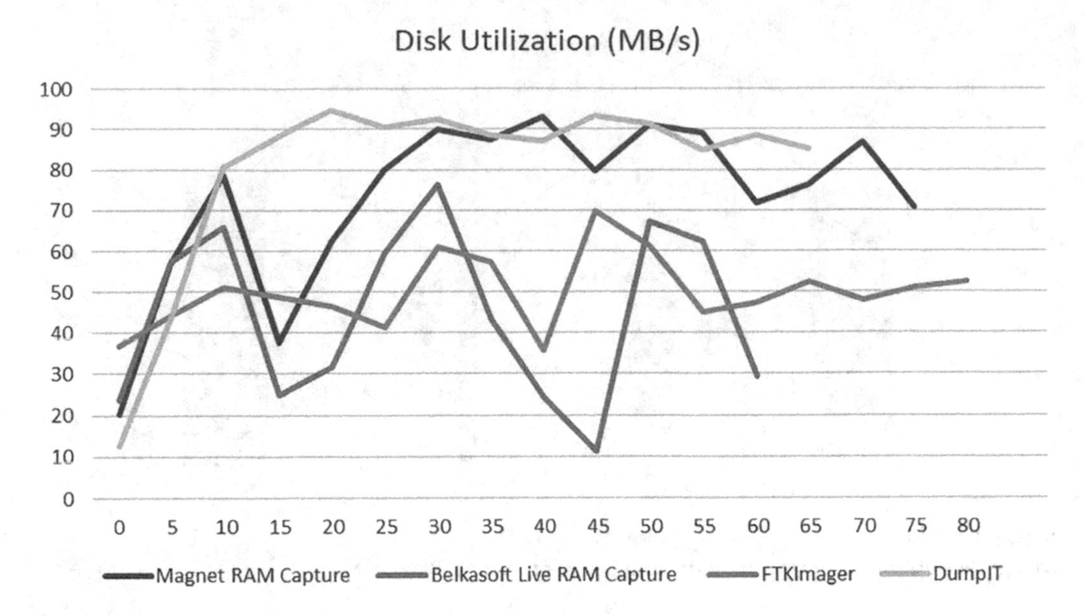

FIGURE 3.9
Disk utilization with respect to time (4GB).

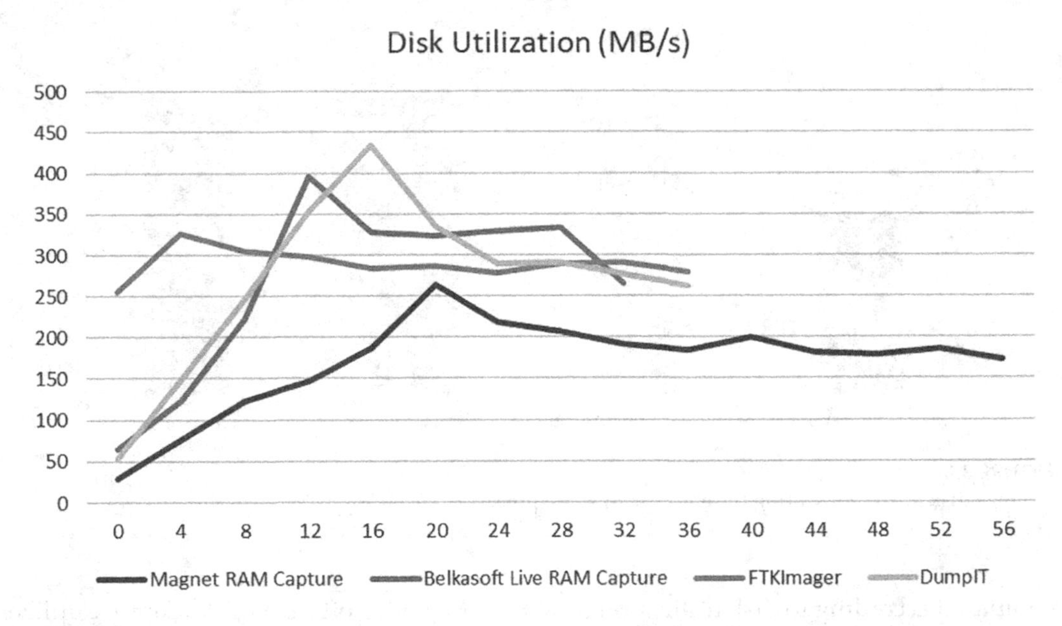

FIGURE 3.10
Disk utilization with respect to time (8GB).

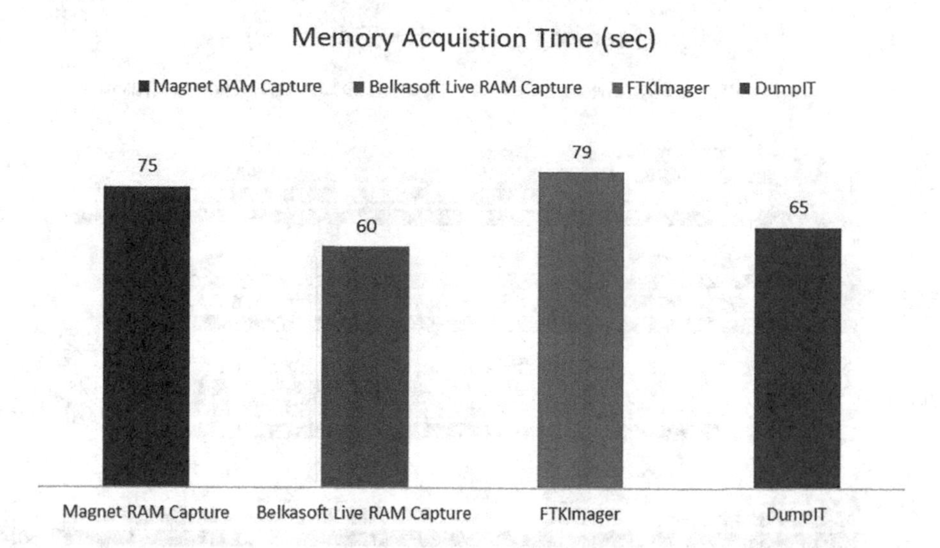

FIGURE 3.11
Average time taken to acquire entire volatile memory (4GB).

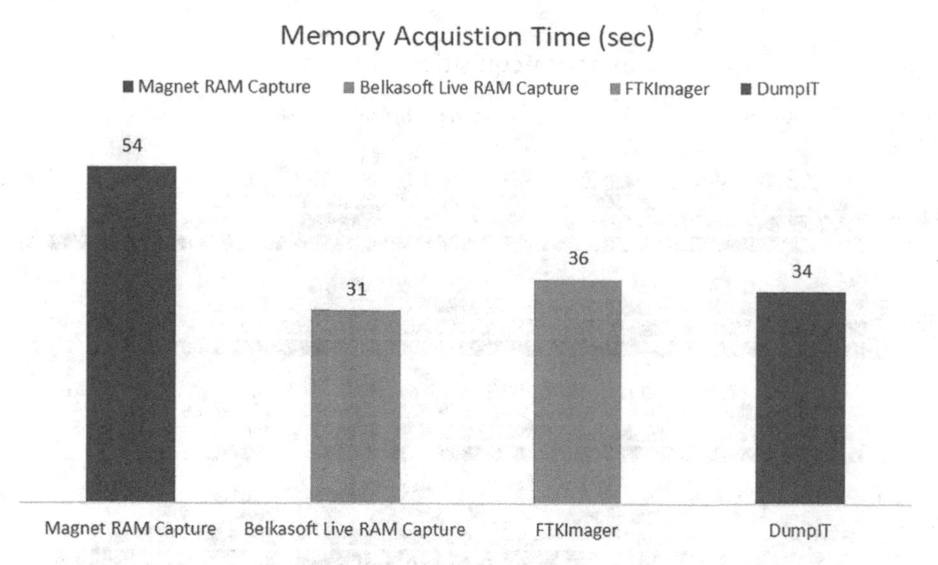

FIGURE 3.12
Average time taken to acquire entire volatile memory (8GB).

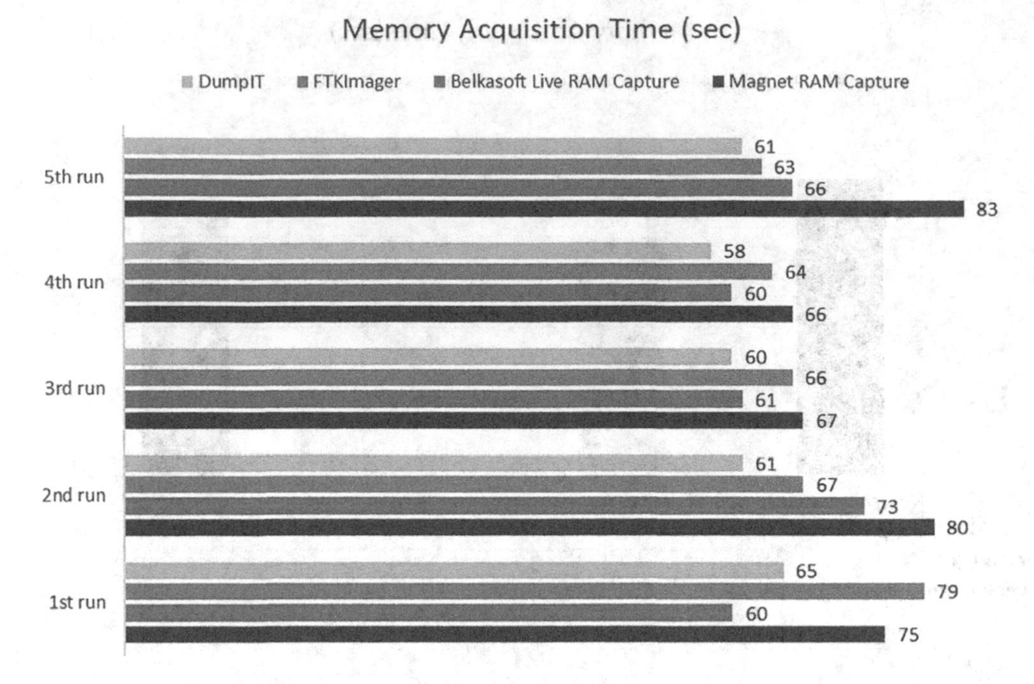

FIGURE 3.13
Time taken by tools in different runs to acquire memory (4GB).

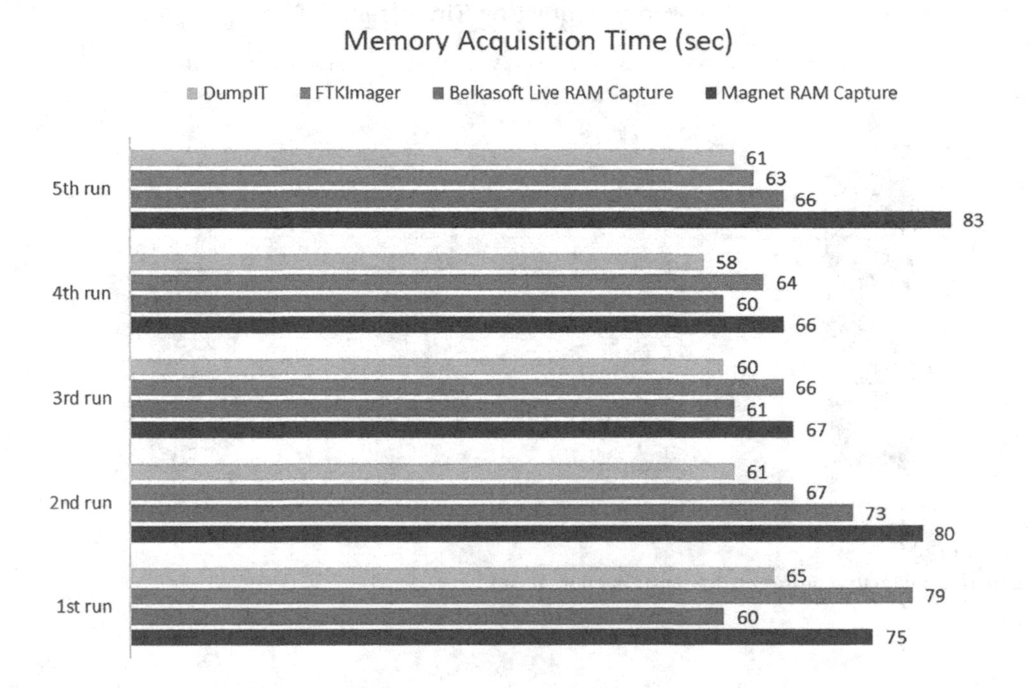

FIGURE 3.14
Time taken by tools in different runs to acquire memory (8GB).

3.9 Conclusion

In this paper, we have reviewed various memory acquisition techniques that include hardware-based, software-based, and firmware-based techniques. Hardware-based techniques include dedicated hardware card, special hardware bus, and cold booting, whereas software-based techniques include user-level, kernel-level, and virtualization-based. Further, we have also defined various memory dump formats that include raw memory dump, windows crash dump, and AFF4 Format. We have selected four memory acquisition tools as they are reliable, easy to use, easily available, and famous in the market. The methodology for each of these tools is outlined in its own section. These four tools were implemented in two Windows systems with different configurations and results obtained were analyzed on the basis of different parameters such as memory acquisition time, processor utilization, and disk utilization. The result depends upon the various factors such as the processes going on while executing the acquisition tools or the windows configurations, as varying configurations will result in varying outcomes. Belkasoft Live RAM Capture appears to be the best in terms of memory acquisition time, but the variation in the result of processor and disk utilization is not constant as it might increase, decrease or remain constant. The outcomes are examined on the basis of various graphs obtained from the data. Hence, various analysis techniques can be used to further analyze the memory dump acquired through these tools, which is out of the scope for this paper.

References

2020_Google_services_outages @ en.wikipedia.org. https://en.wikipedia.org/wiki/2020_Google_services_outages#:~:text=On 14 December 2020%2C another, still accessible with private browsing. The last access/edit date was 25 March, 2022.

Balogh, Š., & Mydlo, M. (2013). New possibilities for memory acquisition by enabling DMA using network card. In *2013 IEEE 7th International Conference on Intelligent Data Acquisition and Advanced Computing Systems (IDAACS)* (Vol. 2, pp. 635–639).

Beebe, N. (2009). Digital forensic research: The good, the bad and the unaddressed. *IFIP Advances in Information and Communication Technology*, *306*, 17–36. https://doi.org/10.1007/978-3-642-04155-6_2

Carrier, B. D., & Grand, J. (2004). A hardware-based memory acquisition procedure for digital investigations. *Digital Investigation*, *1*(1), 50–60. https://doi.org/10.1016/j.diin.2003.12.001

ftkimager3 @ marketing.accessdata.com. (n.d.). https://marketing.accessdata.com/ftkimager3.4.2

Garner, G. M. (2009). Forensic acquisition utilities. http://gmgsystemsinc. com/fau/

Harichandran, V. S., Breitinger, F., Baggili, I., & Marrington, A. (2016). A cyber forensics needs analysis survey: Revisiting the domain's needs a decade later. *Computers and Security*, *57*(March), 1–13. https://doi.org/10.1016/j.cose.2015.10.007

Hilgers, C., Macht, H., Muller, T., & Spreitzenbarth, M. (2014). Post-mortem memory analysis of cold-booted android devices. In *Proceedings -8th International Conference on IT Security Incident Management and IT Forensics, IMF 2014* (pp. 62–75). https://doi.org/10.1109/IMF.2014.8

Klein, T. (2006). Process dumper. http://www.trapkit.de/research/forensic/ pd/index.html

Latzo, T., Palutke, R., & Freiling, F. (2019). A universal taxonomy and survey of forensic memory acquisition techniques. *Digital Investigation*, *28*, 56–69.

McDown, R. J., Varol, C., Carvajal, L., & Chen, L. (2016). In depth analysis of computer memory acquisition software for forensic purposes. *Journal of Forensic Sciences*, *61*, S110–S116.

Palutke, R., Block, F., Reichenberger, P., & Stripeika, D. (2020). Hiding process memory via anti-forensic techniques. *Forensic Science International: Digital Investigation, 33*, 301012. https://doi.org/10.1016/j.fsidi.2020.301012

Parekh, M., & Jani, S. (2018). Memory forensic: Acquisition and analysis of memory and its tools comparison. *International Journal of Engineering Technologies and Management Research, 5*(2), 90–95.

ram-capture-twitter @ go.magnetforensics.com. (n.d.). http://go.magnetforensics.com/ram-capture-twitter

ram-capturer @ belkasoft.com. (n.d.). https://belkasoft.com/ram-capturer

Rathnayaka, C., & Jamdagni, A. (2017). An efficient approach for advanced malware analysis using memory forensic technique. In *Proceedings -16th IEEE International Conference on Trust, Security and Privacy in Computing and Communications, 11th IEEE International Conference on Big Data Science and Engineering and 14th IEEE International Conference on Embedded Software and Systems, Trustcom/BigDataSE/ICESS 2017* (pp. 1145–1150). https://doi.org/10.1109/Trustcom/BigDataSE/ICESS.2017.365

Reust, J., & Friedburg, S. (2006). DFRWS 2005 Workshop Report. Online at http://www.dfrws.org/2005/download. http://www.w1npp.org/ares/RFI/2005FI~1.PDF

Schatz, B. (2007). BodySnatcher: Towards reliable volatile memory acquisition by software. *Digital Investigation, 4*(suppl.), 126–134. https://doi.org/10.1016/j.diin.2007.06.009

Shaw, A. L., Bordbar, B., Saxon, J., Harrison, K., & Dalton, C. I. (2014). Forensic virtual machines: Dynamic defence in the cloud via introspection. In *Proceedings -2014 IEEE International Conference on Cloud Engineering, IC2E 2014* (pp. 303–310). https://doi.org/10.1109/IC2E.2014.59

Sokolov, A. N., Barinov, A. E., Antyasov, I. S., Skurlaev, S. V., Ufimtcev, M. S., & Luzhnov, V. S. (2018). Hardware-based memory acquisition procedure for digital investigations of security incidents in industrial control systems. In *Proceedings -2018 Global Smart Industry Conference, GloSIC 2018* (pp. 1–7). https://doi.org/10.1109/GloSIC.2018.8570109

Stüttgen, J., Vömel, S., & Denzel, M. (2015). Acquisition and analysis of compromised firmware using memory forensics. *Digital Investigation, 12*, S50–S60. https://doi.org/10.1016/j.diin.2015.01.010

Vasisht, V. R., & Lee, H. H. S. (2008). SHARK: Architectural support for autonomic protection against stealth by rootkit exploits. In *Proceedings of the Annual International Symposium on Microarchitecture, MICRO, 2008 PROCEEDINGS* (pp. 106–116). https://doi.org/10.1109/MICRO.2008.4771783

Vidas, T. (2007). The acquisition and analysis of random access memory. *Journal of Digital Forensic Practice, 1*(4), 315–323. https://doi.org/10.1080/15567280701418171

Vömel, S., & Freiling, F. C. (2011). A survey of main memory acquisition and analysis techniques for the windows operating system. *Digital Investigation, 8*(1), 3–22. https://doi.org/10.1016/j.diin.2011.06.002

Wang, J., Zhang, F., Sun, K., & Stavrou, A. (2011). Firmware-assisted memory acquisition and analysis tools for digital forensics. In *2011 Sixth IEEE International Workshop on Systematic Approaches to Digital Forensic Engineering* (pp. 1–5).

Witherden, F. (2010). *Memory Forensics over the IEEE 1394 Interface* (pp. 1–28). https://freddie.witherden.org/pages/ieee-1394-forensics.pdf

4

IoMT and Blockchain-Based Intelligent and Secured System for Smart Healthcare

Hitesh Kumar Sharma, Ravi Tomar, and Preeti
University of Petroleum and Energy Studies

CONTENTS

4.1 Introduction 62
4.2 Application of IoT in Intelligent Systems 62
4.2.1 Smart Homes 62
4.2.2 Smart City 62
4.2.3 Self-Driven Cars 63
4.2.4 IoT in Healthcare 63
4.3 Application of IoT in Healthcare 63
4.3.1 Health Monitoring 63
4.3.2 Remote Sensors 64
4.3.3 Clinical Alert Systems 64
4.4 Advantages of IoMT 64
4.5 Blockchain-Based Security of Intelligent IoMT-Based Smart Healthcare 65
4.6 Application of Blockchain in Intelligent Systems 66
4.6.1 Cash Transfer and Installment Handling 66
4.6.2 Supply Chains Checking 67
4.6.3 Tracking the Supplies 67
4.6.4 Verification of the Team 67
4.6.5 A Focus on Transactions 67
4.7 Advantages of Blockchain in Healthcare 67
4.7.1 Transparency in the Supply Chain 68
4.7.2 Patient-Centric Electronic Health Records 68
4.7.3 IoT Security for Remote Monitoring 69
4.8 Challenges for Blockchain in Healthcare 69
4.8.1 Major Shift 69
4.8.2 Refusal to Share 69
4.8.3 Distributed Healthcare 69
4.9 IoMT with Edge/Fog Computing 69
4.10 Conclusion 72
References 72

DOI: 10.1201/9781003291916-4

4.1 Introduction

IoT market in the medical sector is the fastest growing market and is estimated to be worth nearly $176 billion by 2026 [1]. With such vast applications, IoT is extensively used in healthcare. IoMT is a special IoT-based platform where the patient can meet the doctor, show their report in virtual manner, and take the proper guidance without physically meeting them. Few years ago, diagnosis of diseases in the human body was possible only when the patient physically met the doctor. Sometimes the patient even had to stay in hospital for treatment, which increased the healthcare cost and also stained the health-care facility in rural and remote locations. The technological advancement through these years allowed us for diagnosis of various diseases and health monitoring [2]. In the last decade, there were so many technologies created with the help of which now we can monitor the patient blood pressure, level of glucose, oxygen level, and so on in an easier way, and the best thing is that we do not need to physically meet with doctors to show our reports.

4.2 Application of IoT in Intelligent Systems

4.2.1 Smart Homes

The Internet of things (IoT) is used in homes too to make our homes smart. It can be used in various ways:

- Monitoring the activities.
- Automated lightings.
- Day night sensors.
- Advanced locking systems.

All these uses can be beneficial for everyone. This makes it easy for people by saving time. And due to high technology being used, security concerns are solved. All these uses are carried out after lots of testing to make things reliable [3]. Smart Homes are becoming popular every day. Due to this, the IoT has gained the trust of many people. This technology will keep growing because there's no end to the needs and imagination of humans. Hopefully, this can play a major role in sending people to places other than Earth [4].

4.2.2 Smart City

Every country wants to have many Smart Cities. But why? The answer is simple—because of the ease of everything and time efficiency. A Smart City uses the IoT [5]. Few examples are:

- Solar panels store energy from sun rays and later use it to light up the nights. This method is being adopted by almost every city now.
- Day–night sensors.

- Speed tracking machines.
- Fast tags at toll gates.

There are a few other uses like GPS tracking and internet connectivity that are available everywhere [6]. Many cars use this feature to modernize their products. This makes smart connectivity between everyone. India is also using the IoT to develop. The developments allowed India to form a movement called Digital India [7].

4.2.3 Self-Driven Cars

We have seen Elon Musk use this technology to make an outstanding product, Tesla, where cars can move from one destination to the other on their own. Many other brands have also tried using this technology—Google tried to make its self-driven bicycles and bikes. This can solve many transport issues as not only do they drive on their own, but they also interact with other vehicles to ensure safety of the passenger. This can also overcome human interaction as there wouldn't be any confusion.

4.2.4 IoT in Healthcare

The IoT helps the health industry by its powerful sensory devices and data interactions. Since doctors are mostly busy, self-medication applications are built and improved constantly to ensure the health of an individual. Many sensory devices send real-time information that are then diagnosed online and also provided with the treatments. Advantages of IoT in healthcare:

- Minimal errors.
- Online consultancy which saves time.
- Decreases frequent visits.

Healthcare is using and promoting the IoT to decrease workloads and also ensure the safety of a person.

4.3 Application of IoT in Healthcare

4.3.1 Health Monitoring

Nowadays, wellness groups, pulse-checking sleeves, circulatory strain estimating groups, glucometer, and further developed shrewd medical care gadgets furnish patients with ongoing customized wellbeing status. They remind the patient to be aware of their ordinary calorie utilization, further develop practice schedule, recognize changing circulatory strain pattern, etc. For elderly patients, it is critical to follow their medical issues consistently, and keen medical service gadgets make their way of life simpler by telling them about any conceivable peril ahead of time! It alarms the patient just as to the patient's relatives to caution them about the startling variances in the patient's ailments [8].

4.3.2 Remote Sensors

Remote sensors are used for automation in pathology labs to control the proper environment required for various kinds of lab samples taken from the patients' body. Like proper cooling of blood samples etc. [9].

4.3.3 Clinical Alert Systems

These IoT-based systems are used for alerting the emergency contacts in case of any medical emergency. The person wearing smart health band will send an alert message to his/her emergency contacts saved in his/her mobile phone if any health related mishappen occur.

4.4 Advantages of IoMT

Internet of Medical Things (IoMT) refers to the physical objects or devices which are interconnected through the internet, all collectively taking in and sharing data without requiring human touch or interaction. In the past couple of decades, IoT has spread in every household, connecting our phones, kitchen appliances, and other electrical devices (Figure 4.1).

Similar to our household, IoT also plays a huge part in healthcare. Cases for the same are given below:

- In a hospital, various devices are used to track and monitor patients' heartbeats; their psychological condition, or their movements. All such instruments are connected to each other and provide real-time information to the doctors or nurses.

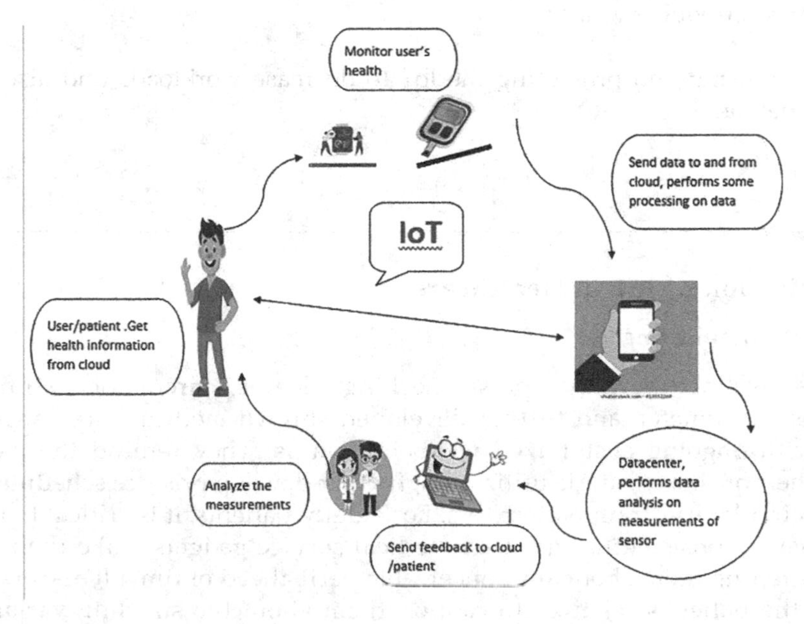

FIGURE 4.1
Data flow in smart healthcare application using IoT.

Such technology helps the staff to keep a check on all the patients at the same time and even reduces hospital stay time [4].

- IoT also has a huge role in our daily lives. Everyday use devices, such as fitness bands, oxymeters, sphygmomanometers, etc., can be tuned/programed to provide daily reports on calories, blood pressure, oxygen levels, and even help in booking appointments. IoT can also help in preventing emergency situations by sending alert signals to close kin and concerned authorities when there is a disturbance in daily patterns or reports of the person.
- Another huge part played by IoT in healthcare is in Healthcare Insurance companies. Such companies make use of large amounts of data and information collected extensively from various healthcare facilities and health-monitoring devices. This data helps them in understanding the most common ailments, injuries, and diseases and coming up with the most helpful and profitable insurance policies for their customers. It also aids in bringing transparency between people and themselves in terms of prices, terms and conditions, etc.

Some examples of IoT-based smart healthcare devices are:

- Remote patient monitoring.
- Glucose monitoring.
- Heart-rate monitoring.
- Depression and mood monitoring.
- Hand hygiene monitoring.

4.5 Blockchain-Based Security of Intelligent IoMT-Based Smart Healthcare

Cyber-attack concerns are very common in the world and nowadays there are serious attacks even in the healthcare sector. The current infrastructure is incapable of ensuring protection against such data breaches, putting patients' health information's privacy and security at jeopardy. The current smart healthcare record models create another door to a problematic situation, namely, patients' data being in the custody of healthcare organizations, putting patients' information at risk, and resulting in inefficient data delivery to patients' healthcare. For instance, if information concerning a patient's health is not communicated from one service provider to the next on time, the patient's treatment may be delayed.

EHR has such limitations practically, which can be overcome by using Blockchain [10]. Blockchain allows data to be collected from a variety of sources and saved in a transaction audit log, which helps maintain track of data accountability and transparency throughout data exchange. As shown in Figure 4.2, Blockchain can provide more secure way of sharing medical data between different users of smart healthcare systems.

Blockchain is an arising innovation stage for creating decentralized applications and information stockpiling, over and past its job as the innovation basis for digital currencies. The fundamental principle of this stage is that it permits one to make a dispersed

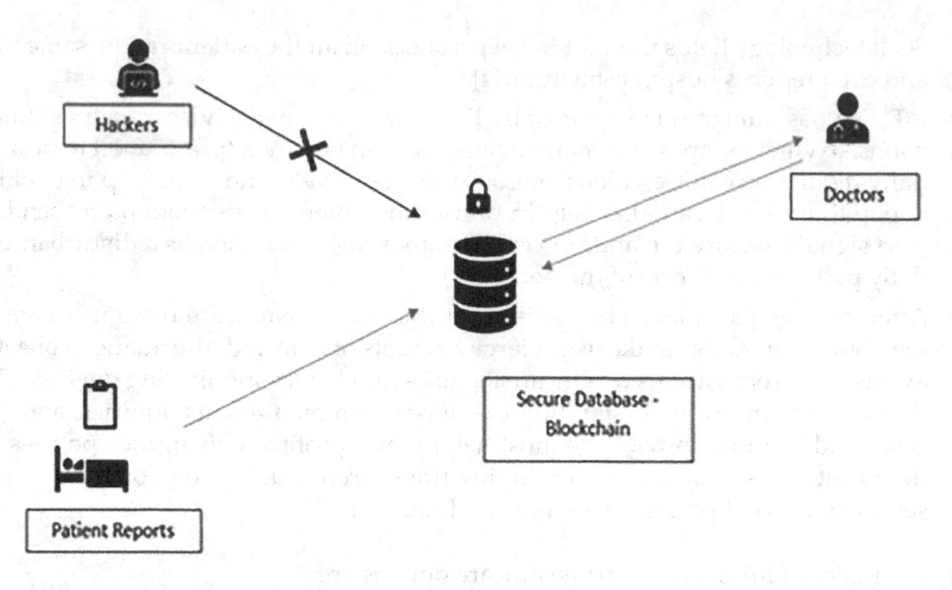

FIGURE 4.2
Role of blockchain in smart healthcare.

and recreated record of occasions, exchanges, and information created through different IT processes with solid cryptographic certifications of alter obstruction, changelessness, and unquestionable status. Public blockchain stages permit us to ensure these properties with overpowering probabilities in any event, when untrusted clients are members of dispersed applications with the capacity to execute on the stage [11]. Despite the fact that blockchain innovation has become prominently known due to its utilization in the execution of Cryptographic forms of money like Bit Coin, Ethereum, and so on, the actual innovation holds significantly more guarantee in different regions, for example, time-stepping, logging of basic occasions in a framework, recording of exchanges, reliable e-administration, and so on. Numerous specialists are chipping away at many such use cases like decentralized public key foundation, self-sovereign personality the executives, library support, wellbeing record the board, decentralized verification, decentralized DNS, and so on. Additionally, organizations, for example, IBM and Microsoft are fostering their own applications in assorted fields like the IoT, and so forth, in any event, empowering blockchain stages on the cloud. Thinking about the need to spread the arising ideas for understudies, we chose to set up another seminar on blockchain innovation stages and applications [12].

4.6 Application of Blockchain in Intelligent Systems

4.6.1 Cash Transfer and Installment Handling

Conceivably the best and most reasonable utilization of blockchain innovation is utilizing it as a way to speed up the exchange of assets starting with one party then onto the next. Most exchanges extended by means of blockchain can be settled inside mere seconds, while banks require 24 hours every day and surprisingly 7 days per week.

4.6.2 Supply Chains Checking

Blockchain innovation is not difficult to apply with regard to checking supply chains. By destroying paper-based preliminaries, ventures can spot shortcomings inside their inventory chains quickly, just as to distinguish things progressively [13]. Blockchain likewise empowers endeavors, and even shoppers, to see how items perform according to a quality-control perspective as they move from their place of beginning to the retailer.

4.6.3 Tracking the Supplies

Besides privacy and security, Blockchain tech can also be used to keep an eye on the supplies. For e.g., if we have ordered something from a medical provider, it can be tracked by the customer to make him aware of the product. MediLedger is one of the examples that uses Blockchain to increase the authenticity and transparency of the goods/supplies.

4.6.4 Verification of the Team

Blockchain can also be used to verify the credentials of the health support. This makes an individual know more about his team. This can lead to better relationships with the team. ProCredEX, an American company, has built an application that verifies the details of the support.

4.6.5 A Focus on Transactions

The validity of transactions will be determined by consensus, and while transactions can be interpreted widely, not all healthcare transactions will be appropriate for this technology. With blockchain, smart contracts are possible, which are downloadable computer code that allows programs, functions, or transactions to be automated. In addition, the technology provides a way to create value for transactions through tokens. Also, cryptocurrency settings have been implemented with tokenization.

4.7 Advantages of Blockchain in Healthcare

DLT can be applied in numerous medical services regions, yet every type of effort inside medical services isn't connected to exchanges. Be that as it may, public blockchains can't be utilized to store private data like recognizing wellbeing information, in light of the fact that the information in them is broadly open. This straightforwardness commands that suppliers consider security issues to guarantee ensured wellbeing data (PHI) [14]. Furthermore, blockchain innovation is powerless against certain sorts of assaults; however, it offers inbuilt security against others [15,16]. The blockchain code exposes it to multiday assaults and bugs, just as friendly designing. In this manner, data security should be given concentrated consideration, particularly when utilized in medical services. DLT can be applied in numerous medical services regions, yet the activity of any kind inside medical care isn't connected to exchanges. Be that as it may, public blockchains can't be utilized to store private data like recognizing wellbeing information, on the grounds that the information in them is generally open. This straightforwardness commands that suppliers consider security issues to guarantee ensured wellbeing data (PHI).

Furthermore, blockchain innovation is defenseless against certain sorts of assaults; however, it offers inbuilt insurance against others. The blockchain code exposes it to multiday assaults and bugs, just as friendly designing. Along these lines, data security should be given escalated consideration, particularly when utilized in medical care.

4.7.1 Transparency in the Supply Chain

Monitoring and security in supply chain is one of the major concerns in supply chain management. Storing the information for each movement of goods in the supply chain at each level can be stored in blocks of a public/private blockchain which cannot be altered and updated in real time.

4.7.2 Patient-Centric Electronic Health Records

Every country and area is struggling with the issue of data segregation, which means that patients and their healthcare professionals have an incomplete picture of their medical history. One possible answer to this issue is to create a blockchain-based record that can be then be joined with existing electronic medical records (EMRs) and serve as a single file.

Any new record, such as a physician's note, a prescription, or a test result, gets attached to this existing blockchain. In this scenario, any change to a patient record, as well as the patient's agreement to disclose a portion of their medical information, is recorded as a transaction on the blockchain. Medicalchain is a notable example of a firm that works with healthcare providers to adopt blockchain-enabled EMRs. In Figure 4.3, we have shown the benefits of Blockchain in smart healthcare systems.

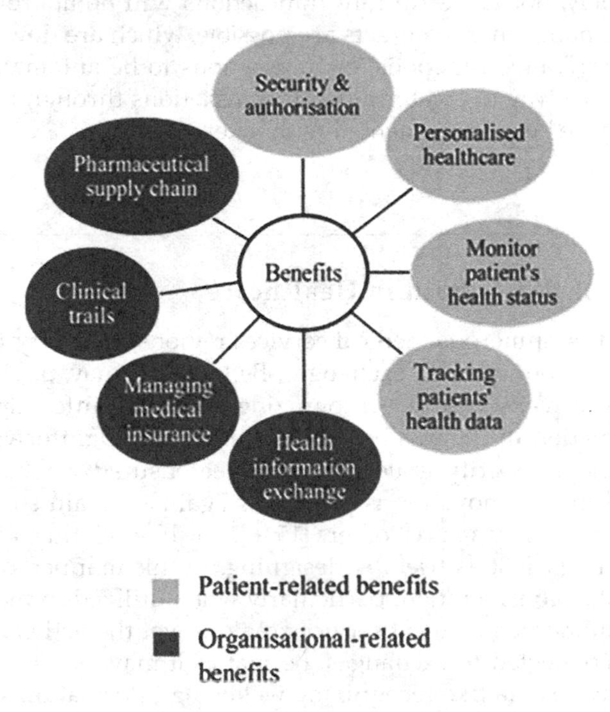

FIGURE 4.3
Benefits of Blockchain in smart healthcare.

4.7.3 IoT Security for Remote Monitoring

Blockchain technology has the potential to impact all recordkeeping processes, including the way transactions are initiated, recorded, authorized, and reported. Blockchain could be used to securely and efficiently transfer user data across platforms and systems All important lost data and information of any such problems are overcome after introduction of Blockchain technology, now the patient's problem looks like it has moved very far away from them as those important reports are stored over the Blockchain through which they can easily access on time of use or in any emergency. Doctors have also got ease in check-up of any patients with the help of Blockchain data as that very patient is either suffering from earlier from any such problems that may lead to this new kind of problem to that very patient.

4.8 Challenges for Blockchain in Healthcare

4.8.1 Major Shift

Many physicians are still stuck on paper at the moment. As a result, convincing them to switch from paper to electronic healthcare records (EHR) utilizing blockchain is a tall order.

4.8.2 Refusal to Share

The active refusal of insurance payers and hospitals to disclose data is a famous illustration of this. Keeping expense data to themselves gives hospitals a competitive advantage. They may receive different fees for different patients if they are required to share with insurance companies.

4.8.3 Distributed Healthcare

In terms of how various organizations handle records, healthcare physician providers and insurance payers, etc. all employ different methods. It would be extremely difficult to get all of these groups together to utilize blockchain as a technology without a simplified approach.

4.9 IoMT with Edge/Fog Computing

Edge/Fog Computing is concerned with high performance of intelligent systems. It is anything other than a continuously virtualized paradigm in which cloud workers operate with, accumulate, and create resources. In addition, the majority of information conveyed by these IoMT contraptions should be updated and rapidly decommissioned in order to increase the amplitude of IoMT requests. Fog Computing looks at, collects a smashing point and interference network up to a specific association edge in order to deal with the rapid issue of IoMT and complete IoMT warranty. Fog Processing provides a complete licensing arrangement to various applications and firms. In the middle of a few IoMT's that mention such as linked cars, the fog can pass persuasive short tranny. The analysis of Fog Computing is the superior solution for applications with reduced time demands such as increased realism, fluctuating press streaming and game playback, etc. The integration of IoMT with Fog Computing has a number of advantages to various IoT applications. Fog processing helps to reduce support times, particularly with IoT requests that are time fragile, by short contact amid IoMT devices. Furthermore, the capacity to restore sensor connections to a huge degree is often among the essential components connected with fog computing. Fog Computing might give various IoMT applications several advantages. This section highlights the distinctive research publications on the interconnection between IoMT and Fog Computing. In this section, we assess the current work that has been discussed in different applications regarding IoMT and Fog Computing joints. The manufacturers have maintained an eye on the motive and characteristics of IoMT applications. It displays data that can be handled by fog effortlessly by IoMT devices. It also conveys the problems of holding up time and block, which can simply be driven by fog. Fog Computing will also help in a wonderful way, regulate decentralized and fast-developing IoMT plans, enhance new businesses inside the network and restrict the results of different programs connected to network facilitator activity, as well as potential customers. The builders began by contemplating many challenges in the construction of IoMT structures and that this is extremely difficult in order to pick issues along with current versions of the business and calculation frameworks. Then we talk about the advantages of the book plan for the structuring of frames, how to cope with them, and the limit. Just how this method may be applied to create amazing possibilities for businesses. In this way, the unique framework analyzes the particular characteristics of Fog processing and helps to reduce some IoMT issues. This pushed upon cellular IoT story plus explored the particular key problems associated with conflict that may have the ability.

This gives the cloud a middleware that foreshadows fog that acts without reservation. As a further alternative to fog cells, cloud-based middleware is investigated, tested, and organised. Furthermore, the manufacturers suggested a structured fog framework. They passed on software with respect to a breeze farm with a shrewd, light-fitting visitor to look at the qualities of their style. Articles on the regulating causes of fog processing in research content are seldom truly investigated. The particular process employs the specific task and unambiguously determines resources for control sources and takes into consideration everything. These people stated that they could better regulate their own job in order to achieve much more specific research and development linked to one another so that they might use Fog Computing plus IoMT to convert to grouped product sales associated with the special impairment of professions. The makers have defined a big improvement problem to prevent the reactivity of available nebulous care. They evaluated their specific design, sometimes simply, by differentiating and active approaches, which results in a fall in inactivity. The manufacturers have evaluated the best way of restoring IoMT devices, which are generally resource-focused barriers. In addition, three motivating situations with WSNs, clever automobiles, and clever organizations, will be presented, showing that Fog's activities may be used for a few useful purposes and organizations. The fog compact utilizes a center area process to create goods possible integrated into a professional slave interface, where expert centers use the planning resources to record information from slave centers. Fog provides different advantages for using IoT, as they may collect and work near the edge of the association. Fog movable rear load changes between the procenters to correlate the slave center points with the stacking master, which updates the

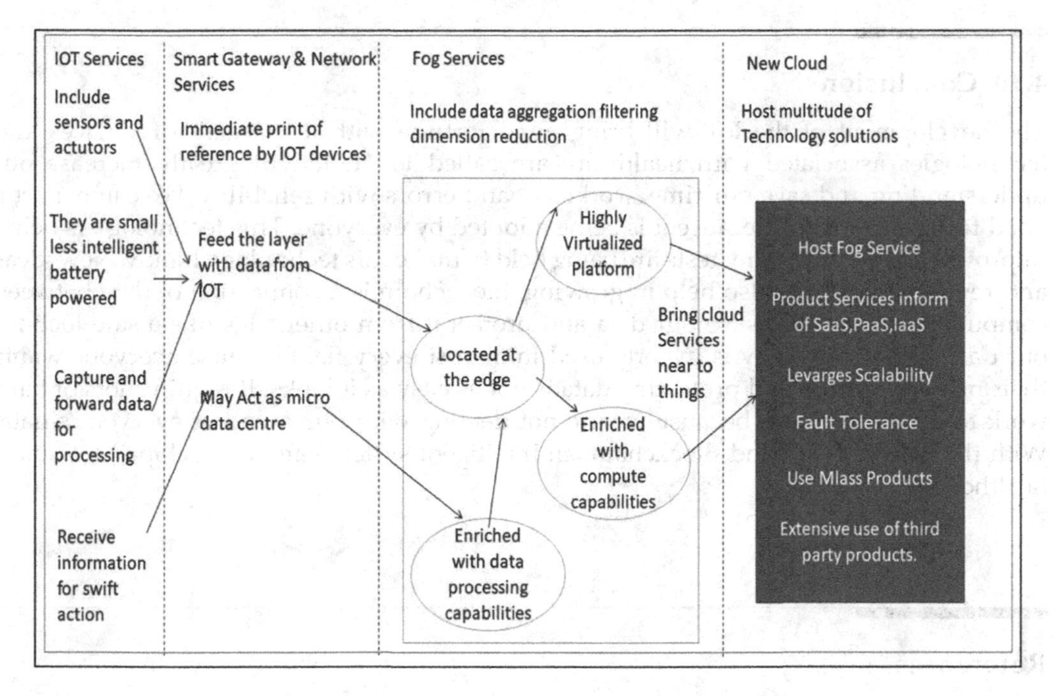

FIGURE 4.4
Edge/Fog computing services for IoMT-based intelligent systems.

whole system's flexibility. The manufacturers have proposed a tonal decreasing approach for Fog-related centers that often reduces IoT center inactivity companies. They proposed a technique of just spreading the local store that would build up Fog to Fog to convey corporate lethargy. With regard to unloading calculations, the specific approach obviously represents simply not just the full line but also a few interest lessons, which otherwise have a distinct estimate time (Figure 4.4).

The unique manufacturer has noticed the particular basic advances for the particular viewpoint of the fog. In addition, a flexible phase for tasks in Fog Computing is provided to start to finalize operations of fog about the requirement that the cycle be determined. Furthermore, these folks presented a case for using IoMT environmental factors to relax the prosperity problem by fog computing in confirmatory reversal data.

They also considered safety measures to be beneficial in ensuring that the IoT fog emphasizes an interest in facilitating the provision of fog computing through the use of diverse protective systems. There are scarcely any research publications considered in automobiles in fog computing. A car strategy reported since the transmission with handling business is known as automotive Fog Digesting. They are simply recommended by physicians to increase the partnership with clients to finish the transmission and preparation, based on vehicle assets. They are optimistic in connection with this building.

Unused approaches are used by fog vehicles to influence fog computing. In the same vein, the manufacturers expect a cross-cutting design of the fog vehicles to depict the dimensions of policy-making and, for instance, the fog centers in different organisations (Figure 4.4).

4.10 Conclusion

The development of the IoT will bring good fortune with it. Special IoT devices and technologies associated with healthcare are called IoMT. It will greatly increase our understanding and save our time, workload, and errors with reliability. We can expect a good future for the IoT because it is being adopted by everyone. This technology is being improved day by day, many tests are being held to make this technology handy. New ideas and reviews by people also help in growing. Blockchain is a connection of data between computers. It is used to save our data and protect it from others. It's like a safe lock for our data. This technology is majorly used in almost every field because everyone wants their privacy. Storing and protecting data is not as easy as it looks. It requires lots of hard work to do these things because we are not dealing with our own but everyone's data. With the help of IoMT and Blockchain, an intelligent system can be developed for smart healthcare.

References

1. Aazam, M., Zeadally, S., Harras, K.A., "Offloading in fog computing for IoT: Review, enabling technologies, and research opportunities", *Future Generation Computer Systems* 87, pp. 278–289 (2018). https://doi.org/10.1016/j.future.2018.04.057
2. https://labs.sogeti.com/iot-vs-edge-vs-fog-computing.
3. https://www.embeddedcomputing.com/technology/iot/wireless-sensor-networks/how-fog-computing-can-solve-the-iot-challenges.
4. Atlam, H.F., Walters, R.J., Wills, G.B., "Fog computing and the internet of things: A review", *Big Data and Cognitive Computing* 2(2), p. 10 (2018). https://doi.org/10.3390/bdcc2020010
5. https://www.rtinsights.com/what-is-fog-computing-open-consortium.
6. https://www.powersystemsdesign.com/articles/five-reasons-why-your-iot-application-needs-fog-computing/140/14857.
7. Alli, A.A, Alam, M.M. "The fog cloud of things: A survey on concepts, architecture, standards, tools, and applications", *Internet Things* 9, p. 100177 (2020).
8. https://labs.sogeti.com/iot-vs-edge-vs-fog-computing/
9. Patni, J.C., Ahlawat, P., Biswas, S.S., "Sensors based smart healthcare framework using internet of things (IoT)", *International Journal of Scientific and Technology Research* 9(2), pp. 1228–1234 (2020).
10. Taneja, S., Ahmed, E., Patni, J.C., "I-Doctor: An IoT based self patient's health monitoring system", In *2019 International Conference on Innovative Sustainable Computational Technologies (CISCT)* (2019).
11. Khanchi, I., Agarwal, N., Seth, P., Ahlawat, P., "Real time activity logger: A user activity detection system", *International Journal of Engineering and Advanced Technology* 9(1), pp. 1991–1994 (2019).
12. Tiwari, R., Sharma, H.K., Upadhyay, S., Sachan, S., Sharma, A., "Automated parking system-cloud and IoT based technique", *International Journal of Engineering and Advanced Technology* 8(4C), pp. 116–123 (2019).
13. Kshitiz, K., Shailendra, "NLP and machine learning techniques for detecting insulting comments on social networking platforms", In *Proceedings on 2018 International Conference on Advances in Computing and Communication Engineering (ICACCE)*, pp. 265–272 (2018).

14. https://medium.com/yeello-digital-marketing-platform/what-is-fog-computing-why-fog-computing-trending-now-7a6bdfd73ef.
15. Gupta, M., Benson, J., Patwa, F., Sandhu, R., "Dynamic groups and attribute-based access control for next-generation smart cars", In *Proceedings of the Ninth ACM Conference on Data and Application Security and Privacy*, pp. 61–72 (2019).
16. Sharma, A., Tomar, R., Chilamkurti, N., Kim, B.G. "Blockchain based smart contracts for internet of medical things in e-healthcare", *Electronics* 9, p. 1609 (2020). https://doi.org/10.3390/electronics9101609

5

AI-Enabled Cloud-Based Intelligent System for Telemedicine

Hitesh Kumar Sharma, Ravi Tomar, and Preeti
University of Petroleum and Energy Studies

Prashant Ahlawat
GL Bajaj Institute

CONTENTS

5.1 Introduction....75
5.2 Artificial Intelligence in Intelligent Systems....76
5.2.1 Application of Artificial Intelligence in Telemedicine Systems....76
5.2.1.1 AI and BCI....77
5.2.1.2 AI and Nanotechnology....77
5.2.1.3 AI with Machine Learning in Research....78
5.2.1.4 AI and Drug Creation....78
5.2.2 Challenges of AI in Telemedicine Systems....78
5.3 Cloud Computing in Intelligent Systems....78
5.3.1 Application of Cloud Computing in Telemedicine Systems....80
5.4 Conclusion....83
References....83

5.1 Introduction

In the current generation, Artificial Intelligence or AI is one of the most popular fields in terms of research, applications, jobs, etc. It has a wide range of uses in every sector and domain that also includes healthcare. Artificial Intelligence is the ability of the computer to sense and analyze data provided to it. Healthcare big database [1] is useful for AI to implement machine learning algorithms. It helps in finding out necessary solutions where humans need much more effort to do so. AI can change the Health Infrastructure by using the data provided by IoT devices over cloud and analyzing it in depth [2]. AI helps in detecting diseases, administration of chronic situations, deliver health and security services, and may help in inventing new drugs or discovering drugs by using the existing database [2–4]. AI also can plan resources—depending upon the condition of the patient, it can quickly allocate necessary resources (if required) for them and provide the information to the doctor. In today's world, AI can be used in mobile applications; patient can use these apps to feed data regarding their mood swings, anxiety, and depression; weight, height, and other parameters, and AI could analyze this data and can provide an approximate diagnostic for the symptoms [5]. More often, smart devices track such

DOI: 10.1201/9781003291916-5

information and upload it over the AI database. In surgeries and operations, AI can use its database to provide the exact location of a certain wound or internal bleeding. Surgeons can use this information to carry out the operation/surgery with more accuracy. AI is also being used to detect tumors that may cause cancer in the future. We can also disburden Electronic Health Records (EHR) using AI. EHR developers now use AI to create better processing power and limit the time spent by the user on maintaining EHR by automating the functions.

Similarly, Cloud Computing is the delivery of different services through the internet, including data storage, servers, databases, networking, software, and gaming. It makes it possible to save files to a remote database and retrieve them on demand [6]. Cloud computing in health care can be very beneficial—it will increase the efficiency in sharing of database of every patient with different hospitals without using any extra efforts. It will also need less money as hospitals depend on large storage devices to store patient data, which cost lots of money and their energy consumption is also very high, resulting in high expenses [7]. By using cloud computing, you can get a large amount of storage at very less prices, and it will be more secure in emergency and its accessibility is way higher. Institutes that provide insurances can also get a boost as they can also track the health of their insurance holder as they can now get more accurate status of health and also can amend new policies to increase their profits [8]. By uploading different medical conditions, students all over the world can study different medical conditions and their cure without much effort. It will also develop the healthcare industries.

5.2 Artificial Intelligence in Intelligent Systems

AI is basically a machine mimicking basic human cognitive functions. In laymen's terms, it is a programming machine to think for themselves and make their own decisions without the help of humans [9]. This, when applied in healthcare, can become a big boost to how the whole sector in itself functions. The biggest advantage of using AI in the health sector is its gathering and processing of data and giving a well-defined result to the user/patient. These programs can be used for diagnosis, treatments, drug development and manufacture, patient monitoring, and providing personal medication [10]. Basically, every task done by a doctor can be programmed into or taught to an AI to perform it with perfection.

5.2.1 Application of Artificial Intelligence in Telemedicine Systems

In Figure 5.1, we have shown various roles played by AI in various fields of healthcare. Another upcoming technology that uses Brain–Computer Interfaces (BCI) aims to help patients with neurological diseases and trauma. Patients who are unable to communicate with their bodies and have lost the ability to move or speak can be helped through this technology. Using BCI and AI, neural activities associated with a simple body function (such as waving one's hand) can be analyzed, decoded, and relayed to output devices that carry out the same desired functions [11]. Such technology can drastically improve many people's lives suffering from strokes, spinal cord injuries, or ALS.

Nanotechnology is already a huge part of the healthcare sector with scientists using this technology to target tumors inside the human body. However, operating at such atomic levels is very tricky and dangerous. This is where AI plays a very vital role. AI nanobots

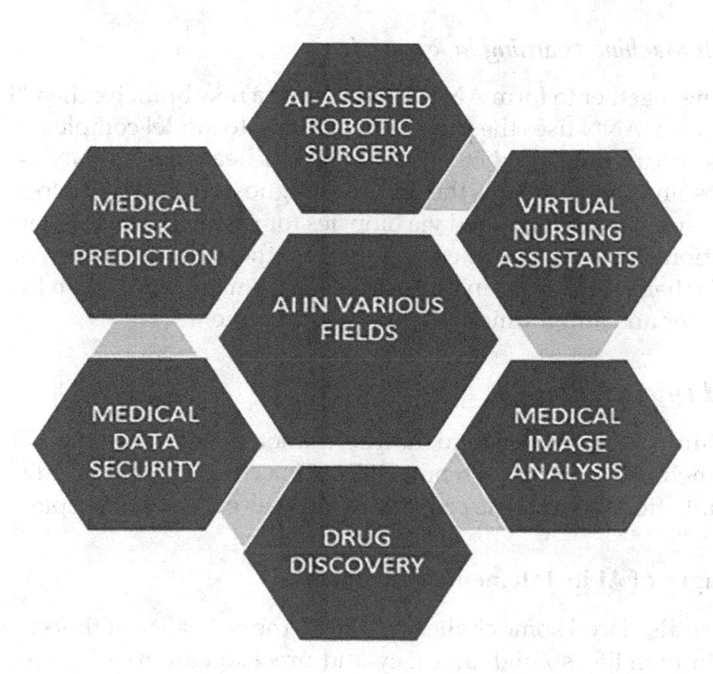

FIGURE 5.1 Role of AI in smart healthcare.

may currently just be a theory but their applications, when manufactured, are endless. These tiny "living robots" could enter our bodies, carry out specific tasks, and even help in curing cancer! AI and ML come together to form Artificial Neural Networks (ANN), which forms a new branch, called "Deep Learning." To give a basic idea, ANN uses the processing of brain to model complex patterns. Through these patterns, it can then predict unseen data. In healthcare, ANN is used to predict future diseases and inform about clinical diagnosis [12].

Some of the popular application are defined in the following subsections.

5.2.1.1 AI and BCI

Another up-and-coming technology that uses BCI aims to help patients with neurological diseases and trauma. Patients who are unable to communicate with their bodies and have lost the ability to move or speak can be helped through this technology. Using BCI and AI, neural activities associated with a simple body function (such as waving one's hand) can be analyzed, decoded, and relayed to output devices that carry out the same desired functions. Such technology can drastically improve many people's lives suffering from strokes, spinal cord injuries, or ALS [13].

5.2.1.2 AI and Nanotechnology

Nanotechnology is already a huge part of the healthcare sector with scientists using this technology to target tumors inside a human body. But operating at such atomic levels is very tricky and dangerous. This is where AI plays a very vital role. AI nanobots may currently just be a theory but their applications, when manufactured, are endless. These tiny "living robots" could enter our bodies and carry out specific tasks and even help in curing cancer!

5.2.1.3 AI with Machine Learning in Research

AI and ML come together to form ANN which forms a new branch called "Deep Learning." To give a basic idea, ANN uses the processing of brain to model complex patterns. Through these patterns, it can then predict unseen data. In healthcare, ANN is used to predict future diseases and inform about the clinical diagnosis. Many radiological departments use physical tissue samples obtained via biopsies for research, but this carries considerable risks for infection. Here, AI can be used to replace these tissues with a new generation of radiology tools that are accurate and detailed. These images could also help to understand cancer cells better and aim for more appropriate treatments [14].

5.2.1.4 AI and Drug Creation

With its own thinking capabilities, and the information from around the world at its disposal, AI can create medicines, and it even has. DSP-1181, a drug made for OCD, was invented by AI through joint efforts from some companies and has even been accepted for human trial.

5.2.2 Challenges of AI in Telemedicine Systems

AI in healthcare also faced some challenges due to the criticality of this system. It is directly connected to human life, so high accuracy and precision are required for using AI-based algorithms in smart healthcare systems.

1. The biggest risk in using AI is that AI systems might be wrong at some points. Let's say in the process of detecting a tumor inside a patient, if AI gets the wrong location, the patient may get fooled and suffer consequences.
2. The next challenge faced by AI is the lack of accurate data. Even we are using Electronic Health Reports and continuously providing data, but the lack of authenticity persists. Even if AI learns the database but if the information provided is not relevant, then there is no point in learning and storing all that data. We first need to ensure that the data, which we gather, is accurate and relevant.
3. Privacy plays a major role in our lives. We need to develop a system that we can trust and ensure that our data is not leaked [15].
4. AI must gather relevant data throughout the world. If only a part of the population is used to analyze the data, the remainder, which might have different traits or disease, may not be accounted for future use.
5. Tools could be mistaken with confidence and algorithms could be misleading. Unsafe AI could harm healthcare providers and patients.
6. Some surgeries are too costly.
7. Due to overdose of some rays can also cause skin or body cells damage.
8. Sometimes the result produced after check-ups have blunder mistakes.

5.3 Cloud Computing in Intelligent Systems

The healthcare industry has historically been behind in the adoption of the bleeding edge of technological developments. Given the gravity of the work this industry deals

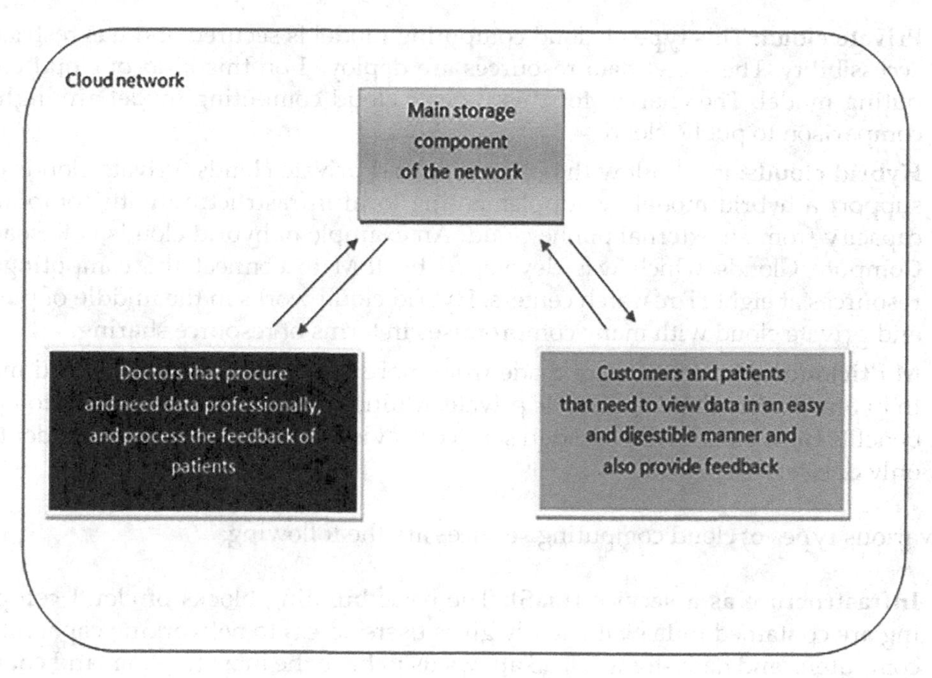

FIGURE 5.2 Cloud storage for medical data.

with, they have to be significantly more careful with any change made and have a smaller margin of error [16]. If they unwittingly adopt a technology that was faulty in ways not obvious in the early stages of widespread implementation of the same, then it can lead to some disastrous situations. Despite this attitude, the healthcare industry has been one of the fastest and most willing at adapting cloud computing technology. From West Monroe Partner's report, 35% of healthcare organizations surveyed held more than 50% of the data or infrastructure in the "cloud," or in such a network, as of 2018. Such monumental moves to adopting this technology have not been seen in any other industry. In Figure 5.2, block diagram is shown for data storage and usage through the cloud computing model.

Cloud computing, in layman speak is the ability to store data and use processing power that isn't exactly "yours," i.e., physically a part of your computer. For example, it can allow for barebones computers to execute highly demanding programs or tasks by simply requesting another computer over some sort of network to do it in its stead, deflecting the demands of processing, data, and such requirements to the other computer instead. This strips the would-be stringent requirements of the computer hardware of the barebones computer this request originated from, making it easier to perform a larger variety of tasks from "anywhere" given a sufficiently well-made network. Cloud computing provides multiple models and multiple services to fulfill the computation and storage requirements as per customers' need [17–22].

The various types of cloud computing models are:

- **Public cloud:** Public clouds are those type of cloud computing models where it is accessible by all. The resources in this model can be accessed by all people who want those resources. This type of model is provided by Google Cloud, AWS, Microsoft Azure. When privacy is not a concern, then an organization or individual can go for this type of cloud computing model.

- **Private cloud:** This type of cloud computing model is secured and has restricted accessibility. The secret data resources are deployed on this kind of cloud computing model. The charges for this type of cloud computing model are high in comparison to public cloud.
- **Hybrid clouds:** It is built with both public and private clouds. Private clouds can support a hybrid model by supplementing local infrastructure with computing capacity from an external public cloud. An example of hybrid clouds is Research Compute Clouds which was developed by IBM to connect the computing IT resources at eight IT research centers. Hybrid cloud works in the middle of public and private cloud with many compromises in terms of resource sharing.
- **Multiclouds:** Multiclouds are made from more than one cloud service and more than one vendor that is public or private. Multicloud helps organizations to avail benefits from more than one cloud service provider instead of being dependent on only one service provider.

The various types of cloud computing services are the following:

- **Infrastructure as a service (IaaS):** The basic building blocks of cloud computing are contained in IaaS. It usually gives users access to networking capabilities, computers, and data storage. IaaS allows us to have the most freedom and control over our IT resources.
- **Platform as a service (PaaS):** This relieves us of the burden of managing the underlying infrastructure, allowing us to concentrate on the deployment and maintenance of our apps. This allows us to be more productive.
- **Software as a service (SaaS):** This gives us s fully functional product that is managed and maintained by the service provider.

The characteristics of cloud computing services are the following:

- **Agility**: The cloud provides us quick access to a wide range of technologies, allowing us to develop more quickly and construct almost anything we can dream.
- **Elasticity**: With cloud computing, we don't need to over-provision resources to accommodate future peal levels of the company activity; instead, we allocate the exact quantity of resources that we require.
- **Cost-savings**: The cloud enables us to swap fixed costs for variable ones, allowing us to pay just for IT. Furthermore, due to economies of scale, the variable expenditures are far lower than what we would spend if we did it ourselves.
- **Deploy globally in minutes**: We may quickly grow to new geographic locations and deploy internationally using the cloud. Taking the example of AWS, which has infrastructure all over the world, allows us to deploy the app in different physical locations in a matter of seconds.

5.3.1 Application of Cloud Computing in Telemedicine Systems

Technology uses internet for storing the data and accessing it through the internet. They are a cost-effective solution for enterprises. For example, Google cloud which includes app engine, Google cloud storage, etc., quality has become an important feature of modern healthcare organizations. When it comes to healthcare, it comes to overcome two industry challenges: increasing cost-effectiveness and building a self-sufficient health ecosystem. The demand and supply fluctuations increased in the pandemic. The technology server is a greater mission to establish a smart healthcare community. It's a huge step toward treating lifestyle-related noncommunicable diseases.

The advantages of cloud computing platform are the following:

1. With the help of cloud computing, the process of data sharing has become easier and simple which can be accessible 24*7.
2. They can access medical reports at anytime and anywhere.
3. It can hold a large amount of information at less amount.
4. It helps the smaller hospitals that have a tight budget to take help from cloud models.
5. You can get real-time updates for all relevant information.
6. It is dependable and consistent.
7. Accurate treatment and decision-making
8. Greater transparency

The disadvantages of adopting cloud computing services are the following:

1. Completely managed by service provider and the organization may have less control over their infrastructure.
2. As there are multiple clients at the same time, it may cause trouble and support challenges.
3. Storing data on third party or external servers may have some risks.
4. Lack of good specialists.

Smart healthcare is a service system that uses technologies such as IoT, internet, etc., to access people related to healthcare and then manages and responds to medical needs in an intelligent manner. The adoption of cloud computing requires time and effort from numerous industries.

As mentioned in the introduction, data storage is one of the primary methods that the industry has put cloud computing to use. It might be slightly unintuitive, given the name, but just this facet of cloud computing has tremendously helped the healthcare industry. The first thing having such an efficient method of data storage means that everyone involved will be able to retrieve it with ease and elegance. This means that people not familiar with the practice, people like the patients, are easily able to see their own medical reports. Things like mobile applications, more interactive and informative websites are now available due to the ease of transfer and organization of data as a result of adopting the cloud network in the medical industry. A mobile application can provide seamless and regular updates to patients about their own health, while also reminding them of things like when and what medicine to take. This has proven to increase the engagement that patients have with their own health and also has had a positive reaction in general [2]. Another potential benefit to cloud technology is the ease at which an insurance company or the workplace of the patient can have the information of the medical proceedings in their hands, alleviating the need for the patient to painstakingly draft up a report. All of these are just the beginning of cloud computing, as its applications like EMRs, and other such implementations are being made daily.

Finally, we come to the miscellaneous benefits of cloud computing in healthcare. One of the most apparent benefits is the flexibility of this system. Most cloud services are handled by third parties who can handle the large amounts of machinery involved. This allows a very scalable program and no growth pains, unlike when the analog systems had to be expanded. Other benefits like the maintenance being handled by the provider. The cloud services are generally very affordable as compared to buying your own cloud system as well. All in all, cloud computing can be said to be a unilaterally positive development for the healthcare industry, and a massive plus for humanity in general.

The centralization of the data storage makes it much easier to manage, and even in such a simple example in which many nuances are ignored, it is clear how streamlined the data flow is under such an arrangement. Cloud computing is a useful tool, and the fundamental reason for its popularity is that businesses, educational institutions, and other organizations can rent storage, processing capacity, and applications from a cloud service provider. The infrastructure and system maintenance are taken care of by the Cloud service provider. For businesses, this means less money spent on computers and fewer employees needed to operate and maintain the system. There are three primary sorts of clouds: public, private, and hybrid. Microsoft Azure, Amazon Web Services, Google Cloud, and other public clouds are the most popular. The provider owns and manages all of the infrastructure and apps in this scenario. The most significant benefit is that the Cloud provider is responsible for all maintenance and management. Private Clouds, on the other hand, are restricted to a single firm or organization. The infrastructure for the Cloud is often managed by an organization's IT department. The advantages of this strategy for major organizations are that their data is secure, and they can deliver all of the benefits of Cloud Computing services inside to their many departments, but at a cost. The best value is available with Hybrid Cloud. The private cloud is used for very sensitive data, whereas the public cloud is used for other services. This gives enterprises more control and security over critical data while yet allowing them to use the public cloud when necessary. Nearly 87% of healthcare firms, for example, choose to employ a hybrid cloud architecture. The private cloud is typically used to host systems that include sensitive patient data, such as imaging and EHR. For storing patients' health files in the cloud, a cloud-based EHR offers a scalable, adaptable, intuitive, and cost-effective solution. We now have the ease of being able to communicate with our doctor on the phone or computer from anywhere in the world, with real-time access to information, anywhere, and compatibility across different systems as a result of this development. Patients' portals, telehealth services, healthcare information exchanges, smart devices, remote monitoring services, and other components of the new healthcare ecosystem are all powered by this. Patients and doctors do not have to be physically present at the same area because of telemedicine and telehealth's Virtual Care platforms. Patients' portals allow patients to access their medical records and contact their doctors safely. Cloud computing is a type of internet-based computing that enables virtual access to a variety of applications and services such as storage, servers, and networking. Cloud computing is a whole new virtualization technology for both individuals and businesses that departs from the old software business model. It is defined as providing end users with remote dynamic access to services and computer resources via the internet. Medical reports and personal information of the patient are available in the online records and can be easily accessed by the physicians, pharmacies, and counselors without charging any extra cost over the diagnosis fees. The healthcare field has a vast expanse and highly complicated ecosystem, including health insurance companies, hospitals, labs, pharmacies/medical stores, patients, and other bodies as well; hence, in order to assure good, effective, and fast functionality of this ecosystem, it is very essential that data is transmitted quickly and securely between the different entities. Cloud computing provides business models to various hospitals and clinic centers to store the information of the patient and update it with his/her every visit. Cloud computing helps in reviewing, exchanging, and sharing the images of MRI, CT-scan, and X-ray reports in a faster and highly secured environment. Uninterrupted services can be provided to various health organizations due to high availability of cloud computing services. The various business models of cloud computing benefit the health organizations in many ways. Cloud computing also provides benefits to the patients like improvement in the quality of services and collaborations between various health organizations. Since human life is priceless and precious and the medical facilities across the world are limited, therefore the healthcare services provided in cloud offer cost-effective concepts so as to benefit both the patients and the health organizations. Thus, we can see that cloud computing has a big scope in terms of healthcare as well.

5.4 Conclusion

Artificial Intelligence and Cloud computing together can help the telemedicare services to perform in a more effective manner. Close by 24*7 availability, clinical consideration providers certainly need to scale the data amassing and association essentials as per the help demands. Cloud development can increase or decrease these limit as indicated by the need of clinical consideration specialists. As cloud development can thoroughly upset your customary methodology of data dealing with, clinical benefit providers need to design the movement cycle well early. Cloud development method diminishes dangers just as it limits the chances of get-away, thwarts information spills, further creates data managing, and builds up security practices.

In this chapter, we have described the role of AI and Cloud computing in an advanced Telemedicine System.

References

1. Kustwar, R. K., & Ray, S. (2020). eHealth and telemedicine in India: An overview on the health care need of the people. *Journal of Multidisciplinary Research in Healthcare*, 6(2), 25–36. doi: 10.15415/jmrh.2020.62004.
2. Saha, A., Amin, R., Kunal, S., Vollala, S., & Dwivedi, S. K. (2019). Review on "Blockchain technology based medical healthcare system with privacy issues". *Security and Privacy*, 2(5), e83. doi: 10.1002/spy2.83.
3. Taneja, S., Ahmed, E., & Patni, J. C. (2019). I-Doctor: An IoT based self patient's health monitoring system. In *2019 International Conference on Innovative Sustainable Computational Technologies (CISCT)*.
4. Sharma, A., Choudhury, T., & Kumar, P. (2018). Health monitoring & management using IoT devices in a Cloud Based Framework. In *2018 International Conference on Advances in Computing and Communication Engineering (ICACCE)*, pp. 219–224.

5. Kamdar, N., & Jalilian, L. (2020). Telemedicine: A digital interface for perioperative anesthetic care. *Anesthesia & Analgesia*, 130(2), 272–275. doi: 10.1213/ANE.0000000000004513.
6. Patni, J. C., Sharma, H. K., Sharma, S., Choudhury, T., Mor, A., Ahmed, M., & Ahlawat, P., (2022) COVID-19 pandemic diagnosis and analysis using clinical decision support systems. In Tavares, J. M. R. S., Dutta, P., Dutta, S., Samanta, D. (eds) *Cyber Intelligence and Information Retrieval. Lecture Notes in Networks and Systems*, vol. 291. Springer, Singapore.
7. Gupta, S., Arya, P., & Sharma, H. K. (2021). User anonymity based secure authentication protocol for telemedical server systems. *International Journal of Information and Computer Security* 1(1), 1.
8. Shailender, C., & Sharma H. K. (2018). Digital cancer diagnosis with counts of adenoma and luminal cells in plemorphic adenoma immunastained healthcare system. *IJRAR* 5(12), 869–874.
9. Joshi, A., Choudhury, T., Sai Sabitha, A., & Srujan Raju, K. (2020). Data mining in healthcare and predicting obesity. In *Proceedings of the Third International Conference on Computational Intelligence and Informatics*, pp. 877–888, Hyderabad, India.
10. Patni, J.C., Ahlawat, P., & Biswas, S.S. (2020). Sensors based smart healthcare framework using internet of things (IoT). *International Journal of Scientific and Technology Research*, 9(2), 1228–1234.
11. Purri, S., Choudhury, T., Kashyap, N., & Kumar, P. (2017). Specialization of IoT applications in health care industries. In *2017 International Conference on Big Data Analytics and Computational Intelligence (ICBDAC)*, pp. 252–256.
12. Krishnan, D. S. R., Gupta, S. C., & Choudhury, T. (2018). An IoT based patient health monitoring system. In *2018 International Conference on Advances in Computing and Communication Engineering (ICACCE)*, pp. 1–7.
13. Shamshad, S., Mahmood, K., Kumari, S., & Chen, C. M. (2020). A secure blockchain-based e-health records storage and sharing scheme. *Journal of Information Security and Applications* 55, 102590.
14. Hussien, H. M., Yasin, S. M., Udzir, S. N. I., Zaidan, A. A., & Zaidan, B. B. (2019). A systematic review for enabling of develop a blockchain technology in healthcare application: Taxonomy, substantially analysis, motivations, challenges, recommendations and future direction. *Journal of Medical Systems*, 43(10). doi: 10.1007/s10916-019-1445-8.
15. Negi, G., Kumar, A., Pant, S., & Ram, M. (2021). Optimization of complex system reliability using hybrid grey wolf optimizer. *Decision Making: Applications in Management and Engineering*, 4(2), 241–256.
16. Koranga, P., Singh, G., Verma, D., Chaube, S., Kumar, A., & Pant, S. (2018). Image denoising based on wavelet transform using visu thresholding technique. *International Journal of Mathematical, Engineering and Management Sciences*, 3(4), 444–449.
17. Fore, V., Khanna, A., Tomar, R., & Mishra, A. (2016). Intelligent supply chain management system. In *2016 International Conference on Advances in Computing and Communication Engineering (ICACCE)*, pp. 296–302. doi: 10.1109/ICACCE.2016.8073764
18. Srivastava, R., Tomar, R., Gupta, M., Yadav, A.K., & Park, J. (2021). Image watermarking approach using a hybrid domain based on performance parameter analysis. *Information*, 12, 310. doi: 10.3390/info12080310
19. Tomar, R., Prateek, M., & Sastry, H. G. (2017). Analysis of beaconing performance in IEEE 802.11p on vehicular ad-hoc environment. In *2017 4th IEEE Uttar Pradesh Section International Conference on Electrical, Computer and Electronics (UPCON)*, pp. 692–696. doi: 10.1109/UPCON.2017.8251133.
20. Gupta, M., Benson, J., Patwa, F., & Sandhu, R. (2019). Dynamic groups and attribute-based access control for next-generation smart cars. In *Proceedings of the Ninth ACM Conference on Data and Application Security and Privacy*, pp. 61–72.
21. Gupta, M., Patwa, F., & Sandhu, R. (2017). Object-tagged RBAC model for the Hadoop ecosystem. In *IFIP Annual Conference on Data and Applications Security and Privacy*, pp. 63–81. Springer, Cham.
22. Gupta, M., Patwa, F., & Sandhu, R. (2018). An attribute-based access control model for secure big data processing in hadoop ecosystem. In *Proceedings of the Third ACM Workshop on Attribute-Based Access Control*, pp. 13–24.

6

Fuzzy Heptagonal DEMATEL Technique and Its Application

A. Felix and PP. Ajeesh
Vellore Institute of Technology

S. Karthik
Vel Tech Rangarajan Dr. Sagunthala R&D Institute of Science and Technology

R. Dinesh Jackson Samuel
Oxford Brookes University

CONTENTS

6.1 Introduction....85
6.2 Preliminaries....86
6.3 Fuzzy Heptagonal DEMATEL Method....90
6.3.1 Numerical Illustration....93
6.4 Conclusion....96
References....97

6.1 Introduction

Multicriteria decision-making (MCDM) method is the distinguished division of decision-making, which creates an interest to the researchers in evaluating, assessing, and ranking alternatives. In the literature, there are about 20 MCDM techniques (Chen and Chen, 2010) that have been designed to solve many real-world decision-making problems. Among them, DEMATEL (decision-making trial and evaluation laboratory) is one of the powerful tools. It was developed in Geneva Research Centre (Fontela and Gabus 1976; Gabus and Fontela, 1972) by the Battelle Memorial Institute. This method is an efficient and practical tool for viewing any difficult structure with matrices/digraphs. The crisp DEMATEL technique involves only the situation wherein the experts provide their opinion on the correlation between the attributes using the crisp values {0,1}. However, in real-life situations, correlations among the attributes involve a lot of uncertainty. Thus, the concept of fuzzy set was introduced to handle the ambiguity and uncertainty of human thought (Zadeh, 1965). Bellmann and Zadeh (1970) introduced decision-making in a fuzzy environment where the membership functions of the fuzzy decision attain maximum values. Zadeh's Fuzzy set theory has paved a way to bring out the fuzzy DEMATEL method (Lin and Wu 2004, 2008; Tseng, 2009a) under the uncertain environment.

When creating a structural model, crisp values are being used to represent human decisions for determining the relationship among factors. However, in most of the

DOI: 10.1201/9781003291916-6

circumstances, human's decisions with favorites seem hard to evaluate by exact values. Instead of assigning numerical values in the problem, linguistic variables can be assigned to take a decision (Zadeh, 1975). In recent times, linguistic terms are widely adopted by researchers in the decision-making problem. Input variables are represented by linguistic terms in decision-making problems and assigned the values by the fuzzy numbers (Karthik et al., 2019, 2020). Fuzzy DEMATEL was extended by Lin and Wu (2008) where the linguistic variables are transformed into triangular fuzzy numbers. Devadoss and Felix (2013a) used DEMATEL with triangular fuzzy numbers to model the cause and effect factor of youth violence. Suo et al. (2012) expanded the DEMATEL method in an uncertain linguistic environment using trapezoidal fuzzy number. Devadoss and Felix (2013a, 2014) extended DEMATEL technique for hexagonal and octagonal fuzzy numbers.

DEMATEL is one of the potent methods to analyze the cause/effect relationship among the criteria (Wu and Lee, 2007). This method supports for analyzing the correlations among criteria in the multifaceted systems (Hori and Shimizu, 1999). Currently, fuzzy DEMATEL method has been used magnificently in various domains such as hospital service quality (Shieh et al., 2010), hotel service quality (Tseng, 2009b), social problems on youth violence (Devadoss and Felix, 2013a), human resource management (Abdullah and Zulkifli, 2015), portfolio of investment projects (Altuntas and Dereli, 2015), business process management (Bai and Sarkis, 2013), green supply chain management (Lin, 2013), global managers competencies (Wu and Lee, 2007), hotel service quality perceptions (Tseng, 2009b), software system design (Hori and Shimizu, 1999), extracting consumer's uneasiness over foods (Tamura et al., 2006), solid waste management (Tseng and Lin, 2009), emotive Music Composition Selection (Aseervatham and Devadoss, 2015), climate change (Felix et al., 2017), solid waste management (Selvaraj et al., 2018; Felix and Dash, 2021), IC fabrication process (Velmathi and Felix, 2019) and bipolar environment (Deva and Felix, 2021). From this review, it is observed that DEMATEL method can be extended using heptagonal fuzzy number under uncertain linguistic environment. To illustrate and demonstrate the heptagonal DEMATEL techniques, the problem on youth aggressive behavior is studied. The structure of this paper is as follows: the important definitions of fuzzy number, linguistic terms, and modified heptagonal fuzzy numbers are given in Section 6.2. In Section 6.3, fuzzy DEMATEL-heptagonal method and its application are presented, and conclusion and direction for future are given in the last section.

6.2 Preliminaries

The important definitions (2.1–2.8) are reviewed and also the modified heptagonal fuzzy number with respect to alpha cut is explored under an uncertain linguistic environment. This heptagonal fuzzy number differs from ordinary heptagonal fuzzy numbers (Rathi and Balamohan, 2014).

Definition 6.2.1

A fuzzy set $\tilde{A} \subseteq X$, in which each element is mapped to [0, 1] by a membership function $\mu_{\tilde{A}}$: $X \to [0,1]$. The membership grade $\mu_{\tilde{A}}(\tau)$ represents the degree of truth that τ is a member of the fuzzy set $\tilde{A}$. Fuzzy set can also be represented by ordered pair by the following manner $\tilde{A} = \{(\tau, \mu_{\tilde{A}}(\tau)) \mid \tau \in X\}$.

Definition 6.2.2

Fuzzy number can be thought as a subset of the real numbers $\mathbb{R}$, which is characterized by the membership function, $\tilde{A}: R \to [0,1]$ holds the subsequent properties,
(i) $\tilde{A}$ is convex.

$$\mu_{\tilde{A}}\left(\lambda\tau_1 + (1-\lambda)\tau_2\right) \geq \min(\mu_{\tilde{A}}(\tau_1), \mu_{\tilde{A}}(\tau_2)),$$

(ii) $\tilde{A}$ is piecewise continuous.
(iii) $\tilde{A}$ is normal, which means the maximum membership value is 1.

Definition 6.2.3

The α-cut of the fuzzy set $\tilde{A}$ of the universal set X is defined as the set of all elements whose membership grade is greater than α, and it is defined as $\tilde{A}_\alpha = \{\tau \in X \,/\, \mu_{\tilde{A}}(\tau) \geq \alpha\}$, where $\alpha \in [0,1]$.

Definition 6.2.4

A Triangular Fuzzy Number $\tilde{T}_r$ can be defined as a triplet (l, m, r) where l, m, and r are real numbers such that $l \leq m \leq r$ and the membership function $\mu_{\tilde{N}}(\tau)$ is defined as

$$\mu_{\tilde{T}_r}(\tau) = \begin{cases} 0 & \tau < l \\ \left(\dfrac{\tau - l}{m - l}\right) & l \leq \tau \leq m \\ \left(\dfrac{r - \tau}{r - m}\right) & m \leq \tau \leq r \\ 0 & \tau > r \end{cases}$$

where l, m, n are real numbers $l \leq m \leq n$.

Definition 6.2.5

A trapezoidal fuzzy number $\tilde{T}_{re}$ can be defined as (l, m_1, m_2, n) and the membership function is defined as

$$\mu_{\tilde{T}_{re}}(\tau) = \begin{cases} \left(\dfrac{\tau - l}{m_1 - l}\right) & l \leq \tau \leq m_1 \\ 1 & m_1 \leq \tau \leq m_2 \\ \left(\dfrac{n - \tau}{n - m_2}\right) & m_2 \leq \tau \leq n \\ 0 & l \leq 0 \,\&\, n \geq 0 \end{cases}$$

where l, m_1, m_2, n are real numbers $l \leq m_1 \leq m_2 \leq n$.

Definition 6.2.6

A heptagonal fuzzy number $\tilde{H}_e$ can be defined as $(a_1, a_2, a_3, a_4, a_5, a_6, a_7)$, and the membership function is defined as $\mu_{\tilde{H}_e}(\tau)$

$$\mu_{\tilde{H}_e}(\tau) = \begin{cases} \dfrac{(\tau - a_1)}{3(a_2 - a_1)}, & a_1 \le \tau \le a_2 \\ \dfrac{(a_3 - a_2) + (\tau - a_2)}{3(a_3 - a_2)}, & a_2 \le \tau \le a_3 \\ \dfrac{2(a_4 - a_3) + (\tau - a_3)}{3(a_4 - a_3)}, & a_3 \le \tau \le a_4 \\ \dfrac{3(a_5 - a_4) - (\tau - a_4)}{3(a_5 - a_4)}, & a_4 \le \tau \le a_5 \\ \dfrac{2(a_6 - a_5) - (\tau - a_5)}{3(a_6 - a_5)}, & a_5 \le \tau \le a_6 \\ \dfrac{(a_7 - \tau)}{3(a_7 - a_6)}, & a_6 \le \tau \le a_7 \end{cases}$$

A Heptagonal fuzzy number $\tilde{H}_e$ can also be defined as $\mu_{\tilde{H}_e}(\tau) = \big(P_1(t), Q_1(u), R_1(v), P_2(t), Q_2(u), R_2(v)\big)$ for $t \in [0, 0.33)$, $u \in [0.33, 0.66)$, and $v \in [0.66, 1]$, where

$$P_1(t) = \frac{(\tau - a_1)}{3(a_2 - a_1)}, Q_1(u) = \frac{(a_3 - a_2) + (\tau - a_2)}{3(a_3 - a_2)}, R_1(v) = \frac{2(a_4 - a_3) + (\tau - a_3)}{3(a_4 - a_3)}$$

$$P_1(w) = \frac{(a_7 - \tau)}{3(a_7 - a_6)}, Q_2(u) = \frac{2(a_6 - a_5) - (\tau - a_5)}{3(a_6 - a_5)}, R_2(v) = \frac{3(a_5 - a_4) - (\tau - a_4)}{3(a_5 - a_4)}$$

Here,

1. $P_1(t), Q_1(u)$, and $R_1(v)$ is bounded and continuous increasing function over [0, 0.33), [0.33, 0.66) and [0.66, 1] respectively.
2. $P_2(t), Q_2(u)$ and $R_2(v)$ is bounded and continuous decreasing function over [0, 0.33), [0.33, 0.66) and [0.66, 1] respectively.

Definition 6.2.7

The α-cut of the fuzzy set $\tilde{H}_e$ of the universe of discourse X is defined as

$$\tilde{H}_e = \{\tau \in X / \mu_{\tilde{H}_e}(\tau) \ge \alpha\} \text{ for } \alpha \in [0,1]$$

$$= \begin{cases} [P_1(\alpha), P_2(\alpha)] & \text{for } \alpha \in [0, 0.33) \\ [Q_1(\alpha), Q_2(\alpha)] & \text{for } \alpha \in [0.33, 0.66) \\ [R_1(\alpha), S_2(\alpha)] & \text{for } \alpha \in [0.66, 1] \end{cases}$$

Definition 6.2.8 (Generalized Heptagonal Fuzzy Number)

A generalized heptagonal fuzzy number is defined as $G\tilde{H} = (a_1, a_2, a_3, a_4, a_5, a_6, a_7; m, n)$ where $m, n \in (0,1)$, and its membership function is defined as (Figure 6.1)

$$\mu_{G\tilde{H}}(x) = \begin{cases} m\left(\dfrac{\tau - a_1}{a_2 - a_1}\right), & a_1 \le \tau \le a_2 \\ m-(m-n)\left(\dfrac{\tau - a_2}{a_3 - a_2}\right), & a_2 \le \tau \le a_3 \\ n-(n-1)\left(\dfrac{\tau - a_3}{a_4 - a_3}\right), & a_3 \le \tau \le a_4 \\ 1, & x = a_4 \\ n-(n-1)\left(\dfrac{a_5 - \tau}{a_5 - a_4}\right), & a_4 \le \tau \le a_5 \\ m-(m-n)\left(\dfrac{a_6 - \tau}{a_6 - a_5}\right), & a_5 \le \tau \le a_6 \\ m\left(\dfrac{a_7 - \tau}{a_7 - a_6}\right), & a_6 \le \tau \le a_7 \\ 0, & \tau \le a_1 \text{ and } \tau \ge a_7 \end{cases}$$

Definition 6.2.9

If $\tilde{A} = (a_1, a_2, a_3, a_4, a_5, a_6, a_7)$ and $\tilde{B} = (b_1, b_2, b_3, b_4, b_5, b_6, b_7)$ are the Heptagonal Fuzzy Numbers, then the addition, subtraction, multiplication, and division are defined as

- $\tilde{A} \oplus \tilde{B} = (a_1 + b_1, a_2 + b_2, \cdots, a_7 + b_7)$

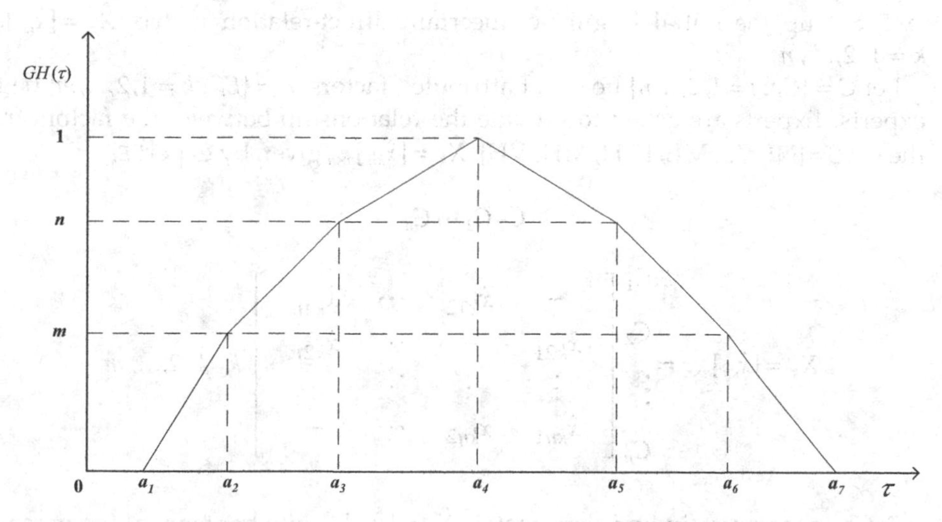

FIGURE 6.1
Heptagonal fuzzy number.

TABLE 6.1

Heptagonal Linguistic Values and Terms

Linguistic Terms	Linguistic Values
No influence (NI)	(0, 0, 0, 0, 0.05, 0.1, 0.15)
Very low influence (VL)	(0, 0.05, 0.1, 0.15, 0.2, 0.25, 0.3)
Medium low influence (ML)	(0.15, 0.2, 0.25, 0.3, 0.35, 0.4, 0.45)
Low influence (L)	(0.3, 0.35, 0.4, 0.45 0.5, 0.55, 0.6)
High influence (H)	(0.45 0.5, 0.55, 0.6, 0.65, 0.7, 0.75)
Medium high influence (MH)	(0.6, 0.65, 0.7, 0.75, 0.8, 0.85, 0.9)
Very high influence (VH)	0.75, 0.85, 0.9, 0.95, 1, 1, 1)

- $\tilde{A}\Theta\tilde{B} = (a_1 - b_7,\ a_2 - b_6, \ldots, a_7 - b_1)$
- $\tilde{A} \otimes \tilde{B} = (a_1b_1,\ a_2b_2,\ a_3b_3,\ \ldots, a_7b_7)$
- $\tilde{A} \div \tilde{B} = \left(\frac{a_1}{b_7}, \frac{a_2}{b_6}, \frac{a_3}{b_5}, \ldots, \frac{a_7}{b_1}\right)$

$\tilde{A} \oplus \tilde{B}, \tilde{A}\Theta\tilde{B}, \tilde{A} \otimes \tilde{B}, \tilde{A} \div \tilde{B}$ are also Heptagonal Fuzzy Numbers.

Next, the linguistic variables are represented by heptagonal fuzzy numbers in the following Table 6.1.

6.3 Fuzzy Heptagonal DEMATEL Method

In this section, the DEMATEL method is extended through heptagonal fuzzy numbers.

Step 1: Set up the initial linguistic uncertain direct-relation matrix $\hat{X}_k = [\hat{x}_{kij}]_{n\times n}$, $k = 1, 2, \ldots, n$

Let $C = \{C_i : i = 1,2,\ldots n\}$ be set of attributes/factors. $E = \{E_k : k = 1,2,\ldots,m\}$ be the experts. Experts are asked to provide the relationship between the factors from the set S={NI, VL, ML, L, H, MH, VH}. $\hat{X}_k = [\hat{x}_{kij}]_{n\times n}$ given by expert E_k

$$\hat{X}_k = [\hat{x}_{kij}]_{n\times n} = \begin{array}{c} \\ C_1 \\ C_2 \\ \vdots \\ C_n \end{array} \begin{array}{c} C_1\ C_2 \cdots C_n \\ \begin{bmatrix} - & \hat{x}_{k12} & \cdots & \hat{x}_{k1n} \\ \hat{x}_{k21} & - & \cdots & \hat{x}_{k2n} \\ \vdots & \vdots & \ddots & \vdots \\ \hat{x}_{kn1} & \hat{x}_{kn2} & \cdots & - \end{bmatrix} \end{array}, k=1, 2,\ldots, m$$

Step 2: Change uncertain linguistic matrix $\hat{X}_k = [\hat{x}_{kij}]_{n\times n}$ into heptagonal fuzzy matrix $\tilde{X}_k = [\tilde{x}_{kij}]_{n\times n}$.

The initial uncertain linguistic direct-relation matrices are transformed into heptagonal fuzzy numbers. i.e.,

$\hat{X}_k = [\hat{x}_{kij}]_{n\times n}$ is transformed into $\tilde{X}_k = [\tilde{x}_{kij}]_{n\times n}$. $\hat{x}_{kij} = (x^1_{kij}, x^2_{kij}, x^3_{kij}, x^4_{kij}, x^5_{kij}, x^6_{kij}, x^7_{kij})$, k=1, 2,..., m and i, j=1, 2,..., n. Particularly, $\hat{x}_{kij} = '-'$ is transformed into $\hat{x}_{kij} = (0,0,0,0,0,0,0)$.

Step 3: Design the group uncertain direct-relation matrix $\tilde{X}_k = [\tilde{x}_{kij}]_{n\times n}$.

All the uncertain direct-relation matrices $\tilde{X}_1, \tilde{X}_2, ..., \tilde{X}_m$ are aggregated into a group uncertain direct-relation matrix $\tilde{X}_k = [\tilde{x}_{kij}]_{n\times n}$.

It is denoted as $\tilde{x}_{kij} = (x^1_{kij}, x^2_{kij}, x^3_{kij}, x^4_{kij}, x^5_{kij}, x^6_{kij}, x^7_{kij})$ where

$$x^1_{ij} = \frac{1}{m}\sum_{k=1}^{m} x^1_{kij}, \quad x^2_{ij} = \frac{1}{m}\sum_{k=1}^{m} x^2_{kij}, \quad x^3_{ij} = \frac{1}{m}\sum_{k=1}^{m} x^3_{kij}, \quad x^4_{ij} = \frac{1}{m}\sum_{k=1}^{m} x^4_{kij},$$

$$x^5_{ij} = \frac{1}{m}\sum_{k=1}^{m} x^5_{kij}, \quad x^6_{ij} = \frac{1}{m}\sum_{k=1}^{m} x^6_{kij}, \quad x^7_{ij} = \frac{1}{m}\sum_{k=1}^{m} x^7_{kij} \quad i, j = 1,2,..., n,$$

Step 4: Frame normalized uncertain relation matrix $\tilde{Z}_k = [\tilde{z}_{kij}]_{n\times n}$.

$\tilde{X}_k = [\tilde{x}_{kij}]_{n\times n}$ is changed into $\tilde{Z}_k = [\tilde{z}_{kij}]_{n\times n}$.

It is given as $\tilde{z}_{ij} = (z^1_{ij}, z^2_{ij}, z^3_{ij}, z^4_{ij}, z^5_{ij}, z^6_{ij}, z^7_{ij})$,

where

$$z^1_{ij} = x^1_{ij} \Big/ \max_{1\le i\le n}\left\{\sum_{j=1}^{n} x^7_{ij}\right\}, \; z^2_{ij} = x^2_{ij} \Big/ \max_{1\le i\le n}\left\{\sum_{j=1}^{n} x^7_{ij}\right\}, \; ...., \; z^7_{ij} = x^7_{ij} \Big/ \max_{1\le i\le n}\left\{\sum_{j=1}^{n} x^7_{ij}\right\},$$

Here $\max_{1\le i\le n}\left\{\sum_{j=1}^{n} x^7_{ij}\right\} \neq 0$ and $0 \le z^1_{ij} \le z^2_{ij} \le z^3_{ij} \le ... \le x^7_{ij} < 1.$,

$\tilde{Z}$ is factorized into seven crisp value matrices $Z^1, Z^2, Z^3 ..., Z^7$

$$Z^1 = \begin{bmatrix} 0 & z^1_{12} & \cdots & z^1_{1n} \\ z^1_{21} & 0 & \cdots & z^1_{2n} \\ \vdots & \vdots & \ddots & \vdots \\ z^1_{n1} & z^1_{n2} & \cdots & 0 \end{bmatrix}, \; Z^2 = \begin{bmatrix} 0 & z^2_{12} & \cdots & z^2_{1n} \\ z^2_{21} & 0 & \cdots & z^2_{2n} \\ \vdots & \vdots & \ddots & \vdots \\ z^2_{n1} & z^2_{n2} & \cdots & 0 \end{bmatrix}, ...,$$

$$Z^7 = \begin{bmatrix} 0 & z^7_{12} & \cdots & z^7_{1n} \\ z^7_{21} & 0 & \cdots & z^7_{2n} \\ \vdots & \vdots & \ddots & \vdots \\ z^7_{n1} & z^7_{n2} & \cdots & 0 \end{bmatrix}$$

By the crisp DEMATEL method, the uncertain overall-relation matrix $\tilde{T}$ is defined as $\tilde{T} = \lim_{\tau\to\infty}\left((\tilde{Z})^1 + (\tilde{Z})^2 + ... + (\tilde{Z})^\tau\right)$

Step 5: Frame uncertain overall-relation matrix $\tilde{T} = [\tilde{t}_{ij}]_{n\times n}$

Let $\tilde{T} = \begin{bmatrix} \tilde{t}_{11} & \tilde{t}_{12} & \cdots & \tilde{t}_{1n} \\ \tilde{t}_{21} & \tilde{t}_{22} & \cdots & \tilde{t}_{21} \\ \vdots & \vdots & \ddots & \vdots \\ \tilde{t}_{n1} & \tilde{t}_{n1} & \cdots & \tilde{t}_{nn} \end{bmatrix}$, where $\tilde{t}_{ij} = (t_{ij}^1, t_{ij}^2, t_{ij}^3, t_{ij}^4, t_{ij}^5, t_{ij}^6, t_{ij}^7)$ Then

$$[t_{ij}^1]_{n\times n} = Z^1(I - Z^1)^{-1},\ i,j = 1,2,\ldots,n$$

$$[t_{ij}^2]_{n\times n} = Z^2(I - Z^2)^{-1},\ i,j = 1,2,\ldots,n$$

$$\ldots$$

$$[t_{ij}^7]_{n\times n} = Z^7(I - Z^7)^{-1},\ i,j = 1,2,\ldots,n$$

Step 6: Derive the total intensities of influencing and influenced correlation of factors C_i, $\tilde{c}_i$ and $\tilde{h}_i$, $i = 1,2,\ldots,n$,

Let $\tilde{c}_i$ be the overall intensity factor C_i influencing others. It is denoted by $\tilde{c}_i = \left(c_i^1, c_i^2, c_i^3, c_i^4, c_i^5, c_i^6, c_i^7\right)$,

where $c_i^1 = \sum_{j=1}^{n} t_{ij}^1, i = 1,2,\ldots,n,\quad c_i^2 = \sum_{j=1}^{n} t_{ij}^2, i = 1,2,\ldots,n,\quad c_i^3 = \sum_{j=1}^{n} t_{ij}^3, i = 1,2,\ldots,n,$ $c_i^4 = \sum_{j=1}^{n} t_{ij}^4, i = 1,2,\ldots,n,\quad \ldots,\ c_i^7 = \sum_{j=1}^{n} t_{ij}^7, i = 1,2,\ldots,n$

Let $\tilde{h}_j$ be the overall intensity factor C_j which is influenced by others. It is represented as $\tilde{h}_j = \left(h_j^1, h_j^2, h_j^3, h_j^4, h_j^5, h_j^6, h_j^7\right)$, where $h_j^1 = \sum_{i=1}^{n} t_{ij}^1, j = 1,2,\ldots,n,\quad h_j^2 = \sum_{i=1}^{n} t_{ij}^2, j = 1,2,\ldots,n,\quad h_j^3 = \sum_{i=1}^{n} t_{ij}^3, j = 1,2,\ldots,n,\quad h_j^4 = \sum_{i=1}^{n} t_{ij}^4, j = 1,2,\ldots,n,\ \ldots,\ h_j^7 = \sum_{i=1}^{n} t_{ij}^7, j = 1,2,\ldots,n,$

Step 7: Derive the uncertain prominence and relation of each factor $\tilde{p}_i$ and $\tilde{r}_i$

Let $\tilde{p}_i$ be the uncertain prominence of factor C_i. It is denoted as $\tilde{p}_i = \left(p_i^1, p_i^2, p_i^3, p_i^4, p_i^5, p_i^6, p_i^7\right)$, where $p_i^1 = c_i^1 + h_i^1$, $p_i^2 = c_i^2 + h_i^2$, $p_i^3 = c_i^3 + h_i^3$, ..., $p_i^7 = c_i^7 + h_i^7$, $i = 1,2,\ldots,n,$

Let $\tilde{r}_i$ be the uncertain relation of factor C_i. It is denoted as $\tilde{r}_i = \left(r_i^1, r_i^2, r_i^3, r_i^4, r_i^5, r_i^6, r_i^7\right)$, where $r_i^1 = c_i^1 - h_i^1$, $r_i^2 = c_i^2 - h_i^2$, $r_i^3 = c_i^3 - h_i^3$, ..., $r_i^7 = c_i^7 - h_i^7$, $i = 1,2,\ldots,n$

Step 8: Find the crisp prominence and relation between each factor p_i and r_i.

The crisp values of prominence (p_i) and relation (r_i) of factor C_i are determined as,

$$p_i = \frac{1}{7}\left(p_i^1 + p_i^2 + p_i^3 + p_i^4 + p_i^5 + p_i^6 + p_i^7\right),\quad i = 1,2,\ldots,n.$$

$$r_i = \frac{1}{7}\left(r_i^1 + r_i^2 + r_i^3 + r_i^4 + r_i^5 + r_i^6 + r_i^7\right),\quad i = 1,2,\ldots,n$$

Step 9: Construct the causal diagram based on p_i and r_i.

The importance and classification of factors are determined through the value of p_i and r_i. A causal diagram can be depicted based on p_i and r_i.

6.3.1 Numerical Illustration

Violence is a common event occurring everywhere, which disturbs our life in numerous ways. Although human beings are taking a lot of effort to form civilized societies, we are unable to stop the influence of violence and aggression. Therefore, it is our concern to analyze the cause and effects of violence and the people who are responsible for it. It is reported by WHO that in most cases of violence, the active participants are the youth. Also, WHO stated that youth violence is the fourth leading cause of death among youngsters worldwide, and due to this, two lakhs people die every year (WHO, 2016). Aggressiveness among the youth is one of the unhealthy developmental problems in many families. The anger of youth is an easy target and vulnerable section of the society to be misused by anyone. The terrorist acts, for instance, are motivated by two factors, viz., social and political injustice, and hence, violence as the only tool for change. Violence includes a range of actions from bullying, physical fighting, severe sexual, physical assault to homicide, and other atrocities. Youth engaging themselves in violence and aggressive behavior is unpredictable because the factors that influence may vary and they belong to the category of uncertain factors. It is extremely difficult to predict the behavior of the male youth in comparison to their female counterpart. Therefore, the study aims to analyze the cause and effect of youth aggressiveness.

We have collected the following attributes which are related to the youth aggressiveness through the unsupervised method by interviewing the 100 youths from Chennai, Tamil Nadu. From their statement, the key purposes for youth violence have been selected as the attributes. C_1—Poor monitoring and supervision of children by parents, C_2—Academic failure/dropping out of school, C_3—Delinquent peers/Gang membership, C_4—Addiction to drugs and alcohol, C_5—Poverty/unemployment, C_6—Involvement in other forms of antisocial behavior, C_7—Aggressive behavior, C_8—Parental substance or criminality, C_9—Depression, C_{10}—Castisem/inequality. Then, three experts—a sociologist, a psychiatrist, and victims—were interviewed to give opinions on the existence and intensities of the correlation among the factors from S={No Influence, Very Low, Medium Low, Low, High, Medium High, Very High}.

To illustrate the proposed heptagonal DEMATEL technique, the proposed technique has been applied to solve the problem of youth violence, and the computational procedure is summarized as follows:

Step 1: At the initial stage, with the three expert's view, the initial linguistic uncertain direct-relation matrices $\hat{X}_k$, $k = 1,2,3$ are given below in Tables 6.2–6.4.

Step 2 and 3: After transforming initial linguistic uncertain direct-relation matrices into heptagonal fuzzy numbers, the group uncertain direct-relation matrix $\tilde{X}_k = [\tilde{x}_{kij}]_{n\times n}$. is obtained by aggregating all three matrices, which is shown in Table 6.5.

Step 4: The uncertain overall-relation matrix $\tilde{Z}_k$ is obtained from the group uncertain direct-relation matrix $\tilde{X}_k$.

Step 5: The uncertain overall-relation matrix $\tilde{T}$ is obtained.

Step 6–8: The overall intensities of influencing and influenced correlation of factors and prominence ($\tilde{p}_i$) and relation ($\tilde{r}_i$) are obtained and the crisp prominence (p_i) and relation (r_i) are also obtained, which are shown in Table 6.6. To view the complicated (10 by 10) matrices in a simple manner, the causal diagram is depicted using p_i and r_i. The causal diagram divides the factors into causes and effect group.

TABLE 6.2
Initial Uncertain Direct-Relation Matrix $\hat{X}_1$ Provided by E_1

	C_1	C_2	C_3	C_4	C_5	C_6	C_7	C_8	C_9	C_{10}
C_1	–	VH	VH	H	H	VH	H	NI	L	NI
C_2	NI	–	VH	H	VH	VH	H	NI	L	H
C_3	H	VH	–	L	H	VH	H	L	H	H
C_4	L	H	H	–	VL	H	H	VL	L	H
C_5	VH	H	VH	L	–	VL	VH	VL	VH	NI
C_6	VH	H	VH	VH	NI	–	H	H	L	H
C_7	H	L	L	H	H	VL	–	H	VH	VH
C_8	NI	VH	VH	H	H	VL	L	–	VH	NI
C_9	H	NI	VL	H	H	H	VH	H	–	VH
C_{10}	NI	H	VH	VH	L	H	VH	NI	H	–

TABLE 6.3
Initial Uncertain Direct-Relation Matrix $\hat{X}_2$ Provided by E_2

	C_1	C_2	C_3	C_4	C_5	C_6	C_7	C_8	C_9	C_{10}
C_1	–	H	VH	H	H	H	L	VL	VL	NI
C_2	VL	–	VH	H	VH	H	L	NI	VL	L
C_3	L	H	–	L	VH	VH	H	VL	L	VL
C_4	L	H	H	–	VL	H	VH	L	H	L
C_5	H	L	VH	H	–	H	VH	H	VH	L
C_6	H	H	H	VH	L	–	H	L	H	VH
C_7	H	L	VH	H	VH	H	–	H	VH	H
C_8	NI	VH	H	VH	H	VH	H	–	VH	H
C_9	H	L	H	H	VH	L	VL	H	–	H
C_{10}	NI	H	VH	H	L	H	H	NI	VH	–

TABLE 6.4
Initial Uncertain Direct-Relation Matrix $\hat{X}_3$ Provided by E_3

	C_1	C_2	C_3	C_4	C_5	C_6	C_7	C_8	C_9	C_{10}
C_1	–	VH	VH	H	H	L	VL	NI	L	NI
C_2	VL	–	H	VH	L	NI	VL	H	L	VL
C_3	L	H	–	H	VH	VH	H	VH	NI	L
C_4	VL	VH	H	–	VL	VH	H	L	H	VH
C_5	VH	L	VH	H	–	H	VH	H	H	VH
C_6	H	H	VH	VH	VL	–	H	L	VL	H
C_7	H	VH	VH	H	H	VL	–	H	VH	H
C_8	NI	H	VH	H	VH	VH	H	–	VH	NI
C_9	VL	H	VL	H	VH	H	H	H	–	H
C_{10}	NI	VL	H	H	VH	VH	H	NI	H	–

TABLE 6.5

Group Uncertain Direct-Relation Matrix $\tilde{X}_k$

	C_1	C_2	C_3	C_4
C_1	(0, 0, 0, 0, 0, 0, 0)	(0.6, 0.7, 0.7, 0.8, 0.8, 0.9, 0.9)	(0.7, 0.8, 0.8, 0.9, 0.9, 1, 1)	(0.4, 0.5, 0.5, 0.6, 0.6, 0.7, 0.7)
C_2	(0.05, 0.08, 0.11, 0.2, 0.25, 0.3)	(0, 0, 0, 0, 0, 0, 0)	(0.6, 0.7, 0.7, 0.8, 0.8, 0.9, 0.9)	(0.5, 0.6, 0.6, 0.7, 0.7, 0.8, 0.8)
C_3	(0.3, 0.4, 0.4, 0.5, 0.5, 0.6, 0.6)	(0.5, 0.6, 0.5, 0.7, 0.7, 0.8, 0.8)	(0, 0, 0, 0, 0, 0, 0)	(0.3, 0.4, 0.4, 0.5, 0.5, 0.6, 0.6)
C_4	(0.2, 0.2, 0.3, 0.3, 0.4, 0.4, 0.5)	(0.5, 0.6, 0.5, 0.7, 0.7, 0.8, 0.8)	(0.4, 0.5, 0.5, 0.6, 0.6, 0.7, 0.7)	(0, 0, 0, 0, 0, 0, 0)
C_5	(0.6, 0.7, 0.7, 0.8, 0.8, 0.9, 0.9)	(0.3, 0.4, 0.4, 0.5, 0.5, 0.6, 0.6)	(0.7, 0.8, 0.8, 0.9, 0.9, 1, 1)	(0.4, 0.4, 0.5, 0.5, 0.6, 0.6, 0.7)
C_6	(0.5, 0.6, 0.6, 0.7, 0.7, 0.8, 0.8)	(0.4, 0.5, 0.5, 0.6, 0.6, 0.7, 0.7)	(0.6, 0.7, 0.7, 0.8, 0.8, 0.9, 0.9)	(0.7, 0.8, 0.8, 0.9, 0.9, 1, 1)
C_7	(0.4, 0.5, 0.5, 0.6, 0.6, 0.7, 0.7)	(0.3, 0.4, 0.4, 0.5, 0.5, 0.6, 0.6)	(0.6, 0.6, 0.7, 0.7, 0.8, 0.8, 0.8)	(0.4, 0.5, 0.5, 0.6, 0.6, 0.7, 0.7)
C_8	(, 0, 0, 0, 0, 0.0, 0.1, 0.1)	(0.6, 0.7, 0.7, 0.8, 0.8, 0.9, 0.9)	(0.6, 0.6, 0.7, 0.7, 0.8, 0.8, 0.8)	(0.5, 0.6, 0.6, 0.7, 0.7, 0.8, 0.8)
C_9	(0.3, 0.3, 0.4, 0.4, 0.5, 0.5, 0.6)	(0.2, 0.2, 0 3, 0.3, 0.4, 0.4, 0.5)	(0.2, 0.2, 0.3, 0.3, 0.4, 0.4, 0.5)	(0.4, 0.5, 0.5, 0.6, 0.6, 0.7, 0.7)
C_{10}	(0, 0, 0, 0, 0.05, 0.1, 0.15)	(0.3, 0.3, 0.4, 0.4, 0.5, 0.5, 0.6)	(0.6, 0.6, 0.7, 0.7, 0.8, 0.8, 0.8)	(0.5, 0.6, 0.6, 0.7, 0.7, 0.8, 0.8)

	C_5	C_6	C_7	C_8
C_1	(0.4, 0.5, 0.5, 0.6, 0.6, 0.7, 0.7)	(0.5, 0.5, 0.6, 0.6, 0.7, 0.7, 0.7)	(0.2, 0.3, 0.3, 0.4, 0.4, 0.5, 0.5)	(0, 0.0, 0.0, 0.1, 0.1, 0.2, 0.2)
C_2	(0.6, 0.6, 0.7, 0.7, 0.8, 0.8, 0.8)	(0.4, 0.4, 0.4, 0.5, 0.5, 0.6, 0.6)	(0.2, 0.3, 0.3, 0.4, 0.4, 0.5, 0.5)	(0.1, 0.1, 0.1, 0.2, 0.2, 0.3, 0.3)
C_3	(0.6, 0.7, 0.7, 0.8, 0.8, 0.9, 0.9)	(0.7, 0.8, 0.8, 0.9, 0.9, 1, 1)	(0.4, 0.5, 0.5, 0.6, 0.6, 0.7, 0.7)	(0.3, 0.4, 0.4, 0.5, 0.5, 0.6, 0.6)
C_4	(0, 0.05, 0.1, 0.1, 0.2, 0.2, 0.3)	(0.5, 0.6, C.6, 0.7, 0.7, 0.8, 0.8)	(0.5, 0.6, 0.6,0.7, 0.7, 0.8, 0.8)	(0.2, 0.2, 0.3, 0.3, 0.4, 0.4, 0.5)
C_5	(0, 0, 0, 0, 0, 0, 0)	(0.3, 0.3, 0.4, 0.4, 0.5, 0.5, 0.6)	(0.7, 0.8, 0.8, 0.9, 0.9, 1, 1)	(0.3, 0.3, 0.4, 0.4, 0.5, 0.5, 0.6)
C_6	(0.1, 0.1, 0.1, 0.2, 0.2, 0.3, 0.3)	(0, 0, 0, 0, 0, 0, 0)	(0.4, 0.5, 0.5, 0.6, 0.6, 0.7, 0.7)	(0.3, 0.4, 0.4, 0.5, 0.5, 0.6, 0.6)
C_7	(0.5, 0.6, 0.6, 0.7, 0.7, 0.8, 0.8)	(0.1, 0.2, 0.2, 0.3, 0.3, 0.4, 0.4)	(0, 0, 0, 0, 0, 0, 0)	(0.4, 0.5, 0.5, 0.6, 0.6, 0.7, 0.7)
C_8	(0.5, 0.6, 0.6, 0.7, 0.7, 0.8, 0.8)	(0.5, 0.5, 0.6, 0.6, 0.7, 0.7, 0.7)	(0.4, 0.4, 0.5, 0.5, 0.6, 0.6, 0.7)	(0, 0, 0, 0, 0, 0, 0)
C_9	(0.6, 0.7, 0.7, 0.8, 0.8, 0.9, 0.9)	(0.4, 0.4, 0.5, 0.5, 0.6, 0.6, 0.7)	(0.4, 0.4, 0.5, 0.5, 0.6, 0.6, 0.6)	(0.4, 0.5, 0.5, 0.6, 0.6, 0.7, 0.7)
C_{10}	(0.4, 0.5, 0.5, 0.6, 0.6, 0.7, 0.7)	(0.5, 0.6, 0.6, 0.7, 0.7, 0.8, 0.8)	(0.5, 0.6, 0.6, 0.7, 0.7, 0.8, 0.8)	(0, 0, 0, 0, 0.0, 0.1, 0.1)

	C_9	C_{10}
C_1	(0.2, 0.2, 0.3, 0.3, 0.4, 0.4, 0.5)	(0, 0, 0, 0, 0.0, 0.1, 0.1)
C_2	(0.2, 0.2, 0.3, 0.3, 0.4, 0.4, 0.5)	(0.2, 0.3, 0.3, 0.4, 0.4, 0.5, 0.5)
C_3	(0.2, 0.2, 0.3, 0.3, 0.4, 0.4, 0.5)	(0.2, 0.3, 0.3, 0.4, 0.4, 0.5, 0.5)
C_4	(0.4, 0.4, 0.5, 0.5, 0.6, 0.6, 0.7)	(0.5, 0.5, 0.6, 0.6, 0.7, 0.7, 0.7)
C_5	(0.6, 0.7, 0.7, 0.8, 0.8, 0.9, 0.9)	(0.3, 0.3, 0.4, 0.4, 0.5, 0.5, 0.5)
C_6	(0.2, 0.3, 0.3, 0.4, 0.4, 0.5, 0.5)	(0.5, 0.6, 0.6, 0.7, 0.7, 0.8, 0.8)
C_7	(0.7, 0.8, 0.8, 0.9, 0.9, 1, 1)	(0.5, 0.6, 0.6, 0.7, 0.7, 0.8, 0.8)
C_8	(0.7, 0.8, 0.8, 0.9, 0.9, 1, 1)	(0.1, 0.1, 0.1, 0.2, 0.2, 0.3, 0.3)
C_9	(0, 0, 0, 0, 0, 0, 0)	(0.5, 0.6, 0.6, 0.7, 0.7, 0.8, 0.8)
C_{10}	(0.5, 0.6, 0.6, 0.7, 0.7, 0.8, 0.8)	(0, 0, 0, 0, 0, 0, 0)

TABLE 6.6

Computational Results

	$\tilde{p}_i$	$\tilde{r}_i$	p_i	r_i
C_1	(1.5316, 1.9317, 2.4410, 3.1110, 4.1666, 5.7683, 7.6674)	(0.1469, 0.1741, 0.2086, 0.2537, 0.3043, 0.3812, 0.4465)	3.8025	0.2736
C_2	(1.8578, 2.3160, 2.8985, 3.6639, 4.8294, 6.5980, 8.6646)	(−0.2290, −0.2666, −0.3143, −0.3766, −0.4515, −0.5652, −0.7042)	4.4040	−0.4153
C_3	(2.3550, 2.8825, 3.5528, 4.4333, 5.7514, 7.7516, 9.9661)	(−0.3063, −0.3394, −0.3816, −0.4371, −0.5240, −0.6559, −0.7233)	5.2418	−0.4811
C_4	(2.0175, 2.5160, 3.1495, 3.9814, 5.2100, 7.0744, 9.2744)	(−0.2688, −0.2986, −0.3366, −0.3866, −0.4632, −0.5796, −0.6904)	4.7462	−0.4320
C_5	(2.1761, 2.6837, 3.3288, 4.1762, 5.4433, 7.3661, 9.5157)	(0.1301, 0.1462, 0.1668, 0.1938, 0.2321, 0.2903,0.3385)	4.9557	0.2140
C_6	(2.0976, 2.5951, 3.2273, 4.0577, 5.3014, 7.1886, 9.3467)	(−0.0186, −0.0202, −0.0222, −0.0249, −0.0296, −0.0369, −0.0163)	4.8306	−0.0241
C_7	(2.1532, 2.6674, 3.3206, 4.1786, 5.4460, 7.3693, 9.5942)	(0.0529, 0.0583, 0.0652, 0.0744, 0.0892, 0.1117, 0.1291)	4.9613	0.0830
C_8	(1.6839, 2.0998, 2.6292, 3.3254, 4.4234, 6.0897, 8.0307)	(0.4588, 0.5150, 0.5864, 0.6802, 0.8151, 1.0200, 1.1509)	4.0403	0.7466
C_9	(1.9758, 2.4555, 3.0652, 3.8661, 5.0718, 6.9013, 9.0279)	(−0.0487, −0.0577, −0.0690, −0.0838, −0.1007, −0.1264, −0.1137)	4.6234	−0.0857
C_{10}	(1.7698, 2.1933, 2.7323, 3.4414, 4.5623, 6.2631, 8.2424)	(0.0827, 0.0889,0.0967, 0.1069, 0.1283, 0.1608, 0.1829)	4.1721	0.1210

Here, p_i shows the degree of importance that factor C_i acts in the entire system. On the contrary, the r_i portrays the effect that factor C_i pays to the system. Precisely, if r_i is +ve, factor C_i is a cause group, while factor C_i is an effect group if r_i is –ve.

To view the computational results in (Table 6.6), a causal diagram is provided in Figure 6.2. From the causal diagram, it is observed that the evaluation criteria were virtually divided in to the cause group, including C_1—Poor monitoring and supervision of children by parents, C_5—Poverty/unemployment, C_7—Aggressive behavior, C_8—Parental substance or criminality, C_{10}—Castisem/inequality, while the effect group was composed of criteria C_2—Academic failure/dropping out of school, C_3—Delinquent peers/Gang membership, C_4—Addiction to drugs and alcohol, C_6—Involvement in other forms of antisocial behavior, and C_9—Depression. Moreover, by only considering the relation of factors, a ranking order $C_8 \succ C_1 \succ C_5 \succ C_{10} \succ C_7 \succ C_6 \succ C_9 \succ C_4 \succ C_2 \succ C_3$ is obtained for making good decision. Consequently, C_1—Poor monitoring and supervision of children by parents and C_8—Parental substance or criminality are the major causes of youths are involving in violence. Therefore, parent's role is very important in the behavior of the youth and their growth.

6.4 Conclusion

This present study has proposed an extension of DEMATEL method using heptagonal fuzzy numbers for analyzing the relationship among factors. With the aid of this technique,

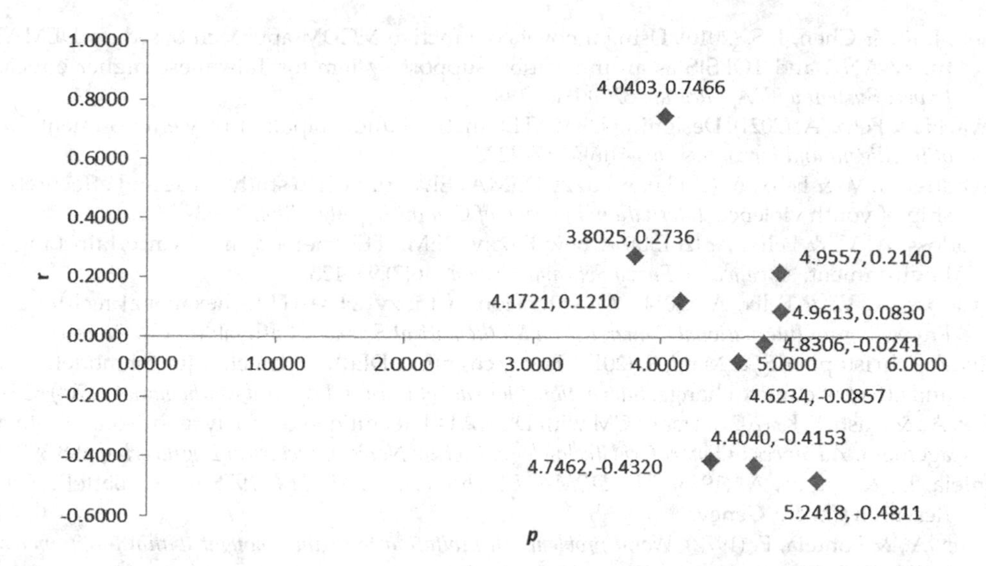

FIGURE 6.2
Causal diagram.

the complex interaction between factors and digraphs can be converted into an observable model. The merit of this technique lies in catching the complexity of any problem through which an insightful decision can be taken by the decision-maker. Compared with the other existing DEMATEL methods, the unique characteristics of this new extended DEMATEL method are the following: (a) appropriate to solve the problem of interrelated factor analysis in fuzzy environment when the judgments and decisions are uncertain in nature and (b) this technique acts better to reduce the uncertainty instead of using triangular and trapezoidal fuzzy number. This proposed technique can also be applied in all decision-making fields such as Engineering, Data Science, Medical, and Social Science. Further research, linguistic variables may be represented by nonagonal and decagonal fuzzy number to extend the DEMATEL method.

References

Abdullah, L., & Zulkifli, N. (2015). Integration of fuzzy AHP and interval type-2 fuzzy DEMATEL: An application to human resource management. *Expert Systems with Applications* 42(9):4397–4409.

Altuntas, S., & Dereli, T. (2015). A novel approach based on DEMATEL method and patent citation analysis for prioritizing a portfolio of investment projects. *Expert Systems with Applications* 42(3):1003–1012.

Aseervatham, S., & Devadoss A. V. (2015). Analysis on criteria based emotive music composition selection using a new trapezoidal fuzzy DEMATEL - TOPSIS hybrid technique. *Journal of Fuzzy Set Valued Analysis* 2:122–133.

Bai, C., & Sarkis, J. (2013). A grey-based DEMATEL model for evaluating business process management critical success factors. *International Journal of Production Economics* 146(1):281–292.

Bellmann, R. E., & Zadeh, L. A. (1970). Decision making in fuzzy environment. *Management Science* 17(4):141–164.

Chen, J. K., & Chen, I. S. (2010). Using a novel conjunctive MCDM approach based on DEMATEL, fuzzy ANP, and TOPSIS as an innovation support system for Taiwanese higher education. *Expert System with Application* 37(3):1981–1990.

Deva, N., & Felix, A. (2021). Designing DEMATEL method under bipolar fuzzy environment. *Journal of Intelligent and Fuzzy Systems* 41(6):7257–7273.

Devadoss, A. V., & Felix, A. (2013a). A Fuzzy DEMATEL approach to study cause and effect relationship of youth violence. *International Journal of Computing Algorithm* 2:363–372.

Devadoss, A. V., & Felix, A. (2013b). A new Fuzzy DEMATEL method in an Uncertain Linguistic Environment. *Advances in Fuzzy Sets and Systems* 16(2):93–123.

Devadoss, A. V., & Felix, A. (2014). An extension of fuzzy DEMATEL- hexagonal techniques. In *Proceedings of International Conference on Mathematical Sciences*, Sathyabama University.

Felix, A., Christopher, S., & Mani, R. (2017). Fuzzy cognitive DEMATEL technique for modeling cause and effect of climate change. *International Journal of Pure and Applied Mathematics* 117(4):655–661.

Felix, A., & Dash, S. K. (2021). Haar FCM with DEMATEL techniques to analyze the solid waste management. *Advances in Smart Grid Technology, Lecture Notes in Electrical Engineering* 688:393–402.

Fontela, E., & Gabus, A. (1976). *The DEMATEL observer, DEMATEL 1976 report*. Battelle Geneva Research Centre, Geneva.

Gabus, A., & Fontela, E. (1972). *World problems an invitation to further thought within the framework of DEMATEL*. Battelle Geneva Research Centre, Geneva.

Hori, S., & Shimizu, Y. (1999). Designing methods of human interface for supervisory control systems. *Control Engineering Practice* 7(11):1413–1419.

Karthik, S., Sarojkumar, D., & Punithavelan, N. (2019). Haar ranking of linear and non-linear heptagonal fuzzy number and its application. *International Journal of Innovative Technology and Exploring Engineering* 8(6):1212–1220.

Karthik, S., Sarojkumar, D., & Punithavelan, N. (2020). A fuzzy decision-making system for the impact of pesticides applied in agricultural fields on human health. *International Journal of Fuzzy System Applications* 9(3):42–62.

Lin, R. J. (2013). Using fuzzy DEMATEL to evaluate the green supply chain management practices. *Journal of Cleaner Production* 40:32–39.

Lin, C. J., & Wu, W. W. (2004). A fuzzy extension of the DEMATEL method for group decision making. *European Journal of Operational Research* 156(1):445-455.

Lin, C. J., & Wu, W. W. (2008). A causal analytical method for decision making under fuzzy environment. *Expert System with Applications* 34(1):205–213.

Rathi, K., & Balamohan, S. (2014). Representation and ranking of fuzzy numbers with heptagonal membership function using value and ambiguity index. *Applied Mathematical Sciences* 8(87):4309–4321.

Selvaraj, A., Dash, S. K., & Punithavelan, N. (2018). Hexagonal fuzzy DEMATEL approach to analyze the solid waste management. *International Journal of Pure and Applied Mathematics* 118(5): 475–492.

Shieh, J. I., Wu, H. H., & Huang, K. K. (2010). A DEMATEL method in identifying key success factors of hospital service quality. *Knowledge-Based System* 23(3):277–282.

Suo, W. L., Feng, B., & Fan, Z. P. (2012). Extension of the DEMATEL method in an uncertain linguistic environment. *Soft Computing* 16(3):471–483.

Tamura, H., Okanishi, H., & Akazawa, K. (2006). Decision support for extracting and dissolving consumer's uneasiness over foods using stochastic DEMATEL. *Journal of Telecommunication Information and Technology* 4:91–95.

Tseng, M. L. (2009a). A causal and effect decision making model of service quality expectation using grey-fuzzy DEMATEL approach. *Expert System with Applications* 36(4):7738–7748.

Tseng, M. L. (2009b). Using the extension of DEMATEL to integrate hotel service quality perceptions into a cause-effect model in uncertainty. *Expert Systems with Applications* 36(5):9015–9023.

Tseng, M. L., & Lin, H. Y. (2009). Application of fuzzy DEMATEL to develop a cause and effect model of municipal solid waste management in Metro Manila. *Environmental and Monitoring Assessment* 158:519–533.

Velmathi, G., & Felix, A. (2019). IC fabrication process steps analysis using fuzzy DEMATEL method. *International Journal of Innovative Technology and Exploring Engineering* 9(1):1968–1973.

WHO (2016). https://www.who.int/news-room/fact-sheets/detail/youth-violence

Wu, W. W., & Lee, Y. T. (2007). Developing global managers competencies using the fuzzy DEMATEL method. *Expert Systems with Applications* 32:499–507.

Zadeh, L. A. (1965). Fuzzy Sets. *Information and Control* 8(2):338–353.

Zadeh, L. A. (1975). The concept of a linguistic variable and its application to approximate reasoning (Part II). *Information Science* 8:301–357.

7

A Comparative Study of Intrapersonal and Interpersonal Influencing Factors on the Academic Performance of Technical and Nontechnical Students

Deepti Sharma, Rishi Asthana, and Vaishali Sharma
BML Munjal University

CONTENTS

7.1 Introduction 101
7.1.1 Intrapersonal Factors 102
7.1.2 Interpersonal Influencing Factors 103
7.2 Aim of the Research 105
7.3 Methodology 106
7.4 Data Analysis 107
7.5 Results and Discussions 108
7.6 Conclusion 114
References 115

7.1 Introduction

Students' varying academic temperaments often surprise the academicians. Some students' enthusiasm to embrace academic challenges delights the academicians, whereas others' sheer reluctance and disengagement toward academics disheartens their endeavors. Students' diverse and miscellaneous academic behavior thus intrigues the researchers to unravel the possible factors that help them understand their students' academic performance. Mere assessment of some assignments and examinations cannot give an elaborative picture of students' behavior and attitude toward their academic work and success in their formal education

A lot of research has already been done to study the different factors which affect students' performance (Saa et al., 2019; Cavilla, 2017; Anderson & Good, 2016; Gibson, 1982) but we have categorized the various factors into intrapersonal and interpersonal influencing factors to see how they impact students' academic performance.

The cultural diversity of India makes it unique and remarkable in every sense. It is quite challenging for universities to accept and embrace the multidimensional aspects of diversity (Cooke & Saini, 2010). The students come from diverse backgrounds/cultures, which to some extent propel their ways of formulating an opinion and also influence their intrapersonal and interpersonal influencing factors. Hence, the varied backgrounds of the students fascinated us to study how different factors navigate the mindsets of the students which affect their academic performance culture is "the way of life especially the general

DOI: 10.1201/9781003291916-7

customs and beliefs, of a particular group of people at a particular time" (The Cambridge English Dictionary, 2015).

Intrapersonal as well as interpersonal influencing factors consciously and sometimes unconsciously affect an individual's behavior. Kassarnig et al. (2018) state that several behavioral patterns are closely attached to academic performance. They can influence social ties and other activities in which students participate. These factors also shape the individual's mindset. Mindset is interpreted as a collection of our thoughts and beliefs, which culminates our inner and outer world. Many researchers have shared their valuable insights to define Mindset. Dweck and Yeager (2019) state, Mindsets help to organize multiple variables such as goals, attributions, helplessness, etc. to provide them a clear meaning, which further adds a significant value of right efforts to achieve better results.

Various factors such as leadership (Deng et al., 2020), parenting style (Yang & Zhao, 2020) personality traits (Credê & Kuncel, 2008), or stress (Elias et al. 2011) collectively play a significant role to affect the mindset of the students which consciously or unconsciously impacts their academic performance. The cumulative effect of different factors paves a route to consolidate different thoughts. People with fixed mindset believe that intelligence is fixed and cannot be changed, whereas people with growth mindset feel that intelligence can be developed and improved with the right efforts.

Various intrapersonal and interpersonal influencing factors pave our way of thinking and sometimes motivate or demotivate us in different situations. It is hence difficult to ascertain which factor has an edge over the other. Wentzel and Wigfield (1998) stated that a comprehensive study of students' academic behavior is possible when both social and academic factors are sincerely taken into consideration.

7.1.1 Intrapersonal Factors

Intrapersonal factors are the individual's beliefs such as a sense of self, awareness to restrict behavior in extreme states, recognition of the power of expression, etc. Tatnell et al. (2014) state that among intrapersonal factors, emotion regulation and self-esteem show the strongest role in behavior. These factors have tremendous scope of bringing exponential growth in life provided the individual is aware of his abilities. Rampullo et al. (2015) state intrapersonal factors play a significant role; they can act as a source of support or a barrier. They are the talents and abilities which reinforce the individual's competence to face and accomplish the challenges and goals. Flavian (2016) states that students' self-fulfillment is based on realistic self-awareness. Research shows that factors including students' awareness about themselves, their goals to achieve, and their expectations from themselves and external world play a pivotal role in their academic performance and mental well-being (Gu et al., 2015). Bandura, an eminent Social Cognitive Psychologist, stated that self-efficacy or confidence is one of the pivotal factors which guides an individual to develop an optimistic view about his goal. It is self-belief in one's abilities that ascertains the successful accomplishment of the task. They are the predictors of the guaranteed completion of the task (Bandura, 1997; Bandura et al., 1999).

To hone intrapersonal skills, language determines a significant role. Language is the most potent vehicle of expression. English is the most common language used in education and professional world. Nonnative English learners may feel uncomfortable and anxious while expressing their opinions. MacIntyre et al. (1998) describe foreign language anxiety as "the worry and negative emotional reaction aroused when learning or using a second language." It is perceived that students suffer foreign language learning anxiety and it reflects on their academic and peer performance: "Language anxiety can originate from learner's own sense of self, their self-related cognitions, language learning difficulties, differences in learners" (Hashemi, 2011). Students who suffer from language anxiety usually avoid eye contact and feel uncomfortable in expressing their thoughts in the classroom and gradually they start avoiding the class discussions.

Each student has a unique way to respond to setbacks and challenges. Some accept the hurdles to improvise their performance and some may get intimidated and see them as insurmountable blocks.

Anxiety, nervousness, and stress are the immediate responses, which occur when students face some unexpected challenges. It is interesting to study to what extent anxiety obstructs academic performance. Research shows different results, in a few cases, it was found that anxiety impedes success (Seipp, 1991; Keogh et al. 2004; Dobson 2012; Macher et al., 2012; Owens et al., 2012) On the contrary, in a few cases, anxiety helped the students to perform better (Alkhalaf, 2018).

Joining university is a huge transition in students' life. A major shift takes place in their physical, mental, psychological, and social arena. The sudden shift may overwhelm some students to adapt to the new changes. As mentioned earlier, individuals with growth mindset will take this as a huge opportunity as they enter a new phase of life. On the contrary, those who feel little intimidated by the change may get affected by some of their preconceived notions or feedback they receive about the teachers. Students who frequently miss the mark become fearful of new tasks and distraught by setbacks (Dweck, 2006). Prejudices, preconceived notions, and biasedness can have a major influence on somebody's behavior if he lacks the vision to pierce through them to see the reality (Duckitt, 1992; Olson, 2009; Rattan & Dweck, 2010).

Some students are eager to take feedback from their seniors and peers about the university, curriculum, faculty, other activities, etc. (Owen, 2016). They are not affected by the feedback they receive as they wish to make their own opinions as their academic journey progresses. They don't allow their preconceived notions to affect their thinking. Cognitive biases can impair the decision-making process and can disrupt collective outcomes (Caviola et al. 2014).

Correct opinions are formed after an exhaustive analysis of the situation. Self-driven individuals do not form opinions easily. They critically evaluate the situation, spend time in thorough analysis, observe how others respond in the same situation (Moussaïd et al., 2013), and then derive their conclusions. Conscious opinions help in taking wiser decisions (Shahsavarani & Abadi, 2015).

7.1.2 Interpersonal Influencing Factors

Interpersonal influencing factors are the external factors that students face while coming in contact with their external surroundings. In the book *The Psychology of Interpersonal Relationships,* the authors quoted the work of social psychologists Baumeister and Leary who stated that since evolution, humans showed a strong inclination toward interpersonal attachments as they provided them a need to belong to a certain group (Berscheid & Rega, 2005).

Interpersonal influencing factors comprise different external entities which come into our contact on daily basis. They consist of different social groups which are an integral part of a personality development (Lamb & Bornstein, 2011).

To study the effect of interpersonal influencing factors on students we have extensively concentrated on the teacher's verbal and nonverbal skills, pedagogy, inclusion of technology, and socio-cultural diversity that enhance or affect student engagement. The ever engaging and active bond of the teacher and student always strengthens the overall learning which takes place in the class (Gablinske, 2014). The teacher not only provides academic support but also looks after the emotional and social being of the students (Varga, 2017).

A well-informed and conscious teacher knows that his/her knowledge and skills have a direct and indirect effect on the students. Many studies show that the teacher's communication style has an indelible effect on the students. Right paralinguistic features such as voice modulation, pitch, rate of speech and tone, determine the effectiveness of communication. Positive and efficient communication motivates the students significantly. On the contrary, weak and fragile communication not only demotivates the students but also pushes them to lose interest in a particular subject. It further affects students' communication style due to which they might struggle to voice their opinion in public (Ataunal, 2003).

Effective teaching depends on the teacher's effective communication. A teacher with effective communication skills understands the needs of the students and uses their communication skills to make learning an enjoyable and fruitful experience. Amadi and Paul (2017) state that a teacher's efficient communication skills have a direct positive effect on a students' academic performance.

Besides verbal, a teacher's nonverbal communication also leaves a remarkable impression on the students' attentiveness. Nonverbal communication includes kinesics, proxemics, chronemics, and paralinguistic features. Kinesics includes correct postures, appropriate gesticulation, positive body language, right facial expressions, and correct eye contact. Proxemics helps us to use the space while communicating with others. Bambaeeroo and Shokrpour (2017) found that nonverbal cues should complement verbal communication, and balanced communication enhances students learning and improves academic performance. Sutiyatno (2018) also states that verbal and nonverbal communication has a significant positive impact on a students' learning process. Zeki (2009) indicated that the teacher's nonverbal communication creates a comfortable environment for learning and motivates students to participate in group discussions.

A major transformation can be seen in teaching methodology due to the advancement of technology. Technology has opened multiple platforms where students and teachers find a constant inflow of knowledge (Bruenjes, 2002). Burton (2003) also supports that incorporating technology in the class also helps a teacher to build better rapport with the students. In order to meet the pace, the teaching pedagogies are also evolving. Teachers are experimenting with blended learning to improvise their teaching style. They are making a fine balance by adopting multiple approaches to deal with different topics (Muema et al., 2018). The conventional way of teaching is used, especially to explain theoretical concepts. The interactive mode of teaching promotes an active and participative environment in the class. The interactive mode of teaching enhances collaborative learning. An amalgamation of novel techniques in teaching bridges the gap between the teacher and the student (Jamian & Baharom, 2012). It fosters innovation and creativity, enhances the ability to ask the right questions, gives an opportunity to everyone to participate in the tasks more efficiently, and pushes the individuals to break their boundaries and open up to new challenges (Dexter et al, 1999; Loveless & Ellis, 2001).

University education is always progressive and encompasses diversity to spread its arena. The inclusion of students from different backgrounds has always been a welcoming aspect of university education. Cultural diversity opens new avenues for the students to practically learn various things about different cultures. Each culture possesses a multitude of different beliefs, values, assumptions, attitudes to respond in different situations. It also affects our responses and decision-making and also influences the development of personalities (Traindis & Suh, 2002; McCullers & Plant, 1964; Sears, 1948) Most of the time we judge people as per our own beliefs, which are to an extent influenced by our cultural influences.

This diversity makes a university a vibrant place where people from diverse places promote new language skills and novel ways of thinking by sharing their experiences. Teaching ethnically diverse classrooms starts from embracing views and employing experiential learning that can transmute the differences into positivity (Civitillo et al., 2018).

Students are often receptive to assimilate and accept each other without giving much consideration to cultural differences; however, they may face some challenges in the beginning, such as perceived discrimination, social isolation, and cultural adjustment (Wu et al., 2015; Wright & Schartner, 2013; Russell et al. 2010). A study conducted by Newsome and Cooper, in 2016, pointed the difficulties such as isolation, loneliness, racial discrimination, and economic exploitation faced by international students in adjusting and accommodating to a new environment. Although these differences do not impact the students' mindset significantly, they may leave certain marks of their presence in the initial days. These barriers can be overcome with the passage of time.

7.2 Aim of the Research

This chapter is a comparative analysis of technical and nontechnical students to study and discern the various factors which affect their academic performance. Human behavior cannot be studied in isolation as both internal and external worlds and intra- and interpersonal factors play a combined role to shape behavior. Through our topic we wanted to study the different perspectives and responses of the intrapersonal and interpersonal influencing factors on the academic performance of the technical and nontechnical students. The authors have used the terms intrapersonal and interpersonal influencing factors to observe their effects on students' academic performance. The term technical refers to engineering undergraduates and nontechnical refers to Management and Law students. To analyze intrapersonal factors, we focused on different behavioral patterns such as anxiety, influence of preconceived notions, feedback from the seniors regarding the teachers, and the relation of students' language skills on their academic performance. For interpersonal influencing factors, we have included the effects of teachers' verbal and nonverbal communication, teaching pedagogy: conventional and blended, social factors such as cultural influences. Through our secondary research, we studied papers that either focused on intrapersonal factors such as behavioral issues or interpersonal factors such as a teacher's influence on the students. Above all, we could not find any paper which made a comparative study of technical and nontechnical students to observe the impact of the mentioned factors. Our literature review research propelled us to take a new approach to make a comprehensive and inclusive study of both the factors—intrapersonal and interpersonal influencing factors to see their impact on academic performance. Hence, we took both the factors and prepared broad questions which could possibly answer the important issues—which we thought are very important and significantly impact the academic performance. We also believe that the program structure of both technical and nontechnical groups incites the students to develop a certain thought process and consequently they behave and respond differently in different situations. Most of the technical courses are experiment-based and challenge an inquiry-based learning; on the other hand, the subjective style of the nontechnical courses encourages a more detailed and subjective approach that provokes a discussion-based learning. An adequate understanding of the intrapersonal and interpersonal influencing factors will always support the facilitator to understand and bridge the academic and nonacademic gaps that otherwise impede the teaching learning process, if remain unaddressed.

7.3 Methodology

The population of the study consisted of undergraduate students of an Indian university, located in Haryana. The sample population consisted of 310 undergraduate students, out of which 258 students were in a technical group and 52 students were in a nontechnical group. The term technical refers to engineering undergraduates and nontechnical refers to management and law undergraduate students. The researchers prepared a questionnaire using a four-point Likert scale. The sample population of three schools of the same university: School of Engineering and Technology, School of Management, and School of Law filled the questionnaire shared through a Google form. As seen in Tables 7.1 and 7.2, the Questionnaire had ten questions that were framed keeping the Intrapersonal and Interpersonal influencing factors in mind.

As seen in Table 7.3, a four-point Likert scale has been used in this study ranging from Always, Sometimes, Rarely, Never. We have transformed this to numeric values in the following ways: Always=4, Sometimes=3, Rarely=2, Never=1.

TABLE 7.1

Questionnaire—Intrapersonal Factors

S. No.	Intrapersonal Factors
1	Which specific language skill do you feel is the cause of poor academic performance? (due to your lack of understanding or efficiency of that skill)
2	Does lack of language proficiency impact the overall understanding and performance of core engineering/management
3	To what extent/how frequently does stress or anxiety impact academic performance?
4	Do the preconceived notions, feedback from the seniors about the teacher affect the overall interest/performance in any particular subject?

TABLE 7.2

Questionnaire -Interpersonal Influencing Factors

S. No.	Interpersonal Influencing Factors
1	Does teacher's accent or pronunciation affect your understanding of any subject?
2	Does cultural difference affect the academic performance?
3	To what extent does the usage of digital teaching aids like power point presentations, online courses, models, videos etc., help in improving the understanding of the subject?
4	Does conventional teaching hinder the overall understanding of the subject?
5	Does the interactive mode of teaching help in better understanding of the subject?
6	Do nonverbal communication skills: Body language, postures, gestures, facial expressions and eye contact of the teacher effect in generating interest in any particular subject?

TABLE 7.3

Likert Scale and Weightage Used

Likert Scale Agreement	Weightage
Always	4
Sometimes	3
Rarely	2
Never	1

The value of Cronbach's alpha is 0.53. A possible reason for this Cronbach's alpha value could be the different types of questions that were not similar and consistent in nature. Since this chapter aimed to see the impact of the intrapersonal and interpersonal factors, we decided to take the first response of the students to avoid the biasedness of the responses.

7.4 Data Analysis

The first question of the questionnaire was the following: Which specific language skill do you feel is the cause of poor academic performance? (Due to your lack of understanding or efficiency of that skill). The students were asked to write any one skill from Listening, Speaking, Reading and Writing. Many authors (Nan, 2018; Evans, 2018; Sadiku, 2015; Wallace et al., 2004) have focused on LSRW—Listening, Speaking, Reading and Writing—the four skills that are essentially required while learning a language and known as language foundation skills. Since this question asked them to write only one skill and did not provide options such as Always, Sometimes, Rarely and Never, the analysis of this question has been done separately.

From Figures 7.1 and 7.2, we can see that both technical and nontechnical groups think that their listening skill is the weakest among all the language skills. Although both the groups acknowledge that listening is the weakest skill, still the percentage of listening as the weakest skill is more in the technical group as compared with the nontechnical. There could be a number of reasons for weak listening skills. The majority of the students feel that listening comes naturally. They interpret listening and hearing as the same and don't put in the required efforts in listening.

FIGURE 7.1
Shows the language skills of the technical group.

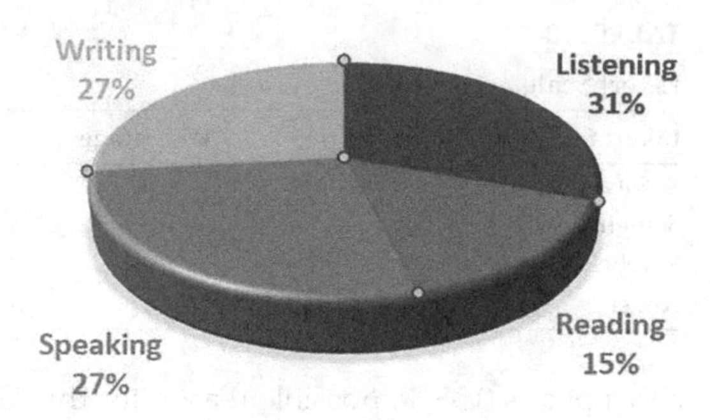

FIGURE 7.2
Shows the language skills of the nontechnical group.

The students get distracted easily due to their short span of concentration. Although students give various reasons for their inept listening skills, at the same time they also realize the barriers to listening can be overcome with their consistent practice. One of the ways to improvise listening skills is active participation, which motivates the students to stay focused while listening. The pie charts also present some interesting facts besides the weakest skill. The technical group sees listening as the weakest skill and speaking as the second weakest. The difference between the two skills is significant. Although in the nontechnical group, the students feel listening is their weakest skill, they find writing and speaking skills equally challenging after listening. This may happen because the nontechnical students often write subjective articles/essays and present them frequently. Due to that they may feel their writing and speaking skills are constantly challenged.

Rest all questions were analyzed following the Likert scale. Each of the questions had four options: Always, Sometimes, Rarely, Never. We have transformed this to numeric values (refer Table 7.3) in the following ways: Always=4, Sometimes=3, Rarely=2, Never=1. We have taken the opinion of students from the following two groups: technical and nontechnical. There were 258 students in the technical group and 52 students in the nontechnical group. We have compared their average (mean) opinion on each question using t-distribution for two samples of technical and nontechnical groups. In this case, we have tested null hypothesis: mean of two populations are same vs alternative hypothesis: mean of two populations are not same. We have also compared the variance of two groups using F-distribution. In this case, we have tested null hypothesis: variances of two populations are same vs. alternative hypothesis: Variances of two populations are not same. Here we have discussed results related to each of these nine questions.

7.5 Results and Discussions

As shown in Table 7.4, we can see that there is no difference in average opinion of the populations of two groups (Technical and nontechnical groups) to measure/observe the impact of lack of language proficiency on subject understanding and anxiety impact on the academic performance. The average opinion of both the samples for this intrapersonal factor is close to 3. The mean results indicate that on average both groups think that sometimes lack of language proficiency impacts the overall understanding and performance of core

TABLE 7.4

Intrapersonal Factors—t Test

Intrapersonal Factors	Lack of Language Proficiency on Subject Understanding	Anxiety Impact on the Academic Performance	Preconceived Notions, Feedback about the Teacher Affect the Interest/Performance in Subject
Mean of technical group	2.810	3.100	2.759
Mean of nontechnical group	2.846	2.923	2.5
Variance of technical group	0.714	0.581	0.665
Variance of nontechnical group	0.995	0.660	0.725
Hypothesized mean difference	0	0	0
t Stat	–0.243	1.452	2.019
P(T≤t) two-tail	0.808	0.150	0.047
t critical two-tail	1.996	1.994	1.993
Conclusion	Do not reject null hypothesis.	Do not reject null hypothesis.	Reject the null hypothesis.

engineering/management subjects. Lack of language proficiency may hinder the students' freedom of expressing their point of view, especially for those who come from semiurban or remote areas where English is not used as the first language to teach different subjects. This may affect their oral and written skills. Technical students may feel confident while writing their core courses exams or assignments as their content is standardized around the technical concepts, but they feel uneasy during presentations.

Nontechnical students generally deal with more subjective topics, and lack of language proficiency impedes their written as well as oral communication and restricts their opinions due to limited vocabulary.

The average opinion of both the samples for anxiety impact on the academic performance is also close to 3. We can conclude that on average, both groups think that sometimes stress or anxiety influences academic performance. Based on the results shown in Table 7.4, it can be ascertained that anxiety to a certain degree affects academic performance. The reasons for distress may vary in technical and nontechnical groups due to the differing nature and assessment patterns of the technical and nontechnical courses. Technical courses are more scientific and empirical in nature. The assessments are also designed to evaluate the scientific and technical aptitude of the students. Besides theoretical, most of the assessments are lab-based where the students check or prove the validity of the concepts, which demand robust technical skill sets. The meticulous and ordered pattern of the course and assessment may create anxiety in several students, especially in those who lack confidence in performing such assessments. Nontechnical subjects are more theoretical and interactive in nature. The assessments are also flexible as compared with the technical courses. Since nontechnical courses have more presentations-based assessments, the students who have low confidence and are conscious because of inept presentation skills may feel somewhat anxious while expressing their opinions.

Means for two samples of technical and nontechnical groups are 2.759 and 2.5, respectively, for interpersonal factor preconceived notions, feedback about the teacher affect the interest/performance in subject. Since there is a difference in average opinion of two populations, we can conclude that both groups think differently on how the preconceived notions, for example, feedback from the seniors about the teacher, affect the overall interest/

TABLE 7.5

Intrapersonal Factors-F Test

Intrapersonal Factors	Lack of Language Proficiency on Subject Understanding	Anxiety Impact on the Academic Performance	Preconceived Notions, Feedback about the Teacher Affect the Interest/Performance in Subject
Variance of technical group	0.714	0.581	0.802
Variance of nontechnical group	0.995	0.660	0.685
Hypothesis	Variance of two populations are same	Variance of two populations are same	Variance of two population are same
F	1.392	1.136	1.169
F critical	1.397	1.397	1.468
Conclusion	Do not reject null Hypothesis.	Do not reject null Hypothesis.	Do not reject the null hypothesis

performance in any subject. Majority of students in the technical group in comparison to the nontechnical group believe that sometimes it affects the overall inderest/performance in any subject. It is observed that Engineering students have very high expectations from their teachers. They always feel that the course is challenging, and it requires tremendous potential to perform well to attain high technical skills. Engineering students are more conscious about their performance, so they may feel a little overwhelmed if they receive any feedback about the teacher from their seniors. This can work in both ways: if they hear positive feedback, then it may help them to increase their interest in that subject. On the contrary, if they receive negative feedback, then it may increase their nervousness as they are already a little daunted due to the complexity of the course. Nontechnical students may also get affected by the feedback, but it does not affect their performance much as their courses focus more on the soft skills where they are more interested to hone their interpersonal skills and hence are not much bothered by the feedback they receive from the seniors.

As shown in Table 7.5, we have compared the variability of the opinion of the two groups (technical and nontechnical groups) for three interpersonal factors. Although there is a difference in the variance of two groups for lack of language proficiency on subject understanding, we can conclude that there is not much variability in the opinion of the two populations. Similarly, we can conclude for Anxiety Impact on the Academic Performance that there is not much variability in the opinion of two populations. There is not much difference in the sample variances of technical and nontechnical groups for preconceived notions, feedback about the teacher affecting the interest/performance in a subject. As it is evident from Table 7.5, we can conclude that there is no variability in the opinion of two populations for this interpersonal factor.

As shown in Table 7.6, we have compared the average opinion of two groups (Technical and nontechnical groups) for different interpersonal factors. We can see that there is no difference in average opinion of the populations of two groups (technical and nontechnical groups) for the factors: teachers' accent/pronunciation affect subject understanding, integrating digital teaching aids to improve the subject understanding, interactive mode of teaching helps in better subject understanding, nonverbal communication of the teacher affects in generating interest in the subject, cultural difference affects the academic performance and there is a difference in average opinion of the two groups for the factor: conventional teaching hinders the subject understanding.

TABLE 7.6

Interpersonal Influencing Factors-t Test

Interpersonal Factors	Teachers Accent/ Pronunciation Affect Subject Understanding	Integrating Digital Teaching Aids Improving the Subject Understanding	Conventional Teaching Hinder the Subject Understanding	Interactive Mode of Teaching Help in Better Subject understanding	Nonverbal Communication of the Teacher Affect in Generating Interest in Subject	Cultural Difference Affect the Academic Performance
Mean of technical group	2.717	3.431	2.751	3.759	3.399	1.964
Mean of nontechnical group	2.5	3.442	2.480	3.711	3.557	2.096
Variance of technical group	0.966	0.535	0.802	0.237	0.684	0.947
Variance of nontechnical group	1	0.486	0.685	0.483	0.526	0.951
Hypothesized mean difference	0	0	0	0	0	0
t Stat	1.431	−0.097	2.124	0.476	−1.402	−0.884
P(≤) two-tail	0.156	0.461	0.036	0.635	0.164	0.379
t critical two-tail	1.993	1.992	1.991	1.999	1.990	1.992
Conclusion	Do not reject null hypothesis.	Do not reject the null hypothesis.	Reject the null hypothesis.	Do not reject the null hypothesis.	Do not reject the null hypothesis.	Do not reject the null hypothesis.

Means for two samples of technical and nontechnical groups are 2.717 and 2.5, respectively, for the factors: teachers' accent/pronunciation affects subject understanding. Since there is no difference in average opinion of two populations, we can conclude that both groups think that the teacher's accent or pronunciation rarely or sometimes affect their understanding of any subject. Many of them believe it sometimes influences their understanding. The teacher's paralinguistic skills play a vital role in overall students' learning process. For technical students, communication skills predominantly affect the students in their skills and perspective courses, where they need good communication skills to perform well. If some students struggle to understand the teacher's accent, then they may feel difficult to cope with the course.

On the other hand, if the students find the teacher's accent unclear or faulty, they struggle to understand the course content. Consequently, it may hinder their understanding of the core engineering courses because these courses are full of technical jargon. Nontechnical students don't use too much jargon and can enjoy the flexibility of subjective terms and expressions.

If some students are already apprehensive about their language skills and constantly strive to understand things, then they may take time to understand the teacher's accent. On the other hand, if the students don't appreciate the teacher's accent but understand the concept taught by him/her, then it may not have much impact on their academic performance. The result also supports/highlights this fact.

Means for two samples of technical and nontechnical groups are 3.431 and 3.442, respectively, for the factor: integrating digital teaching aids improve the subject understanding.

Since there is no difference in average opinion of two populations (in this case sample means are very close as well), we can conclude that both groups think that usage of digital teaching aids like ppts, online courses, models, and videos, sometimes or always help in improving the understanding of the subject. Majority of them believe that it always helps in improving the understanding of the subject. For this question, average opinions of two groups are very similar, and this fact is also verified with high p-value. This clearly shows that the students in both groups feel that technology expands their horizon of learning. Technology strengthens their problem-solving aptitude. It not only helps them to identify the gaps of learning but also offers various platforms to bridge those gaps. Through advanced technology, they can get access to various new courses. Online learning is one such example. The result shows that both technical and nontechnical students welcome and embrace blended learning that amalgamates technology with theoretical concepts.

Means for two samples of technical and nontechnical groups are 2.751 and 2.480, respectively, for the factor: conventional teaching hinders the subject understanding. Since there is a difference in average opinion of the two populations, we can conclude that both groups think differently about conventional teaching. Many students of the technical group in comparison to the nontechnical group believe that sometimes it hinders the overall understanding of the subject. Conventional teaching is interpreted as long descriptive lectures with minimum or no usage of technology. Technical courses are more focused on experiments. The technical students are always inquisitive to validate the concepts that are taught in the class. It can be deduced from the results that technical students may like to give more importance to task-oriented lectures that promote experiential learning instead of listening to long descriptive lectures. That does not deny the value of theoretical lectures but the majority of them welcome blended learning over conventional learning. Nontechnical subjects are more subjective and descriptive such as subjects taught to law students. The students understand the nature of the course and hence are prepared to listen to the lectures. This could be one of the reasons that the majority of nontechnical students are comfortable with conventional methods of teaching.

Means for two samples of technical and nontechnical group are 3.759 and 3.711, respectively, for the factor: interactive mode of teaching help in better subject understanding. Since there is no difference in average opinion of two populations (in this case sample means are very close as well), we can conclude that both groups think that the interactive mode of teaching always helps in better understanding of the subject as two samples means are very close to four. For this question, the average opinions of the two groups are very similar and this fact is also verified with high p-value. Through interactive teaching, a teacher actively involves the students in their learning process by way of regular teacher–student interaction, student–student interaction, use of audio–visuals, and hands-on demonstrations. The students act as active contributors and learn significantly through such interactions. It is clearly demonstrated in the result that both groups hail and acknowledge the advantages of interactive mode of teaching.

Means for two samples of technical and nontechnical groups are 3.399 and 3.557, respectively, for the factor: nonverbal communication of the teacher affects in generating interest in the subject. Since there is no difference in average opinion of two populations, we can conclude that both groups think that nonverbal communication skills that include body language, postures, gestures, facial expressions, and eye contact of the teacher sometimes or always affect in generating interest in any subject. The majority of them believe that nonverbal communication skills: body language, postures, gestures, facial expressions, and eye contact of the teacher always promote interest in any subject. Both the groups feel that nonverbal communication improves a person's ability to relate, engage, and establish meaningful interactions in everyday life and leads to developing a good rapport with the teacher, which helps them to develop more interest and participation in the subject.

The means for two samples of the technical and nontechnical group are 1.964980545 and 2.096153846, respectively, for the factor: cultural difference affects academic performance. Since there is no difference in average opinion of two populations (in this case sample means are very close as well), we can conclude that both groups think that cultural differences rarely affect academic performance as 2 samples means are very close to 2. For this question, the average opinions of the two groups are very similar and this fact is also verified with high p-value. The students in both groups feel that cultural diversity does not have any tangible effect on their performance.

As shown in Table 7.7, we have compared the variability of the opinion of two groups (technical and nontechnical groups) for the questions addressing different interpersonal influencing factors.

As it is evident from Table 7.7, there is not much difference in the sample variances of technical and nontechnical groups for the interpersonal influencing factors discussed in the table. We can conclude that there is no variability in the opinion of two populations for the factors: teacher's accent/pronunciation affects subject understanding, integrating digital teaching improves the subject understanding, conventional teaching hinders the subject understanding, teacher's nonverbal communication yields in generating interest in the subject and, cultural differences do affect the academic performance. There is significant variability in the opinions of two populations for the factor: interactive mode of teaching promotes better subject understanding, although there is no difference in the average opinion of two populations for this factor. In this case, we can conclude that there is variability in opinion of two groups that the interactive mode of teaching helps in better understanding of the subject. This variation is more for nontechnical group in comparison to the technical group as evident from sample variances of technical and nontechnical groups.

TABLE 7.7

Interpersonal Influencing Factors-F Test

Interpersonal Factors	Teachers Accent/ Pronunciation Affect Subject Understanding	Integrating Digital Teaching Aids Improving the Subject Understanding	Conventional Teaching Hinder the Subject Understanding	Interactive Mode of Teaching Help in Better Subject Understanding	NVC of the Teacher Affect in Generating Interest in Subject	Cultural Difference Affect the Academic Performance
Variance of technical group	0.966	0.535	0.665	0.237	0.684	0.947
Variance of nontechnical group	1	0.486	0.725	0.483	0.526	0.951
Hypothesis	Variance of two population are same	Variance of two population are same	Variance of two population are same	Variance of two population are same	Variance of two population are same	Variance of two population are same
F	1.034	1.098	1.089	2.034	1.301	1.003
F critical	1.397	1.468	1.397	1.397	1.468	1.397
Conclusion	Do not reject the null hypothesis	Do not reject the null hypothesis	Do not reject the null hypothesis	reject the null hypothesis	Do not reject the null hypothesis	Do not reject the null hypothesis

7.6 Conclusion

Coping with students' academic challenges is a complex and rigorous experience for academicians. Motivating students to channelize their intellect and emotions in the right direction requires robust and consistent efforts. The present study presents the impact of intrapersonal and interpersonal influencing factors on the academic performance of technical and nontechnical students and highlights how different factors differently affect each group's academic performance. The responses to the questionnaire helped us to understand how these factors play their roles in two different groups. We could deduce some interesting results through the varied responses of each group. For a few factors, the groups showed similar opinions: as the fondness of digital learning in both groups, the impact of lack of proficiency in language skills on their academic performance, nonverbal communication of the teachers and their effect on their interest in that subject, the impact of anxiety on academic performance, least effect of cultural diversity on academic performance. Although the average opinion of the two groups toward an interactive mode of teaching is similar, there is variability in the opinion of the two groups. Besides similarities, we also found some differences in the opinions of the technical and nontechnical groups. The technical group believes that conventional teaching may hinder their performance but the nontechnical group shows no such indifference toward conventional teaching, the technical group believes that the teacher's accent and paralinguistic skills can have an effect on their understanding of the subject, but the nontechnical group seems less affected from this. The technical group indicates the possibility of the effect of the preconceived notions or the feedback they receive about the teacher from their seniors on their academic performance, but the nontechnical group does not agree to this thought much. It is an enriching, scholastic, and edifying experience to see the differences in the responses of the technical and nontechnical students. The study of the intrapersonal and interpersonal influencing factors on two different groups has broadened our view as teachers to understand the effect of different factors on the students' academic performance. The reasons for these differences are explained in the result analysis section, and we feel that one of the major reasons for the difference of opinions in the two groups could be the nature of the courses that the technical and nontechnical students pursue. We are very hopeful that the results of this study will be promising for the academicians to fathom the academic challenges of the technical and nontechnical students and provide them their timely assistance to boost their academic performance. A future study on overcoming the factors can be pursued to see the direct advantages on academic performance. Factors can be undertaken individually to see their vast implication in the academic realm. These factors can also be studied to see their effects on postgraduate students. Following are the limitations of this research:

- We conducted the survey only once to avoid the biasedness of the responses. Had the survey been conducted more than once, we could have applied test–retest reliability.
- We could only collect the sample from one university from different courses that the university offers. To obtain better results, the sample size could have been varied and vast.
- We had decided the influencing factors—intrapersonal and interpersonal—before drafting the questionnaire and adhered to those factors only. In the future, we would like to explore factors by conducting factor analysis.

References

Alkhalaf AM (2018) Positive and negative affect, anxiety, and academic achievement among medical students in Saudi Arabia. *Inter Journal of Emergency Mental Health and Human Resilience* 20(2): 397.

Amadi G, Paul AK (2017) Influence of student-teacher communication on students' academic achievement for effective teaching and learning. *American Journal of Educational Research* 5(10): 1102–1107.

Anderson AS, Good, DJ (2016) Increased body weight affects academic performance in university students. *Preventive Medicine Reports* 5: 220–223.

Ataunal A (2003) *Niçin ve nasıl bir öğretmen? (Why and what kind of teacher?).* Ankara: Milli Eğitim Vakfı Yayınları.

Bambaeeroo F, Shokrpour N (2017) The impact of the teachers' non-verbal communication on success in teaching. *Journal of Advances in Medical Education and Professionalism* 5(2): 51–59.

Bandura A (1997) *Self-efficacy: The exercise of control.* New York: W H Freeman/Times Books/Henry Holt & Co.; MacMillan Publisher.

Bandura A, Freeman WH, Lightsey R (1999) Self-efficacy: The exercise of control. *Journal of Cognitive Psychotherapy* 13(2): 158–166.

Berscheid ES, Rega PC (2005) *The psychology of interpersonal relationships.* Upper Saddle River, NJ: Pearson Prentice Hall.

Bruenjes LS (2002) A multi-case study investigating the disposition of faculty use of technology as a teaching and learning tool in the higher education classroom. Doctoral dissertation University of Massachusetts, Lowell. UMI 3041377.

Burton DB (2003) Technology professional development: A case study, *Academic Exchange Quarterly* 7(2): 2378–2381.

Cambridge English Dictionary (2015) Meaning of culture- Archived from the original on August 15, 2015. Retrieved July 26, 2015.

Cavilla D (2017) The effects of student reflection on academic performance and motivation. *Sage Open* 7(3), 2158244017733790.

Caviola L, Mannino A, Savulescu J, Faulmüller N (2014) Cognitive biases can affect moral intuitions about cognitive enhancement. *Frontiers in Systems Neuroscience.* https://www.frontiersin.org/articles/10.3389/fnsys.2014.00195/full

Civitillo S, Juanga LP, Schachner MK (2018) Challenging beliefs about cultural diversity in education: A synthesis and critical review of trainings with pre-service teachers. *Educational Research Review* 24: 67–83.

Cooke FL, Saini DS (2010) Diversity management in India: A study of organizations in different ownership forms and industrial sectors. *HR Science Forum* 40(3): 477–500.

Credê M, Kuncel NR (2008) Study habits, skills, and attitudes: The third pillar supporting collegiate academic performance. *Perspectives on Psychological Science* 3(6): 425–453.

Deng W, Li X, Wu H, Xu G (2020) Student leadership and academic performance. *China Economic Review 60*

Dexter SL, Anderson RE, Becker HJ (1999) Teachers' views of computers as catalysts for changes in their teaching practice. *Journal of Research on Computing in Education* 31(3): 221–239.

Dobson C (2012) Academic anxiety and coping with anxiety effects of academic anxiety on the performance of students with and without learning disabilities and how students can cope with anxiety at school, Master Thesis, Northern Michigan University, https://www.nmu.edu/education/sites/DrupalEducation/files/UserFiles/Dobson_Cassie_MP.pdf

Duckitt J (1992) Prejudice and behavior: A review. *Current Psychology* 11(4): 291–307.

Dweck CS (2006) *Mindset: The new psychology of success.* New York, NY: Random House.

Dweck CS, Yeager DS (2019) Mindsets: A view from two eras. *Perspectives on Psychological Science* 14(3), 481–496.

Elias H, Ping WS, Abdullah MC (2011) Stress and academic achievement among undergraduate students in Universiti Putra Malaysia. *Procedia-Social and Behavioral Sciences* 29: 646–655.

Evans MB (2018) The Integration of Reading, Writing, Speaking, and Listening Skills in the Middle School Social Studies Classroom in the Middle School Social Studies Classroom. Utah State University Utah State University, All Graduate Theses and Dissertations. 7157. https://digitalcommons.usu.edu/cgi/viewcontent..cgi?article=8259&context=etd

Flavian H (2016) Towards teaching and beyond: Strengthening education by understanding students' self-awareness development. *Power and Education* 8(1), 88–100.

Gablinske PB (2014) A case study of student and teacher relationships and the effect on student learning and the effect on student learning, Doctorate Thesis, University of Rhode Island University. https://digitalcommons.uri.edu/cgi/viewcontent..cgi?article=1284&context=oa_diss

Gibson MA (1982) Reputation and respectability: How competing cultural systems affect students' performance in school, *Anthropology & Education Quarterly*. https://anthrosource.onlinelibrary.wiley.com/doi/abs/10.1525/aeq.1982.13.1.04x0462t

Gu J, Strauss C, Bond R, Cavanagh K (2015) How do mindfulness-based cognitive therapy and mindfulness-based stress reduction improve mental health and wellbeing? A systematic review and meta-analysis of mediation studies. *Clinical Psychology Review* 37: 1–12.

Hashemi M (2011) Language stress and anxiety among the English language learners. *Procedia-Social and Behavioral Sciences* 30: 1811–1816.

Jamian AR, Baharom R (2012) The application of teaching aids and school supportive factors in learning reading skill among the remedial students in under enrolment schools. *Procedia-Social and Behavioral Sciences* 35:187–194

Kassarnig V, Mones E, Nielsen AB, Sapiezynski P, Lassen DD, Lehmann S (2018) Academic performance and behavioral patterns. *EPJ Data Science* 7: 1–16.

Keogh E, Bond FW, French CC, Richards A, Davis RE (2004) Test anxiety, Susceptibility to distraction and examination performance. *J of Anxiety, Stress and Coping* 17(3): 241–252.

Lamb ME, Bornstein MH (2011) *Social and Personality Development: An Advanced Textbook*, Psychology Press Taylor & Francis Group, New York.

Loveless A, Ellis V (2001) *ICT, Pedagogy and the Curriculum*. Taylor and Francis Group. https://teknologipendidikankritis.files.wordpress.com/2011/11/loveless-ellis_ict-pedagogy-and-curriculum.pdf

Macher D, Paechter M, Papousek I, Ruggeri K (2012) Statistics anxiety, trait anxiety, learning behavior, and academic performance. *European Journal of Psychology of Education* 27: 483–498.

MacIntyre PD, Dörnyei Z, Clément R, Noels KA (1998) Conceptualizing willingness to communicate in a L2: A situational model of L2 confidence and affiliation, *Journal of The Modern Language* 82(4): 545–562.

McCullers JC, Plant WT (1964) Review of educational research growth. *Development, and Learning* 34(5): 599–610.

Moussaïd M, Kämmer JE, Analytis PP, Neth H (2013) Social influence and the collective dynamics of opinion formation. *PLoS One*, 8(11): e78433.

Muema JS, Mulwa DM, Mailu, SN (2018) Relationship between teaching method and students. performance in mathematics in public secondary schools in Dadaab Sub County, Garissa County; Kenya. *IOSR Journal of Research & Method in Education* 8(5): 59–63.

Nan C (2018) Implications of Interrelationship among four language skills for high school english teaching, *Journal of Language Teaching and Research* 9(2) 418–423.

Newsome LK, Cooper P (2016) International students' cultural and social experiences in a British University: "Such a hard life [it] is here". *Journal of International Students* 6(1): 195–215.

Olson MA (2009) Measures of prejudice. In: Nelson TD, editor. *Handbook of prejudice, stereotyping, and discrimination*. New York: Psychology Press 367–386.

Owen L (2016) The impact of feedback as formative assessment on student performance. *Inter J of Teaching and Learning in Higher Education* 28(2): 168–175.

Owens M, Stevenson J, Hadwin JA, Norgate R (2012) Anxiety and depression in academic performance: An exploration of the mediating factors of worry and working memory. *School Psychology International* 33: 433–449.

Rampullo A, Licciardello O, Castiglione C (2015) Intrapersonal factors effects on professional orientation and environmental representations. *Procedia-Social and Behavioral Sciences* 205(9):422–428.

Rattan A, Dweck CS (2010) Who confronts prejudice? The role of implicit theories in the motivation to confront prejudice. *Psychological Science* 21(7): 952–959.

Russell J, Rosenthal D, Thomson G (2010) The international student experience: Three styles of adaptation. *Higher Education*, 60(2): 235–249.

Saa AA, Emran MA, Shaalan K (2019) Factors affecting students' performance in higher education: a systematic review of predictive data mining techniques. *Technology, Knowledge and Learning* 24: 567–598.

Sadiku LM (2015) The importance of four skills reading, speaking, writing, listening in a lesson hour. *European Journal of Language and Literature Studies* 1(1): 29–31.

Sears RR (1948) Personality development in contemporary culture. *Proceedings of the American Philosophical Society* 92(5): 363–370.

Seipp B (1991) Anxiety and academic performance: A meta-analysis of findings" *Journal of Anxiety Research* 4(1): 27–41.

Shahsavarani AM, Abadi EAM (2015) The bases, principles, and methods of decision-making: A review of literature. *International Journal of Medical Reviews* 2(1): 214–225.

Sutiyatno S (2018) The effect of teacher's verbal communication and non-verbal communication on students english achievement. *Journal of Language Teaching and Research* 9(2): 430–437.

Tatnell R, Kelada L, Hasking P, Martin G (2014) Longitudinal analysis of adolescent NSSI: The role of intrapersonal and interpersonal factors. *Journal of Abnormal Child Psychology* 42(6): 885–896.

Traindis H, Suh EM (2002) Cultural influences on personality. *Annual Review of Psychology* 5(1): 133–160.

Varga M (2017) The effect of teacher-student relationships on the academic engagement of students. Master Thesis, Graduate Programs in Education, Goucher College.

Wallace T, Stariha E, Walberg, HJ (2004) *Teaching, speaking, listening and writing. Educational practices series-14*. International Academy of Education, International Bureau of Education. https://files.eric.ed.gov/fulltext/ED495377.pdf

Wentzel KR, Wigfield A (1998) Academic and Social Motivational Influences on Students' Academic Performance. *Educational Psychology Review* 10:155–175.

Wright C, Schartner A (2013) "I can't...I won't?" International students at the threshold of social interaction. *Journal of Research in International Education* 12(2):113–128.

Wu HP, Garza E, Guzman N (2015) International student's challenge and adjustment to college. *Education Research International* 2015: 202753. https://www.hindawi.com/journals/edri/2015/202753/

Yang J, Zhao X (2020) Parenting styles and children's academic performance: Evidence from middle schools in China. *Children and Youth Services Review* 113: 105017.

Zeki CP (2009) The importance of non-verbal communication in classroom management. *Procedia-Social and Behavioral Sciences* 1(1): 1443–1449.

8

PIN Solar Cell Characteristics: Fundamental Physics

Aditya N. Roy Choudhury

Techno India University

CONTENTS

8.1 Introduction 119
8.2 Working of the PIN Solar Cell 120
8.3 Dark Current 123
8.4 Photocurrent 127
8.5 Conclusion 131
References 131

8.1 Introduction

Solar cells are prevalent in our lives today. Roadside lamp posts use solar cells, solar vehicles are roaming in the city, and space stations up in the sky are using solar panels for powering up. Some natural questions arise: what are solar cells? How do they function? Well, a short answer is that solar cells are power sources: they convert sunlight into electricity. They act like batteries when they are illuminated. That is, when there is light, a solar cell produces a voltage.

Like many other devices, solar cells have two terminals. Under illumination, if the two terminals are shorted, a short-circuit current flows. To understand solar cell action, further a current–voltage study is needed; questions need to be addressed like, under illumination, and subject to a voltage, what happens to a solar cell's current? Under a known intensity of light, if a certain voltage is applied against a solar cell, can we know how much current will flow in the circuit? This chapter ventures out to answer this question, i.e., it works toward figuring out the current–voltage–light phase space of a solar cell. For that, we would first go through the operations of a solar cell in dark. We would figure out the solar cell's dark current–voltage characteristics. Then we would illuminate the solar cell and would derive the current–voltage characteristics under illumination. Details of the dark characteristics can be found in Refs. [1–4]. The illuminated characteristics were discussed in Ref. [5].

From various types of solar cells that are getting researched at present, we would pick the PIN type. The PIN solar cell has three layers of semiconductors—a P type semiconductor, an I type (intrinsic) semiconductor, and an N type semiconductor. The P layer is doped with negatively charged acceptor ions and has more holes than electrons. Similarly, the N layer is doped with positively charged donor ions and has more electrons than holes. The I layer is intrinsic; it has an equal number of holes and electrons. The I layer can also instead be lightly doped (P or N type) but the doping density should ideally be less than that of the P and N layers.

DOI: 10.1201/9781003291916-8

8.2 Working of the PIN Solar Cell

Under illumination, the I layer acts as an absorber of light. Light is absorbed and excitons are produced inside the semiconductor. These excitons are then dissociated into free electrons and holes. As the P layer is negatively charged (as a result of the PIN structure formation, the P layer has more holes than I layer, in the dark, and so some holes diffuse out of the P layer into the I layer) the positively charged holes in the I layer, under illumination, feel an attraction toward the P layer and move to it. Similarly, the negatively charged electrons feel an attraction toward the positively charged N layer and move to it. These movements of electrons and holes from the absorber to the P and N layers cause an electric current inside the solar cell. If one electrically connects the P and the N layers with a wire, one should be able to measure a current under illumination. This is the simplest understanding of a PIN solar cell. Figure 8.1 shows the above operations schematically. The short-circuit current density flowing through a solar cell is symbolized as J_{SC}. Similarly, if the P and the N sides of the illuminated solar cell in Figure 8.1b is not connected (directly or through a resistor), but kept open, an open-circuit voltage develops across the cell and the cell behaves, in principle, like a battery. This voltage is symbolized as V_{oc}.

For deriving the current–voltage characteristics, an understanding of the electron energy band diagram is required. A semiconductor is characterized by its energy band diagram. It is a potential energy landscape for electrons: if the band tilts or bends down electrons find it easy to move downhill. In the other case if the band bends up electrons find it difficult to climb that energy barrier. In the diagram, the vertical axis is potential energy, and the horizontal axis is space. A straight line in the energy band diagram, parallel to the X-axis, refers to a potential energy level. For the entire semiconductor, as there are many atoms and, therefore, many levels, energy levels merge to form a band. An electron, therefore, can be anywhere within a band. Besides this, a semiconductor is always characterized by a bandgap. The bandgap is a forbidden zone for electrons, no electron can be inside the gap unless, of course, there are trapping levels induced by defects. The bandgap has its upper level as the conduction band edge E_C. For an electron at any energy $E>E_C$, i.e., above E_C, horizontal movement across the band (without any voltage or force from outside) is decided by the slope of E_C. The down level of the gap is the valence band edge E_V. Thus, electrons can be anywhere above E_C or below E_V. Also, in the valence band holes reside. Once a bound electron is pumped up, from the valence band to the conduction band, to become free, it leaves behind a free hole in the valence band. The Fermi level E_F decides the total number of electrons and holes present in a semiconductor. Closer the Fermi level is to the conduction band edge, more electrons (than holes) the semiconductor has (i.e., it is n type), and closer the E_F is to the valence band, likewise, more holes than electrons the semiconductor has (it is *p* type). The electron and the hole concentrations are expressed

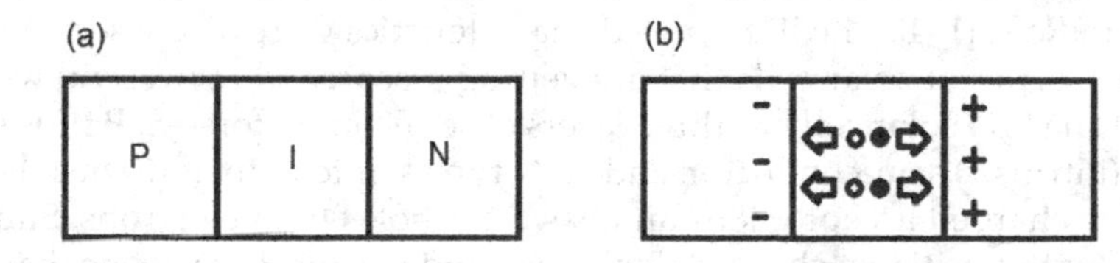

FIGURE 8.1
(a) A PIN solar cell showing the three layers—the P, the I, and the N. (b) Under illumination the I layer absorbs light and produces free electron and hole pairs. These opposite charge carriers travel in opposite directions under the effect of the built-in electric field present in the PIN structure. This causes an electric current from the P side to the N side, or an open-circuit voltage when the cell is left open.

by equations (8.1) and (8.2). N_C and N_V are the effective conduction and the valence band density of states, k_B is the Boltzmann constant, and T is temperature.

$$n = N_C \, exp\left[\frac{E_F - E_C}{k_B T}\right] \tag{8.1}$$

$$p = N_V \, exp\left[\frac{E_V - E_F}{k_B T}\right] \tag{8.2}$$

The electron-hole concentration product, i.e., *np* product is always equal to the semiconductor's intrinsic carrier density squared. The intrinsic carrier concentration n_i is the (equal) number of electrons and holes present in the undoped semiconductor due to thermal activation. n_i is expressed in equation (8.3). E_G is the bandgap.

$$n_i^2 = np = N_C N_V \, exp\left(\frac{-E_G}{k_B T}\right) \tag{8.3}$$

The PIN energy band diagram is drawn keeping in mind what would happen to the P, I, and N layers due to each other's presence. The P layer has more holes than the I layer. This would make holes diffuse out from the P layer into the I layer when the P and the I layer are brought in close proximity. Similarly, the N layer will lose electrons to the I layer. As a result, the I layer's carrier concentration profiles become nontrivial: toward the P side, the I layer will have more holes, and toward the N side, it will have more electrons. This makes the I layer tilt (with respect to the Fermi level) as shown in Figure 8.2a. The P layer and the N layer will not get affected by this tilt. They will just become devoid of respective carriers up to a width $(x_p - w_p)$ and $(w_n - x_n)$.

Electric field inside a semiconductor can be calculated using Poisson's Equation: at any point in space, the gradient of the electric field is given by the total charge density stored at that point divided by the electric permittivity at that point. Applying Poisson's Equation to the PIN structure is done below. N_A and N_D are the doping densities of the P and the N layers (i.e., the total hole and electron concentrations), ε_r is the dielectric constant, and ε_0 is the permittivity of the vacuum.

$$\frac{d\varepsilon}{dx} = \frac{-qN_A}{\varepsilon_{r,P}\,\varepsilon_0} \quad \text{for } w_p < x < x_p. \tag{8.4a}$$

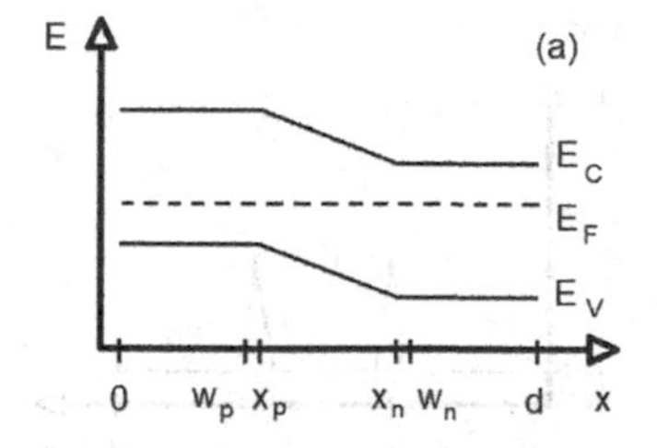

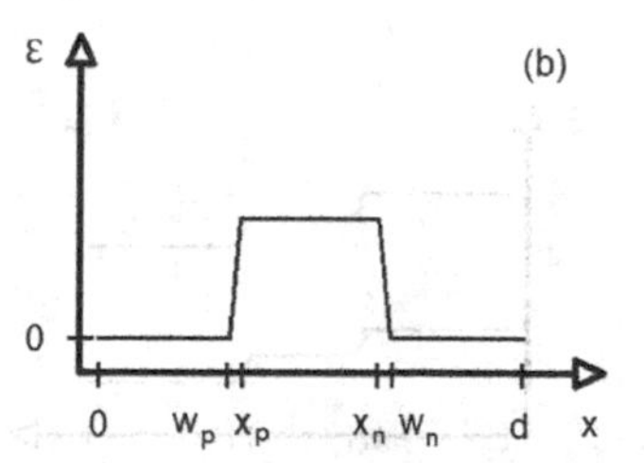

FIGURE 8.2
(a) The PIN energy band diagram. The P semiconductor is x_p units wide. The I layer (absorber of the solar cell) is $(x_n - x_p)$ units wide. The N layer's width is $(d - x_n)$. $x_p - w_p$ is the width of the affected region in the P layer due to the presence of the neighboring I layer, and similarly $(w_n - x_n)$ is the width of the affected portion of the N layer. d is the total width of the solar cell. (b) Electric field in the P, I, and N layers. The absolute value of the field is plotted, i.e., the field $\mathcal{E}$ is actually negative. The field is constant across the I layer and drops to zero in the regions of the P and N layers that are unaffected by the I layer.

$$\frac{d\varepsilon}{dx} = \frac{p(x) - n(x)}{\varepsilon_{r,I}\ \varepsilon_0} \quad \text{for } x_p < x < (x_n - x_p)/2 \tag{8.4b}$$

$$\frac{d\varepsilon}{dx} = -\frac{n(x) - p(x)}{\varepsilon_{r,I}\ \varepsilon_0} \quad \text{for} (x_n - x_p)/2 < x < x_n \tag{8.4c}$$

$$\frac{d\varepsilon}{dx} = \frac{qN_D}{\varepsilon_{r,N}\ \varepsilon_0} \quad \text{for } x_n < x < w_n \tag{8.4d}$$

Solutions of the above Poisson Equations lead to the electric field distribution as shown in Figure 8.2b. Figure 8.2b gives the absolute value of the solved electric field. The real electric field is negative which is easy to find from Figure 8.2a. Down the slope of the tilted I layer, an electron travels from left to right which means the electric field under which the electron is in is negative (the current is negative, and from Ohm's Law, so is the field). This electric field is sufficient to make the PIN structure gain solar cell functionalities. Electrons and holes can follow this field and can travel toward the N and the P side when the I layer is illuminated. This gives rise to a V_{oc}. Thus Figure 8.2 shows the PIN band diagram in dark and gives a rough idea of what can happen when the solar cell is illuminated.

There is a second variant of the PIN band diagram. In reality, having a perfectly intrinsic semiconductor is not possible. So the I layer always remains very lightly doped. In addition, if the I layer is also sufficiently wide, the free carriers inside the I layer can rearrange themselves in a way so as to cancel the electric field in the bulk. The electric field arises due to the immobile charges in the P and N layers' depletion regions, i.e., in the portions (x_p–w_p) and (w_n–x_n). These regions are devoid of carriers and so are negatively and positively charged due to the ionized acceptor and donor atoms. After the field is canceled at the I layer's boundaries, it drops to zero in the I layer's bulk. As a result, the bands in the bulk are flat and don't tilt or bend. The energy band diagram of this type of PIN structure is shown in Figure 8.3.

A natural question, therefore, arises—if the electric field in the I layer's bulk is zero what aids the solar cell function when the device is illuminated? The answer is—diffusion. After the I layer is illuminated, it gets filled with free electrons and holes. Both of these entities tend to diffuse out of the I layer because they are more in number inside the I layer. Due to the band bending present on either sides of the I layer (Figure 8.3a), electrons travel to the right, i.e., toward the N layer (down the band diagram) and holes travel to the P layer (up the band diagram). This ensures solar cell functions when the device is illuminated. Just as it is easy for electrons to move down an electron energy band diagram, it is easy for holes to move up. That is how the band diagram is.

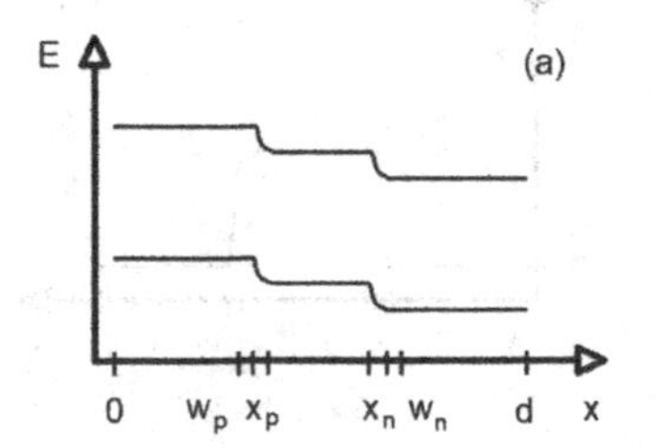

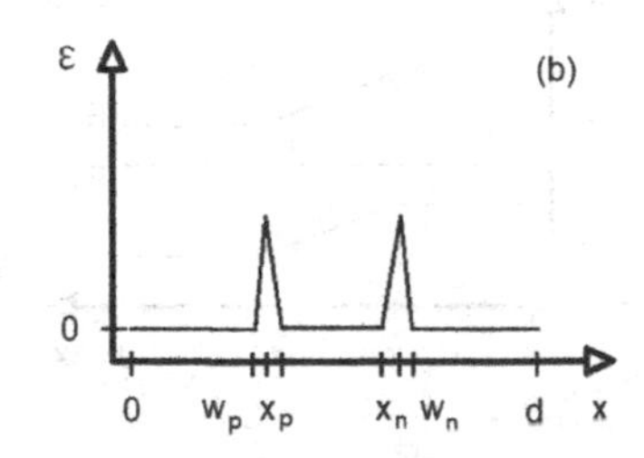

FIGURE 8.3
(a) The energy band diagram of the PIN structure where the I layer is intentionally/unintentionally doped and is also wider. The electric field arising from the P and N layers is canceled at the I layer boundaries making the field in the I layer interior zero. (b) The electric field in this type of PIN structure.

8.3 Dark Current

After getting the fundamentals of the energy band diagram right, one should ponder how will a solar cell's current–voltage curve look in the dark. That is what we will discuss next. Across all the three solar cell layers—P, I, and N—the current–voltage relation will be calculated. In principle, the total current should be independent of space.

There are three kinds of currents involved in semiconductors—drift current, diffusion current, and generation–recombination (RG) current.

$$J_{drift} = \sigma\varepsilon = cq\mu\varepsilon \tag{8.5a}$$

$$J_{diff} = qD\frac{dc}{dx} \tag{8.5b}$$

$$J_{RG} = \int_0^x qMdx \tag{8.5c}$$

The drift current density is given by equation (8.5a). Here, by c we mean either the hole concentration p or the electron concentration n. μ is mobility. By drift, we mean Ohm's Law. Likewise, the diffusion current density is given by Fick's Law in equation (8.5b). D is diffusion coefficient or diffusivity. When a semiconductor is illuminated, after an electron and hole pair is created, a current can flow known as the generation (G) current. Similarly an electron and hole pair can combine back and vanish (i.e., recombine) to give rise to another current component called the recombination (R) current. The generation or recombination current density should go so as to make $1/q\ (dJ_{RG}/dx)=M$, where M stands for G or R. Thus, the G or R current density is given by equation (8.5c).

While deriving currents inside semiconductors, one comes across the Continuity Equations which states that a carrier just cannot appear or disappear at a given point of space. In other words, if there are some electrons, getting created or destroyed per second, for example, inside the P layer (as minority carriers), then those electrons must have arrived at that point of space following the physical processes of either drift, or diffusion, or recombination, or generation. The Continuity Equations are given in equations (8.6a) and (8.6b).

$$\left(\frac{dp}{dt}\right)_{total} = \left(\frac{dp}{dt}\right)_{drift} + \left(\frac{dp}{dt}\right)_{diff} + \left(\frac{dp}{dt}\right)_{G} + \left(\frac{dp}{dt}\right)_{R} \tag{8.6a}$$

$$\left(\frac{dn}{dt}\right)_{total} = \left(\frac{dn}{dt}\right)_{drift} + \left(\frac{dn}{dt}\right)_{diff} + \left(\frac{dn}{dt}\right)_{G} + \left(\frac{dn}{dt}\right)_{R} \tag{8.6b}$$

Further, because there are only two types of currents that are physically possible—the currents due to drift and diffusion—one can write

$$\left(\frac{dp}{dt}\right)_{drift} + \left(\frac{dp}{dt}\right)_{diff} = \frac{1}{q}\frac{dJ_p}{dx} \tag{8.7a}$$

$$\left(\frac{dn}{dt}\right)_{drift} + \left(\frac{dn}{dt}\right)_{diff} = \frac{1}{q}\frac{dJ_n}{dx} \tag{8.7b}$$

In our derivation, we would assume that there is no recombination present in the I layer. As, in addition, it is dark, there is no generation present also. This would mean, in the I layer, $G=R=0$, and, therefore, $J_{RG}=0$ (from equation (8.5c)). Thus, by applying the Continuity Equation in the I layer, one sees that the carriers arrive by drift and leave by diffusion. Therefore,

$$\left(\frac{dp}{dt}\right)_{drift} + \left(\frac{dp}{dt}\right)_{diff} = 0 \tag{8.8a}$$

$$\left(\frac{dn}{dt}\right)_{drift} + \left(\frac{dn}{dt}\right)_{diff} = 0 \tag{8.8b}$$

From equations (8.7) and (8.8) one sees that, in the I layer, both the electron and the hole currents, individually, are constant spatially. We call these values J_{sn} and J_{sp}.

Interestingly, just as drift balances diffusion in the I layer, for the P and N layers, in the quasi-neutral regions (i.e., for $x<w_p$, and for $x>w_n$), it's recombination that balances diffusion. In other words, diffusion brings the carriers and recombination subtracts them. There is no generation or drift present in the quasi-neutral regions. Therefore, the Continuity Equations in the P and N layers go as

$$\left(\frac{dp}{dt}\right)_{diff} + \left(\frac{dp}{dt}\right)_{R} = 0 \tag{8.8c}$$

$$\left(\frac{dn}{dt}\right)_{diff} + \left(\frac{dn}{dt}\right)_{R} = 0 \tag{8.8d}$$

The recombination here follows the monomolecular Shockley Read Hall (SRH) model. In other words, the Continuity Equation terms for recombination becomes

$$\left(\frac{dp}{dt}\right)_{R} = \frac{p-p_0}{\tau_p} = \frac{\Delta p}{\tau_p} \tag{8.9a}$$

$$\left(\frac{dn}{dt}\right)_{R} = \frac{n-n_0}{\tau_n} = \frac{\Delta n}{\tau_n} \tag{8.9b}$$

where p_0 and n_0 are the carrier concentrations that are present in the quasi-neutral regions at zero applied voltage V, i.e., before the applied bias brings in more minority carriers in the P and the N layers. τ is the minority carrier lifetime for holes (τ_p) or electrons (τ_n) accordingly. This means, in the dark and under zero applied voltage, the hole concentration in the N layer for $x>w_n$ is p_0. Likewise, the electron concentration in the P layer for $x<w_p$ is n_0. It is easy to see that $p_0=n_i^2/N_D$, and $n_0=n_i^2/N_A$. As voltage is applied to the PIN structure, with the positive terminal at P and the negative terminal at N, the minority carrier concentrations (i.e., the concentration of holes in the N layer and that of the electrons in the P layer) increase. The recombination is proportional to this increase and inversely proportional to the carrier lifetime τ.

At an applied voltage V the minority carrier concentration, say, increases to p_1 and n_1 in the quasi-neutral regions' edges, i.e., at w_n and w_p.

$$p_1 = \frac{n_i^2}{N_D} exp\left(\frac{qV}{k_BT}\right) \tag{8.10a}$$

$$n_1 = \frac{n_i^2}{N_A} exp\left(\frac{qV}{k_B T}\right) \tag{8.10b}$$

From the fact that $\mathcal{E} = 0$ in the quasi-neutral regions, and from equations (8.5b), (8.7), (8.8c), (8.8d), and (9), we have

$$D_p \frac{d^2(\Delta p)}{dx^2} = \frac{(\Delta p)}{\tau_p} \quad \text{for } x \geq w_n \tag{8.11}$$

Redefining the spatial variable to $x' = (x - w_n)$ we have

$$D_p \frac{d^2(\Delta p)}{dx'^2} = \frac{(\Delta p)}{\tau_p} \quad \text{for } x' \geq 0 \tag{8.12}$$

with boundary conditions

$$\Delta p = p_1 - p_0 \quad \text{at } x' = 0 \tag{8.13a}$$

$$\Delta p = 0 \quad \text{at } x' \to \infty \tag{8.13b}$$

Solving equation (8.12) with the above boundary conditions one arrives at the solution

$$\Delta p\,(x') = (p_1 - p_0)\,\exp\left(-\frac{x'}{L_p}\right) \tag{8.14}$$

where L is called the minority carrier diffusion length. Simplifying the expression one arrives at –

$$\Delta p\,(x') = \frac{n_i^2}{N_D}\left[exp\left(\frac{qV}{k_B T}\right) - 1\right]\exp\left(-\frac{x'}{L_p}\right) \tag{8.15}$$

i.e.,

$$p(x') = \frac{n_i^2}{N_D}\left[exp\left(\frac{qV}{k_B T}\right) - 1\right]\exp\left(-\frac{x'}{L_p}\right) + \frac{n_i^2}{N_D} \tag{8.16}$$

The electron minority carrier concentration will have a similar expression.

After deriving the carrier concentrations, deriving the carrier current densities become trivial. From equations (8.5b) and (8.16) one has

$$J_p(x') = \frac{qn_i^2 D_p}{N_D L_p}\left[exp\left(\frac{qV}{k_B T}\right) - 1\right]\exp\left(-\frac{x'}{L_p}\right) \tag{8.17}$$

One must balance for the minus sign as the hole minority carriers travel from left to right, and, therefore, the current must be positive, and, on the other hand, the slope of $p(x')$, for $x'>0$, is negative.

The electron current will have a similar expression.

From equation (8.17), at $x' = 0$, one can get the expression of the hole current density, i.e.,

$$J_{sp} = \frac{qn_i^2 D_p}{N_D L_p}\left[exp\left(\frac{qV}{k_B T}\right) - 1\right] \tag{8.18}$$

Similarly, for the electron current density

$$J_{sn} = \frac{qn_i^2 D_n}{N_A L_n}\left[exp\left(\frac{qV}{k_B T}\right) - 1\right] \tag{8.19}$$

Keeping in mind that in the I layer the current densities do not change spatially, for the PIN solar cell in dark, the current density–voltage characteristics go as

$$J = J_{sp} + J_{sn} \tag{8.20a}$$

$$J = J_S\left[exp\left(\frac{qV}{k_B T}\right) - 1\right] \tag{8.20b}$$

$$J_S = qn_i^2\left(\frac{D_p}{N_D L_p} + \frac{D_n}{N_A L_n}\right) \tag{8.20c}$$

Strictly speaking equation (8.20b) is valid for a PN junction diode, but it can also hold for a PIN structure when recombination in the I layer is considered negligible. J_S is called the saturation current density. When recombination in the I layer is significant an ideality factor n_{ID} enters in equation (8.20b), but that we do not discuss here. Equation (8.20b) is also called the Ideal Diode Equation or the Shockley Equation. Figures 8.4 and 8.5 show the dark carrier concentrations and the dark current densities of the PIN solar cell. Figure 8.6, finally, shows the current–voltage characteristics that is given by equation (8.20b).

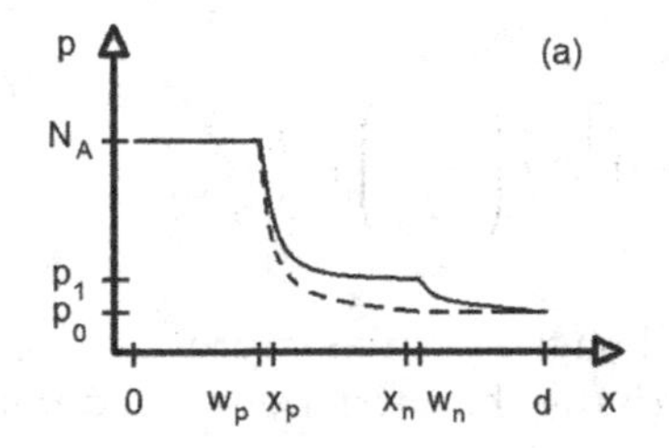

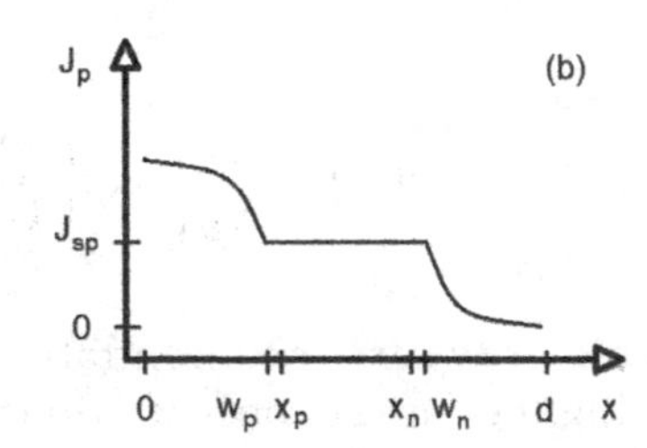

FIGURE 8.4

(a) Hole concentration in the solar cell under dark and at an applied voltage V. The dashed profile is at zero applied voltage. At the N layer edge (more precisely, at the depletion layer edge of the N layer) the hole concentration at $V=0$ is p_0 and becomes p_1 at applied voltage V. (b) The hole current density for the PIN solar cell under dark at applied bias V.

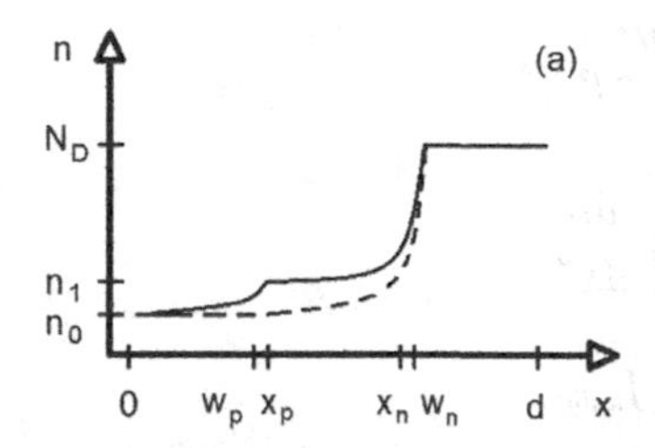

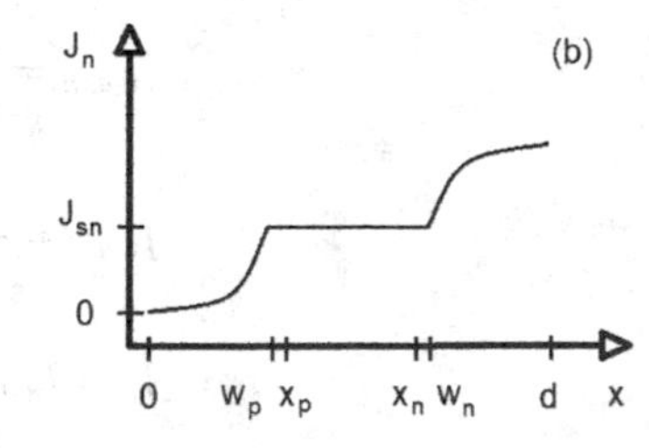

FIGURE 8.5
(a) Likewise, electron concentration in the solar cell under dark and at an applied voltage V. The dashed profile signifies the pattern at zero applied voltage. At the depletion layer edge of the P layer the electron concentration at $V = 0$ is n_0 and becomes n_1 at applied bias V. (b) The electron current density for the PIN solar cell under dark and at applied voltage V.

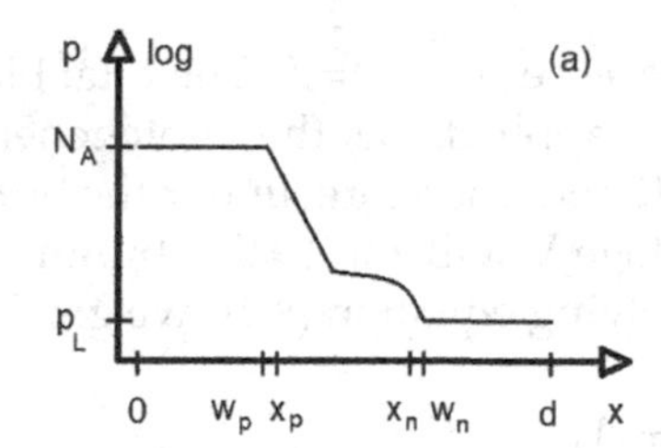

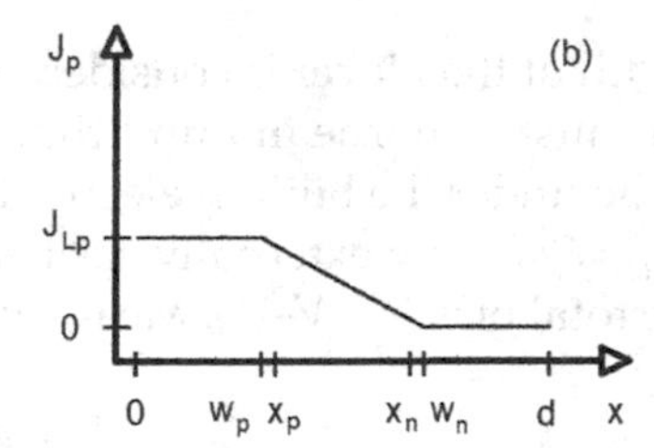

FIGURE 8.6
The dark current–voltage characteristics of the solar cell showed in normal (a) and in natural logarithm (b) scale.

8.4 Photocurrent

After deriving the dark current for our PIN solar cell, the next aim will be to derive the current bias dependence under illumination. For this, we would follow a method similar to what we did for the dark current. We will start with the Continuity Equation, equate the generation current to the current caused by drift and diffusion, and will derive the spatial profiles of carrier concentrations and current densities. The total current density will be independent of space as before. An important point on deriving the photocurrent is that all calculations are done for the I layer, i.e., the absorber. Just as the P and the N layers formed the center of attraction while deriving the dark current, this time, the absorber will take the front seat for getting into photocurrent. In what follows the spatial variable x'' is considered to be 0 at x_p, and to be d_i at x_n.

Following equation (8.5c), the generation term can be written, for both electrons and holes, as

$$\frac{1}{q}\frac{dJ_p}{dx''} = G \tag{8.21a}$$

$$\frac{1}{q}\frac{dJ_n}{dx''} = -G \tag{8.21b}$$

The electron and hole drift and diffusion current densities can be expressed as

$$J_p = J_{p,drift} + J_{p,diff} \tag{8.22a}$$

$$J_{p,drift} = \frac{q\mu_p V'}{d_i} p. \tag{8.22b}$$

$$J_{p,diff} = -qD_p \frac{dp}{dx''} \tag{8.22c}$$

$$J_n = J_{n,drift} + J_{n,diff} \tag{8.23a}$$

$$J_{n,drift} = \frac{q\mu_n V'}{d_i} n \tag{8.23b}$$

$$J_{n,diff} = qD_n \frac{dn}{dx''} \tag{8.23c}$$

Here, the width of the I layer is considered to be d_i, i.e., $(x_n - x_p) = d_i$. The total bias is considered as V'. It must be borne in mind that at zero applied bias the photogenerated charge carriers will be under the built-in electric field $\mathcal{E}$, and, therefore, under the built-in voltage V_{bi}, where $V_{bi} = \mathcal{E}/d_i$. Any externally applied voltage V will add to this built-in voltage and therefore the total bias $V' = V - V_{bi}$. Meanwhile, solving equation (8.21) we get

$$J_p = qGx'' + A_p \tag{8.24a}$$

$$J_n = -qGx'' + A_n \tag{8.24b}$$

By using equations (8.22) and (8.24a) we can solve for the hole spatial profile $p(x'')$

$$p = \frac{qGx'' + A_p}{q\mu_p V' / d_i} + \frac{\left\{ B_p \exp\left(\frac{\mu_p V'}{D_p d_i} x'' \right) + q^2 G D_p \right\}}{\left(q\mu_p V' / d_i \right)^2} \tag{8.25a}$$

Similarly, by using equations (8.23) and (8.24b), we can solve for the electron spatial profile $n(x'')$

$$n = \frac{-qGx'' + A_n}{q\mu_p V' / d_i} + \frac{\left\{ B_n \exp\left(-\frac{\mu_n V'}{D_n d_i} x'' \right) - q^2 G D_n \right\}}{\left(q\mu_n V' / d_i \right)^2} \tag{8.25b}$$

The two unknowns A and B in each of equation (8.25) can be solved by using boundary conditions. Under high-level injection, i.e., when the concentration of photogenerated carriers is higher than the doping densities, it is easy to see that $p = n = 0$ at $x'' = 0$ (i.e., at $x = x_p$) and at $x'' = d_i$ (i.e., at $x = x_n$). In other words, the electron and hole densities in these points will be negligibly small compared to that in the I layer. Under low-level injection, i.e., when the concentration of photogenerated carriers are higher than the minority carrier concentrations in the P and the N layers but lower than the doping densities, these boundary conditions change. This makes the p and n spatial dependencies different, but in either cases the current density graphs and expressions stay approximately the same, and the final bias-dependent photocurrent expression is that which is obtained using the $p=n=0$ boundary conditions at the boundaries of the I layer. The solutions are given as

$$p = \frac{Gd_i^2}{\mu_p V'}\left[\frac{x''}{d_i} - \frac{\left\{exp\left(\frac{qV'}{k_BT}\frac{x''}{d_i}\right) - 1\right\}}{\left\{exp\left(\frac{qV'}{k_BT}\right) - 1\right\}}\right] \tag{8.26a}$$

$$n = -\frac{Gd_i^2}{\mu_n V'}\left[\frac{x''}{d_i} - \frac{\left\{exp\left(-\frac{qV'}{k_BT}\frac{x''}{d_i}\right) - 1\right\}}{\left\{exp\left(-\frac{qV'}{k_BT}\right) - 1\right\}}\right] \tag{8.26b}$$

Following equations (8.22a) and (8.23a) the photocurrent densities can be deduced

$$J_p = J_L\left[\frac{x''}{d_i} - \frac{k_BT}{qV'} + \frac{1}{\left\{exp\left(\frac{qV'}{k_BT}\right) - 1\right\}}\right] \tag{8.27a}$$

$$J_n = -J_L\left[\frac{x''}{d_i} + \frac{k_BT}{qV'} + \frac{1}{\left\{exp\left(\frac{-qV'}{k_BT}\right) - 1\right\}}\right] \tag{8.27b}$$

$$J_L = qGd_i \tag{8.27c}$$

$$J = J_p + J_n = J_L\left[coth\left(\frac{qV'}{2k_BT}\right) - \left(\frac{2k_BT}{qV'}\right)\right] \tag{8.27d}$$

Thus the final bias-dependent photocurrent expression is given by equations (8.27c) and (8.27d).

The photocurrent densities in the P or N layers should be independent of space because both the generation and recombination there are zero. The generation is zero because in our derivation we have considered illumination only in the I layer, i.e., the absorber. The recombination is also zero because for majority carriers in the P and N quasi-neutral regions there is no recombination present.

Figure 8.7 shows the carrier concentration profiles for V' close to zero (but negative). It assumes high-level injection. Figures 8.8 and 8.9 show the carrier profiles under low-level injection, and also the current density profiles. The photocurrents are always negative, however, in the diagrams only the magnitudes are shown.

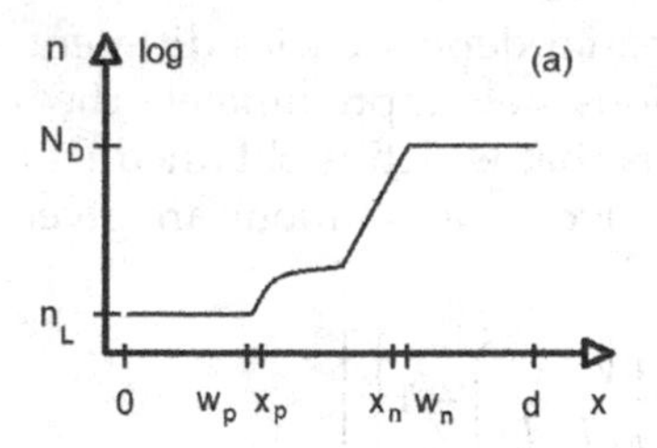

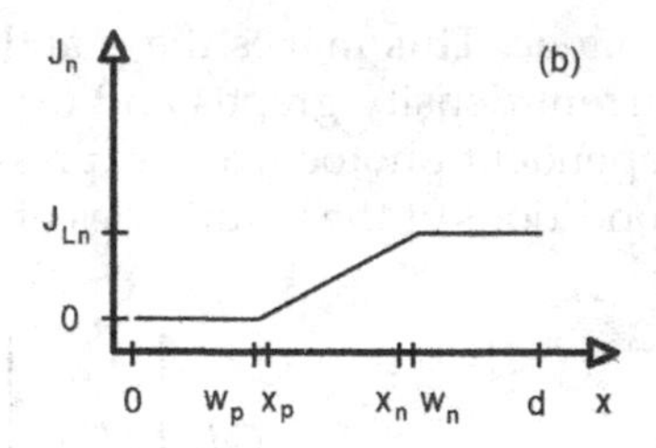

FIGURE 8.7
The solved spatial profiles of holes (a) and electrons (b) in the I layer under the assumption of high-level injection.

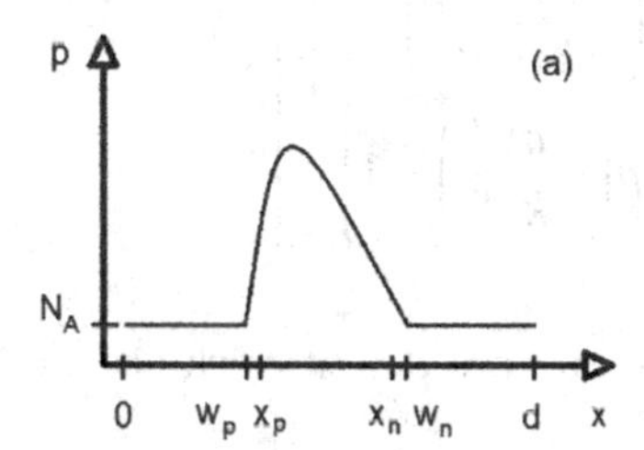

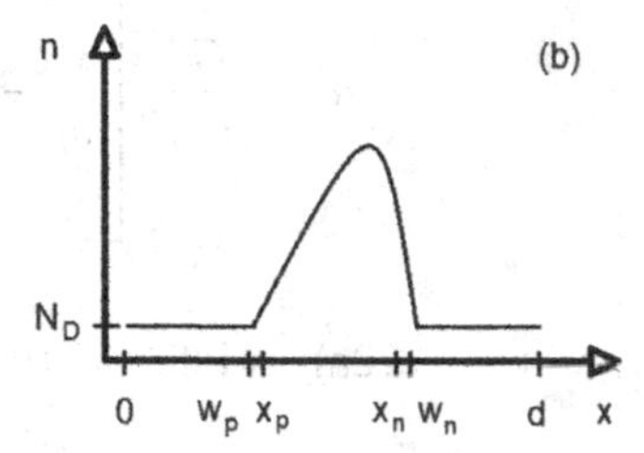

FIGURE 8.8
(a) The hole spatial profile under low-level injection. (b) The hole current density profile. $J_{LP} = J_{Ln} = J_L$. The magnitude of the hole current is drawn.

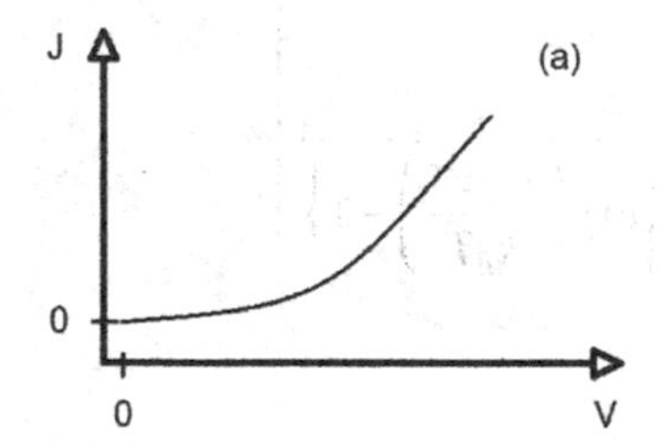

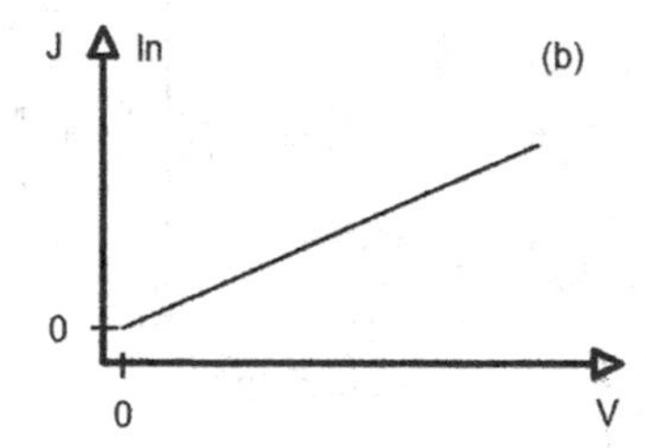

FIGURE 8.9
(a) The electron spatial profile under low-level injection. (b) The electron current density profile. $J_{LP} = J_{Ln} = J_L$. The magnitude of the electron current is drawn.

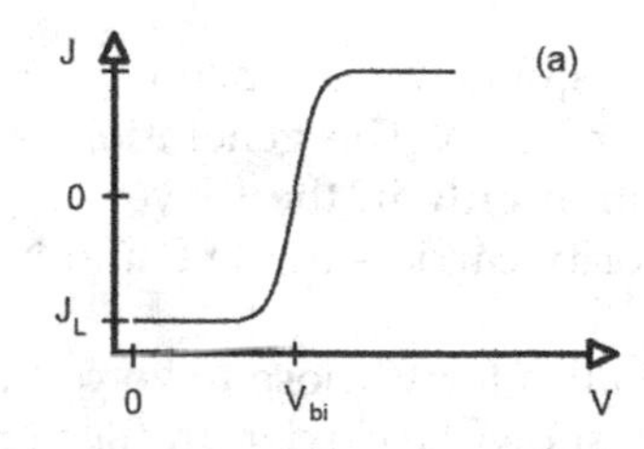

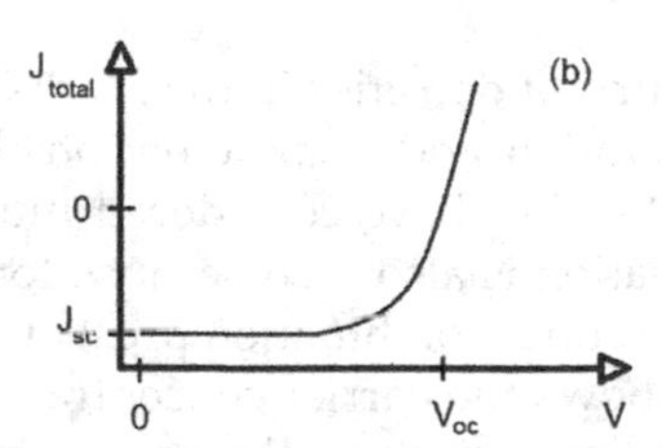

FIGURE 8.10
(a) The bias-dependent photocurrent of the solar cell. (b) The total current–voltage characteristics of the solar cell.

In summary, the bias-dependent photocurrent is shown in Figure 8.10a. It is also often referred to as the S curve. When this photocurrent adds to the solar cell's dark current (shown in Figure 8.6), the total solar cell current looks like what is shown in Figure 8.10b. This is the final current–voltage curve for an illuminated solar cell. The current at zero voltage is called the short-circuit current (J_{sc}) and the voltage at zero current is called the open-circuit voltage (V_{oc}).

8.5 Conclusion

Thus, in summary, in this chapter, we have derived the PIN current–voltage characteristics in dark and under illumination. We have started with the energy band diagram, have introduced the Continuity Equations, and from the Drift Diffusion Equations, we arrived at the fact that for the P and N layers the diffusion will be equal to the recombination and thus the exponential current–voltage curve and the Ideal Diode Equation were derived. Next we assumed high-level injection in the absorber and with those boundary conditions, we equated drift and diffusion to generation in the I layer and arrived at the bias-dependent photocurrent S curve. When both of these characteristics are added, the total current–voltage characteristics of the solar cell are obtained under illumination. With more illumination G increases, and, from Equation 27(c), J_L increases, and thus the current–voltage–light phase space of the solar cell can be chalked out.

References

[1] R. F. Pierret, Semiconductor Device Fundamentals, Addison-Wesley Publishing Company, New York (1996).
[2] D. A. Kleinman, "The forward characteristic of the PIN diode", *Bell Syst. Tech. J.* 35 (1956) 685–706.
[3] A. Herlet, "The forward characteristic of Silicon power rectifiers at high current densities", *Solid State Electron.* 11 (1968) 717–742.
[4] F. Berz, "A simplified theory of the PIN diode", *Solid State Electron.* 20 (1977) 709–714.
[5] R. Sokel and R. C. Hughes, "Numerical analysis of transient photoconductivity in insulators", *J. Appl. Phys.* 53 (1982) 7414.

9

A New Approximation for Conformable Time-Fractional Nonlinear Delayed Differential Equations via Two Efficient Methods

Brajesh Kumar Singh and Saloni Agrawal
Babasaheb Bhimrao Ambedkar University

CONTENTS

9.1 Introduction 133
9.2 Basic Literature 134
9.3 Description of NIDTM 137
9.4 Description of $_0$HANITM 138
  9.4.1 Error Estimation of $_0$HANITM 140
9.5 Test Examples 141
9.6 Conclusion 150
9.7 Result and Discussion 150
Acknowledgments 157
References 157

9.1 Introduction

Delayed differential equations are concerned to be very beneficial in biological and physical sciences, in which the evaluation of the structure confides on both present and former positions. In the literature, vigorous physical/biological models arise in terms of fractional-order systems that could not be handled by any integer order system, for instance, Robertson diffusion model, signal processing, electrical network system, space–time delay models, and so forth [1–4]. The investigation of Burgers' equation assumes an imperative role in understanding distinctive numerical and analytical methods. These models give a premise to the hypothesis of shock waves and turbulence flow, refer [5–7] for more details.

Integro-delay differential equations are generally named pantograph-type delay differential equations or generalized delay equations, which are frequently utilized to demonstrate certain issues with delayed consequence in mechanics and the related logical fields. Numerous models, for example, stress–strain conditions of materials, models of polymer crystallization, movement of rigid bodies are described in Kolmanovskii and Myshkis's monograph [8] and the references therein. In addition, we have considered the time-fractional pantograph differential equation (pantograph: z-formed mechanical connection appended to the top of an electric train/transport to acquire power flexibly by an overhead electric wire) that has a wide range of application in several systems of mechanical/electro-dynamic framework [9].

DOI: 10.1201/9781003291916-9

As we know, it is quite challenging to calculate exact results of fractional differential equations. In order to overcome this problem, an easy and demanding definition of new type of fractional-order derivative "conformable fractional derivative" was first suggested by Khalil et al. [10]. Latterly, the concept of conformable fractional calculus has been applied to compute the approximate solutions of linear and nonlinear partial differential equations (PDEs). Lio [11,12] developed a homotopy analysis method to determine the solution of various nonlinear classical order (PDEs). Thereafter, this technique has been extended to analyze many fractional-order PDEs, see [13–18]. Moreover, q-homotopy analysis method $(qHAM)$ was used to estimate the solution of conformable space–time–fractional Robertson equation with 1-D diffusion [13], generalized Hirota–Satsuma coupled KdV system [17], ERCDTM was implemented to figure out the solution of the nonlinear PDEs with delay argument [19] and so forth.

To the best of our knowledge, no effort has been made in the computation of approximate solution of conformable time-fractional PDEs (CTFPDEs) with proportional delay using $\mathbb{NIDTM}$ and ${}_{0}\mathbb{HANITM}$. In this chapter, we have implemented two reliable techniques for the analytical study of following the autonomous initial valued structure of nonlinear CTFPDEs with proportional delay.

$$\mathcal{N}[{}_{t}\mathcal{T}^{\alpha}\phi(x,t)] - g(x,t) = 0 \qquad \phi(x,0) = f(x), \qquad 0 \le \alpha < 1, x_{i}, t \in [0,1] \tag{9.1}$$

where

$$\mathcal{N}[{}_{t}\mathcal{T}^{\alpha}\phi(x,t)] = {}_{t}\mathcal{T}^{\alpha}\phi(x,t) - \mathcal{F}\left(x, \phi(a_0 x, b_0 t), \phi^{x}(a_1 x, b_1 t), \ldots, \phi^{x^{(m)}}(a_m x, b_m t)\right) \tag{9.2}$$

$\mathcal{N}$ be symbolize the nonlinear operator, ${}_{t}\mathcal{T}^{\alpha}$ symbolize the conformable derivative, $g(x,t)$ be a known function, f be a smooth function, C be a constant and $\phi(x,t)$ is the unknown function to be computed.

9.2 Basic Literature

There are numerous definitions of fractional derivatives proposed in history. Researchers have found some weaknesses in various types of definitions by which few models do not approach toward reality. For instance, Caputo fractional derivative [2] is noncommutative [20]. Keeping this in mind, a new sort of meaning of fractional derivative, i.e., Conformable Fractional Derivative as mentioned above is defined as *f*.

Definition 1 [Conformable Time-fractional derivative (CTFD) [10]]

The CTFD of a function $\phi(x,t) : \mathcal{R} \times (0,\infty) \rightarrow \mathcal{R}$ is described as

$${}_{t}\mathcal{T}^{\alpha}\phi(x,t) = \lim_{h \to 0} \frac{\phi(x, t + ht^{1-\alpha}) - \phi(x,t)}{h}, \qquad \alpha \in (0,1].$$

where α is the fractional order of derivative.

Further, the CTFD (left) of fractional order α of a function $\phi(x,t):\mathcal{R}\times(a,\infty)\to\mathcal{R},\ a\geq 0$ is characterized as below:

$$_t\mathcal{T}^{\alpha}\phi(x,t)=\lim_{h\to 0}\frac{\phi(x,t+h(t-a)^{1-\alpha})-\phi(x,t)}{h},\qquad \alpha\in(0,1].$$

The fundamental conditions of CTFD are given as:

Theorem 9.1 [20]

Suppose $\alpha\in(0,1], C, D$ are arbitrary constants and $\phi(x,t)$ and $\psi(x,t)$ are fractional-order differentiable functions at the given point $(x,t)\in\mathcal{R}\times(a,\infty),\ a\geq 0$. Then

a. $_t\mathcal{T}^{\alpha}(C\ \phi+D\ \psi)=C\ {_t\mathcal{T}^{\alpha}}\phi+D\ {_t\mathcal{T}^{\alpha}}\psi,$

b. $_t\mathcal{T}^{\alpha}(C)=0,$

c. $_t\mathcal{T}^{\alpha}(\phi\ \psi)=\phi\ {_t\mathcal{T}^{\alpha}}\psi+\psi\ {_t\mathcal{T}^{\alpha}}\phi,$

d. $_t\mathcal{T}^{\alpha}(\phi/\psi)=\dfrac{\psi\ {_t\mathcal{T}^{\alpha}}\phi-\psi {_t\mathcal{T}^{\alpha}}\ \phi}{\psi^2},$

e. *For every* $r\in\mathcal{R}$, *i)* $\mathcal{T}^{\alpha}(t^r)=r\ t^{r-\alpha}$, *ii)* $_t\mathcal{T}_a^{\alpha}(t-a)^r=r(t-a)^{r-\alpha}$.
Appropriately, $_t\mathcal{T}_a^{\alpha}(t-a)^{\alpha}=\alpha$.

f. *If* $\phi,\ \psi$ are differentiable functions w.r.t. t, then

$$i)\ {_t\mathcal{T}_a^{\alpha}}\phi=t^{1-\alpha}\frac{\partial\phi}{\partial t},\quad ii)\ {_t\mathcal{T}_a^{\alpha}}\psi=(t-a)^{(1-\alpha)}\frac{\partial\psi}{\partial t},$$

g. $_t\mathcal{T}_a^{\alpha}\left(e^{\lambda\left(\frac{(t-a)^{\alpha}}{\alpha}+x\right)}\right)=\lambda e^{\lambda\left(\frac{(t-a)^{\alpha}}{\alpha}+x\right)}.$

h. *Take* $m<\alpha\leq m+1,\ \beta=\alpha-m$, *if* $\dfrac{\partial^m\psi}{\partial t^m}$ exists, $_t\mathcal{T}_a^{\alpha}\psi={_t\mathcal{T}_a^{\beta}}\left(\dfrac{\partial^m\psi}{\partial t^m}\right)$.

i. Consider $\alpha\in(0,1]$ and $\phi(x,t)$ be $k-$ times differentiable function at $(x,t_0)\in\mathcal{R}\times(0,\infty)$ s.t. $\dfrac{\partial^i\phi}{\partial t^i}|_{(x,t_0)}=0,\ \forall i\in\{1,2\ldots k-1\}$, therefore, the $k-$ times CTFD $\phi(x,t)$ is

$$_t\mathcal{T}_{t_0}^{k\alpha}\phi=(t-t_0)^{k-k\alpha}\left(\frac{\partial^k\phi}{\partial t^k}\right)_{t=t_0},\ \text{where}\ {_t\mathcal{T}_{t_0}^{k\alpha}}\equiv\underbrace{{_t\mathcal{T}_{t_0}^{k\alpha}}\ldots{_t\mathcal{T}_{t_0}^{k\alpha}}}_{\text{k-times}}.$$

The conformable integration of fractional order α as defined in Ref. [21] is given below:

Definition 9.2 [21]

Consider $\phi:\mathbb{R}\to\mathbb{R}$ is a real-valued function. Therefore α^{th} order $(0<\alpha\leq 1)$ "conformable integration" of ϕ is characterized as follows

$$(\mathcal{I}_t^{\alpha}\phi)(t)=\int_0^t\frac{\phi(x)}{x^{1-\alpha}}dx,\quad t>0 \tag{9.3}$$

Moreover if $\alpha\in(n,n+1],\ \beta=\alpha-n,$

if the nth derivative $\phi^n(t)$ of $\phi : \mathbb{R} \to \mathbb{R}$ exists, therefore the conformable integration of ϕ for order α is

$$(\mathcal{I}_t^\alpha \phi)(t) = (\mathcal{I}_t^\beta \phi^n)(t) = \int_0^t \frac{\phi^n(x)}{x^{1-\beta}} dx. \tag{9.4}$$

Definition 9.3 [New integral transform, [22–24]]

The NIT of $\phi(\tau) \in \mathcal{F}$ is defined by

$$\Phi(x,\mu) = \mathcal{K}_\alpha\{\phi(x,t)\} := \frac{1}{\mu}\int_0^\infty \exp\left(\frac{-t^\alpha}{\alpha\mu^2}\right)\phi(x,t)\, t^{\alpha-1} dt, \tag{9.5}$$

Here $\mathcal{F}$ be the class of real-valued exponential order functions:

$$\mathcal{F} = \left\{\phi(x,t) : \exists r_1, r_2 > 0, 0 < K < \infty\, ; |\phi(x,t)| \le K \exp\left(\frac{|t|}{r_\ell^2}\right),\ \text{if } t \in (-1)^\ell \times [0,\infty)\right\},$$

where for given $\phi \in \mathcal{F}$, the constant K must be finite although r_1, r_2 arbitrary.

Some properties of (NIT) required to complete understanding of the present work are reported below:

Theorem 9.2 [Properties of NIT, [22–24]]

a. The NIT of $\frac{\partial^\kappa g(x,t)}{\partial t^\kappa} \in \mathcal{F}$ is defined by

$$\mathcal{K}_\alpha\left\{\frac{\partial^k g(x,t)}{\partial t^k}\right\} = \frac{\mathcal{G}(x,\mu)}{\mu^{2k}} - \sum_{\ell=0}^{k-1} \frac{1}{\mu^{2(k-\ell)-1}} \frac{\partial^\ell g(x,0^+)}{\partial t^\ell}, \quad k \ge 1.$$

b. The NIT of CTFD ${}_t\mathcal{T}^\alpha g(x,t)$ read as follows:

i. $\mathcal{K}_\alpha\{{}_t\mathcal{T}^{-\alpha} g(x,t)\} = \mu^{2\alpha}\mathcal{G}(x,\mu)$,

ii. $\mathcal{K}_\alpha\{{}_t\mathcal{T}^\alpha g(x,t)\} = \frac{\mathcal{G}(x,\mu)}{\mu^{2\alpha}} - \sum_{\ell=0}^{k-1} \frac{1}{\mu^{2(\alpha-\ell)-1}} \frac{\partial^l g(x,0)}{\partial t^l}, \quad k-1 < \alpha \le k \in \mathbb{N}$,

Some conformable NITs of certain functions are listed below:

i. $\mathcal{K}_\alpha\{1\} = \mu$

ii. $\mathcal{K}_\alpha\{t^\lambda\} = \alpha^{\frac{\lambda}{\alpha}} \mu^{\frac{2\lambda}{\alpha}+1} \Gamma(1+\frac{\lambda}{\alpha}),\ \kappa = 0,1,2,\ldots,n$

iii. $\mathcal{K}_\alpha\left\{\exp\left(\frac{t^\alpha}{\alpha}\right)\right\} = \frac{\mu}{1-\mu^2}$

iv. $\mathcal{K}_\alpha\left\{\sin\left(\frac{t^\alpha}{\alpha}\right)\right\} = \frac{\mu^3}{1+\mu^4}$;

v. $\mathcal{K}_\alpha\left\{\cos\left(\frac{t^\alpha}{\alpha}\right)\right\} = \frac{\mu}{1+\mu^4}$

9.3 Description of NIDTM

This section represents the NIDTM [25] for the system of CTFPDEs. Consider a nonlinear PDEs with conformable derivative of the type equation (9.1) whose standard form is given as

$$ {}_t\mathcal{T}^{\alpha}(\phi(x,t)) + \mathcal{L}[\phi(x,t)] + \mathcal{N}[\phi(x,t)] = h(x,t);\ \phi(x,0) = g(x) \tag{9.6} $$

where the $\mathcal{L}$ and $\mathcal{N}$ are linear and nonlinear operators, respectively; $h(x,t)$ is the known source term, ${}_t\mathcal{T}^{\alpha}$ be the conformable fractional operator of order α.

Applying conformable NIT to both sides of equation (6)

$$ \mathcal{K}_{\alpha}\left\{{}_t\mathcal{T}^{\alpha}(\varphi(x,t))\right\} + \mathcal{K}_{\alpha}\left\{\mathcal{L}[\varphi(x,t)] + \mathcal{N}[\varphi(x,t)] - h(x,t)\right\} = 0. \tag{9.7} $$

utilizing the property of conformable NIT operator for ${}_t\mathcal{T}^{k\alpha}(\phi(x,t))$ with initial condition, we obtain

$$ \frac{\mathcal{K}_{\alpha}\{\phi(x,t)\}}{\mu^{2\alpha}} - \sum_{\ell=0}^{k-1} \frac{1}{\mu^{2(\alpha-\ell)-1}} \frac{\partial^{\ell}\phi(x,0)}{\partial t^{\ell}} = -\mathcal{K}_{\alpha}\left\{\mathcal{L}[\phi(x,t)] + \mathcal{N}[\phi(x,t)] - h(x,t)\right\}. \tag{9.8} $$

Now implementing inverse NIT on both sides of equation (9.8), we get

$$ \phi(x,t) = \mathcal{G}(x,t) - \mathcal{K}_{\alpha}^{-1}\left[\mu^{2\alpha}\mathcal{K}_{\alpha}\left\{\mathcal{L}[\phi(x,t)] + \mathcal{N}[\phi(x,t)] - h(x,t)\right\}\right] \tag{9.9} $$

where $\mathcal{G}(x,t)$ represents the source term with initial condition. Formulate linear and nonlinear terms as

$$ \mathcal{L}[\phi(x,t)] = \sum_{k=0}^{\infty} \mathcal{L}[\phi_k(x,t)]; \qquad \mathcal{N}[\phi(x,t)] = \sum_{k=0}^{\infty} \phi_k(x,t). \tag{9.10} $$

where

$$ \mathcal{P}_k\left(\vec{\phi}_k(x,t)\right) = \frac{1}{k!}\frac{\partial^k}{\partial \aleph^k}\left[\mathcal{N}\left(\sum_{k=0}^{\infty} \aleph^k \phi(x,t,\aleph)\right)\right]_{\aleph=0}. $$

Substituting equation (9.10) in equation (9.9), we obtain

$$ \sum_{k=0}^{\infty} \phi_k(x,t) = \mathcal{G}(x,t) - \mathcal{K}_{\alpha}^{-1}\left[\mu^{2\alpha}\mathcal{K}_{\alpha}\left\{\mathcal{L}\left[\sum_{k=0}^{\infty}\phi_k(x,t)\right] + \mathcal{N}\left[\sum_{k=0}^{\infty}\phi_k(x,t)\right] - h(x,t)\right\}\right] \tag{9.11} $$

Comparing the coefficient, we calculate the following terms:

$$ \phi_0(x,t) = \mathcal{G}(x,t) + \mathcal{K}_{\alpha}^{-1}\left[\mu^{2\alpha}\mathcal{K}_{\alpha}h(x,t)\right]; \phi_1(x,t) = -\mathcal{K}_{\alpha}^{-1}\left[\mu^{2\alpha}\mathcal{K}_{\alpha}\left\{\mathcal{L}[\phi_0(x,t)] + \mathcal{N}[\phi_0(x,t)]\right\}\right] $$

In general, for $k \geq 1$:

$$ \phi_{k+1}(x,t) = -\mathcal{K}_{\alpha}^{-1}\left[\mu^{2\alpha}\mathcal{K}_{\alpha}\left\{\mathcal{L}[\phi_k(x,t)] + \mathcal{N}[\phi_k(x,t)]\right\}\right] \tag{9.12} $$

9.4 Description of $_0$HANITM

This section elaborates the $_0$HANITM [13] for the system of CTFPDEs. Consider nonlinear PDEs with conformable derivative of the type equation (9.1) whose standard form is given as

$$ {}_t\mathcal{T}^{\alpha}(\phi(x,t)) + \mathcal{L}[\phi(x,t)] + \mathcal{N}[\phi(x,t)] - h(x,t) = 0 \tag{9.13} $$

where the $\mathcal{L}$ and $\mathcal{N}$ are linear and nonlinear operators, respectively; $h(x,t)$ is the known source term, and ${}_t\mathcal{T}^{k\alpha}$ be the conformable fractional operator of order $k\alpha, k$ be a positive integer.

Applying conformable NIT on both sides of equation (9.13)

$$ \mathcal{K}_{\alpha}\left\{{}_t\mathcal{T}^{\alpha}(\varphi(x,t))\right\} + \mathcal{K}_{\alpha}\left\{\mathcal{L}[\varphi(x,t)] + \mathcal{N}[\varphi(x,t)] - h(x,t)\right\} = 0. \tag{9.14} $$

and utilizing the property of conformable NIT operator for ${}_t\mathcal{T}^{k\alpha}(\phi(x,t))$:

$$ \mathcal{K}_{\alpha}\left\{{}_t\mathcal{T}^{\alpha}\{\phi(x,t)\}\right\} = \frac{1}{\mu^{2\alpha}}\Phi_{\alpha}(x,\mu) - \sum_{\ell=0}^{k-1}\frac{1}{\mu^{2(\alpha-\ell)-1}}\frac{\partial^{\ell}\phi(x,0)}{\partial t^{\ell}}. $$

we get,

$$ \Phi_{\alpha}(x,\mu) - \sum_{\ell=0}^{k-1}\frac{1}{\mu^{-2\ell-1}}\frac{\partial^{\ell}\phi(x,0)}{\partial t^{\ell}} + \mu^{2\alpha}\mathcal{K}_{\alpha}\left\{\mathcal{L}[\phi(x,t)] + \mathcal{N}[\phi(x,t)] - h(x,t)\right\} = 0. $$

Formulate nonlinear operator as follows:

$$ \beth[\varphi(x,t;\aleph)] $$

$$ = \mathcal{K}_{\alpha}\{\varphi(x,t;\aleph)\} - \sum_{\ell=0}^{k-1}\frac{1}{\mu^{2\ell-1}}\frac{\partial^{\ell}\phi(x,0)}{\partial t^{\ell}} + \mu^{2\alpha}\mathcal{K}_{\alpha}\left\{\mathcal{L}[\varphi(x,t;\aleph)] + \mathcal{N}[\varphi(x,t;\aleph)] - h(x,t)\right\}. $$

where $\varphi(x,t;\aleph)$ is real-valued function of $\aleph, x, t$; $\aleph \in [0,1]$ is the standard embedded parameter, $\mathcal{K}_{\alpha}$ stands for conformable NIT operator.

The deformation expression of zeroth-order as in [11] is constructed as

$$ (1-\aleph)\,{}_t\mathcal{T}^{\alpha}\left[\varphi(x,t;\aleph) - \phi_0(x,t)\right] = \aleph\hbar H(x,t)\beth[\varphi(x,t;\aleph)], \tag{9.15} $$

where $\hbar \neq 0, H(x,t)$ be the auxiliary parameter & auxiliary function, respectively; $\phi_0(x,t)$ be the initial value of the function $\phi(x,t)$. It is worth mentioning that $_0$HANITM suggests a huge opportunity for selection of the auxiliary things in proceedings. It is easy to observe that $\varphi(x,t;0) = \phi(x,0)$ *and* $\varphi(x,t;1) = \phi(x,t)$ truly hold, respectively, for $\aleph = 0,1$.

This signifies that as $\aleph$ grows from 0 *to* 1, the solution $\varphi(x,t;\aleph)$ grows simultaneously from the initial guess: $\phi_0(x,t)$ to the exact solution: $\phi(x,t)$.

The function $\varphi(x,t;\aleph)$ can be expanded utilizing Taylor's formula in the form of $\aleph$ as follows:

$$\varphi(x,t;\aleph) = \phi_0(x,t) + \sum_{k=1}^{\infty} \phi_k(x,t)\aleph^k, \tag{9.16}$$

where the parameter $\hbar$ regulates the convergence area of the solution of equation (9.16). The convergence of the solution equation (9.16) at $\aleph = 1$ can be established by suitable selection of the auxiliary parameters: $\hbar, H(x,t) \neq 0$ and the initial guess. Thus,

$$\phi(x,t) = \phi_0(x,t) + \sum_{k=1}^{\infty} \phi_k(x,t). \tag{9.17}$$

where $\phi_k(x,t) = \left.\frac{1}{k!}\frac{\partial^k \varphi}{\partial \aleph^k}\right|_{\aleph=0}$.

$$\vec{\phi}_k(x,t) = (\phi_0(x,t), \phi_1(x,t), \phi_2(x,t), \ldots, \phi_k(x,t)).$$

Subsequently, the deformation equation of order k is evaluated through differentiating k-times to the deformation equation of zeroth-order with respect to the embedded parameter $\aleph at \aleph = 0$, thereafter dividing the resulting expression by $k!$.

In this consequence, the deformation equation of order k is obtained as follows:

$$\mathcal{K}_\alpha\left[\phi_k(x,t) - \chi_k\phi_{k-1}(x,t)\right] = \hbar\aleph H(x,t)\mathcal{P}_k(\vec{\phi}_{k-1}(x,t)), \tag{9.18}$$

where $\chi_k = 0$ if $k \leq 1$ and 1 otherwise. On employing inverse conformable Laplace transform to the above equation (9.19) and setting $\aleph = 1, H(x,t) = 1$, we get

$$\phi_k(x,t) = \chi_k\phi_{k-1}(x,t) + \hbar\mathcal{K}_\alpha^{-1}\left[\mathcal{P}_k(\vec{\phi}_{k-1}(x,t))\right] \tag{9.19}$$

where

$$\mathcal{P}_k(\vec{\varphi}_{k-1}(x,t)) = \left.\frac{1}{(k-1)!}\frac{\partial^{k-1}\mathcal{T}[\varphi(x,t,\aleph)]}{\partial \aleph^{k-1}}\right|_{\aleph=0}.$$

After computing $\phi_k(x,t)$ for $k \geq 1$, Mth-order approximate series solution of equation (9.13) is calculated as follows:

$$S_M(x,t) = \sum_{k=0}^{M} \phi_k(x,t), \tag{9.20}$$

which converges to the exact solution $\phi(x,t)$ of the equation (9.14) accurately as $M \to \infty$.

Theorem 9.3

As far as the series approximation equation (9.17) converges, where $\phi_k(x,t)$ are computed from equation (9.19), then the series approximation equation (9.17) must be the exact solution to the conformable differential equation (9.13).

Let the series approximation of equation (9.13) converges. If we set

$$\Xi(x,t)=\sum_{k=0}^{\infty}\phi_k(x,t),$$

then following condition (*) holds true.

$$(*)\lim_{\kappa\to\infty}\phi_\kappa(x,t)=0.$$

Since, $(**)\phi_\kappa(x,t)=\sum_{\lambda=1}^{\kappa}\left[\phi_k(x,t)-\chi_k\phi_{k-1}(x,t)\right].$

Utilize the condition (*) and (**) in equation (9.20) with property $\aleph\neq 0$, we get

$$\lim_{\kappa\to\infty}\sum_{k=1}^{\kappa}\mathcal{P}_k(\vec{\phi}_{k-1}(x,t))=\sum_{k=1}^{\infty}\mathcal{P}_k(\vec{\phi}_{k-1}(x,t))=0.$$

Therefore,

$$\sum_{k=1}^{\infty}\mathcal{P}_k(\vec{\varphi}_{k-1}(x,t))$$

$$=\sum_{k=1}^{\infty}\left[\mathcal{K}_\alpha[\phi_{k-1}(x,t)]-(1-\chi_k)\left(\sum_{\ell=0}^{k-1}\mu^{2\ell+1}\frac{\partial^\ell\phi(x,0)}{\partial t^\ell}+\mu^{2\alpha}h(x,t)\right)+\mu^{2\alpha}\mathcal{K}_\alpha\left\{\mathcal{L}[\phi_{k-1}(x,t)]+\mathcal{N}[\phi_{k-1}(x,t)]\right\}\right]$$

$$=\mu^{2\alpha}\mathcal{K}_\alpha\left[{}_t\mathcal{T}^\alpha(\phi(x,t))+\mathcal{L}[\phi(x,t)]+\mathcal{N}[\phi(x,t)]-h(x,t)\right]=0,$$

$$\Rightarrow_t\mathcal{T}^\alpha(\phi(x,t))+\mathcal{L}[\phi(\sigma,t)]+\mathcal{N}[\phi(x,t)]=h(x,t)$$

which confirms that the series solution $\phi(x,t)$ in equation (9.18) is the exact solution of the differential equation equation (9.14).

9.4.1 Error Estimation of ${}_0$HANITM

The efficiency and accurateness of the present method can be calculated using root mean square error norm L_2 and maximum error norm (L_∞)

For the problems considered in the previous section. The square residual error [14] in the kth order approximate solution is

$$\Delta_k(\hbar)=\int_0^1\int_0^1\left(\mathcal{S}\left[\sum_{i=0}^{k}\phi_i(x,t)\right]\right)^2 dxdt. \tag{9.21}$$

Here $\mathcal{S}(\phi) = \mathcal{N}[{}_t\mathcal{T}^{\alpha}\phi(x,t)]$, is the residual error. The controlling parameter $\hbar$ appeared in equation (9.17) plays a significant character in governing the convergence area and rate of the convergence for the proposed results. As mentioned in equation (9.16) of Section 9.4, q-HAM yields a family of solutions in view of $\hbar$, and so, $\Delta_k(\hbar)$ treated as the function of $\hbar$. More rapidly $\Delta_k(\hbar)$ decreases to zero, the corresponding q-HAM solution converges faster to the exact solution (i.e., accelerate the convergence of the solution). More precisely, the acceptable interval of $\hbar$ is the line segment in $\hbar$-curve approximately parallel to the horizontal axis. Thus, $\hbar$-curve helps us in an easy form to control & fix the convergence area of the approximate solution equation (9.16) of CTFPDEs with proportional delay. The optimal value of $\hbar$ is defined by the values of $\hbar$ within the $\hbar$-region for which $\Delta_k(\hbar)$ is minimum, i.e., $\frac{d\Delta_k(\hbar)}{d\hbar} = 0$.

As mentioned in [11,14], it is seen that exact square residual error $\Delta_k(\hbar)$ as defined in equation (9.21) takes too much CPU-time in their computation even for small order of approximation, it cause the computation of $\Delta_k(\hbar)$ through equation (9.21) is often not worthy to use. To overcome such type of disadvantage, the authors of Ref. [11] introduced a more efficient formula to compute the residual error, which is known as averaged residual error, and is given in the following

$$\Delta_k(\hbar) = \frac{1}{k_1\ k_2}\sum_{j=0}^{k_1}\sum_{l=0}^{k_2}\left(N\left[\sum_{i=0}^{k}\phi_i(jdx, ldt)\right]\right)^2. \tag{9.22}$$

where $\delta x = \frac{1}{k_1}$ and $\delta t = \frac{1}{k_2}, k_1 = k_2 = 10$.

The bound on the maximum absolute truncated error of the ${}_0\mathbb{HANITM}$, as proved in [14], is reported in the following

Theorem 9.4 [14]

Let $0 < \rho < 1$. If $\|\phi_{i+1}(x,t)\| \le \rho\|\phi_i(x,t)\|$ for each i. In addition, if the truncated series $\sum_{i=0}^{k}\phi_i(x,t)$ represents an approximation for the solution of $\phi(x,t)$.

Then the maximum truncated absolute error can obtain by the following

$$\left\|\phi(x,t) - \sum_{i=0}^{k}\phi_i(x,t)\right\| \le \frac{\rho^{k+1}}{1-\rho}\|\phi_0(x,t)\|. \tag{9.23}$$

9.5 Test Examples

Example 9.1

Firstly, assume nonlinear nonhomogeneous time-fractional pantograph differential equation [26]

$${}_t\mathcal{T}^{\alpha}\phi(t) + 2\phi^2\left(\frac{t}{2}\right) = 1 \qquad \phi(0) = 0, \qquad t \in [0,1], \qquad 0 < \alpha \le 1. \tag{9.24}$$

In particular case when $\alpha = 1$ the exact solution of this equation is $\phi(t) = \sin t$.

Case 1. NIDTM. On implementing NIDTM as described in equation (9.12) on the system of equation (9.24), we get the following recursive equation

$$\phi_0(t) = \mathcal{K}_\alpha^{-1}\left[\mu^{2\alpha}\mathcal{K}_\alpha\{1\}\right]$$

$$\sum_{k=0}^{\infty}\phi_{k+1}(t) = -\mathcal{K}_\alpha^{-1}\left[\mu^{2\alpha}\mathcal{K}_\alpha\left\{2\sum_{j=0}^{k}\phi_j\left(\frac{t}{2}\right)\phi_{k-j}\left(\frac{t}{2}\right)\right\}\right]. \tag{9.25}$$

By solving the recurrence relation equation (9.25), the first three iterations can be computed by using Mathematica software as

$$\phi_0(t) = \frac{t^\alpha}{\alpha};\ \phi_1(t) = -\frac{2^{1-2\alpha}t^{3\alpha}}{3\alpha^3};\ \phi_2(t) = \frac{8^{1-2\alpha}t^{5\alpha}}{15\alpha^5};$$

$$\phi_3(t) = -\frac{2^{3-10\alpha}t^{7\alpha}}{63\alpha^7} - \frac{2^{5-12\alpha}t^{7\alpha}}{105\alpha^7}. \tag{9.26}$$

In this consequence, the kth iterative solutions $\phi_k(t)$ for $k \geq 4$ can be evaluated from equation (9.26). Therefore, sixth-order approximate solution for $\alpha = 1$ is obtained as

$$S_6(t) = \sum_{k=0}^{6}\phi_k(t) = t - \frac{t^3}{3!} + \frac{t^5}{5!} - \frac{t^7}{7!} + \frac{t^9}{9!} - \frac{t^{11}}{11!} + \frac{t^{13}}{13!} \tag{9.27}$$

which is the nearest form of exact result $\phi(t) = \sin t$ at $\alpha = 1$.

Case 2. $_0$HANITM: On implementing $\mathcal{K}_\alpha$ operator on the system of equation (9.24) and using the properties of NIT, we get

$$\mathcal{K}_\alpha\{\phi(t)\} - \mu\phi(0) + \mu^{2\alpha}\mathcal{K}_\alpha\left\{2\phi^2\left(\frac{t}{2}\right) - 1\right\} = 0$$

Construct nonlinear operator as

$$\mathcal{N}\varphi(t,\aleph) = \mathcal{K}_\alpha\{\varphi(t;\aleph)\} + \mu^{2\alpha}\mathcal{K}_\alpha\left\{2\varphi^2\left(\frac{t}{2};\aleph\right) - 1\right\}. \tag{9.28}$$

Utilizing equation (9.28) in equation (9.19), we get the recursive formula to evaluate the term $\phi_k(t)$
for each $k \geq 1$:

$$\phi_k(t) = \chi_k\phi_{k-1}(t) + \hbar\mathcal{K}_\alpha^{-1}\left[\mathcal{P}_k(\vec{\phi}_{k-1}(t))\right] \tag{9.29}$$

where

$$\mathcal{P}_k(\vec{\phi}_{k-1}(t)) = \mathcal{K}_\alpha\{\phi_{k-1}(t)\} - (1-\chi_k)\mu^2 + \mu^{2\alpha}\mathcal{K}_\alpha\left[\left\{2\sum_{j=0}^{k-1}\phi_{k-1-j}\left(\frac{t}{2}\right)\phi_j\left(\frac{t}{2}\right)\right\}\right]$$

With the aid of Mathematica software, solve recurrence relation equation (9.30) for $\phi_k(t)$ for $k \geq 1$ as follows, on solving recurrence equation (9.30), the first three iterations are obtained as:

$$\phi_1(t) = -\frac{\hbar t^\alpha}{\alpha}; \quad \phi_2(t) = -\frac{\hbar^2 t^\alpha}{\alpha} - \frac{\hbar t^\alpha}{\alpha}; \quad \phi_3(t) = \frac{2^{1-2\alpha}\hbar^3 t^{3\alpha}}{3\alpha^3} - \frac{\hbar^3 t^\alpha}{\alpha} - \frac{2\hbar^2 t^\alpha}{\alpha} - \frac{\hbar t^\alpha}{\alpha}$$

In this consequence, the kth iterative solutions $\phi_k(t)$ for $k \geq 4$ can be evaluated from equation (9.29). Thus, sixth-order approximate solution at $\hbar = -1$ is obtained as:

$$S_6(t) = \sum_{k=0}^{6} \phi_k(t) = \frac{t^\alpha}{\alpha} - \frac{2^{1-2\alpha} t^{3\alpha}}{3\alpha^3} + \frac{8^{1-2\alpha} t^{5\alpha}}{15\alpha^5}$$

which is nearest form of exact solution $\phi(t) = \sin t$ at $\alpha = 1$.

Example 9.2

Let us consider the following first-order nonlinear nonhomogeneous pantograph-type integro-differential equation [26]

$$_t\mathcal{T}^\alpha \phi(t) + \left(\frac{t}{2} - 2\right)\phi(t) - 2\int_0^t \phi^2\left(\frac{s}{2}\right) ds = 1, \; 0 < \alpha \leq 1, \; \phi(0) = 0, \; t \in [0,1]. \tag{9.30}$$

In special case when $\alpha = 1$ the exact result is $\phi(t) = te^t$.

Case 1. $\mathbb{NIDTM}$: On implementing $\mathbb{NIDTM}$ as mentioned in equation (9.12) on the system of equation (9.30), we obtain the following recursive equation

$$\phi_0(t) = \mathcal{K}_\alpha^{-1}\left[\mu^{2\alpha}\mathcal{K}_\alpha\{1\}\right]$$

$$\sum_{k=0}^{\infty} \phi_{k+1}(t) = -\mathcal{K}_\alpha^{-1}\left[\mu^{2\alpha}\mathcal{K}_\alpha\left\{\left(\frac{t}{2} - 2\right)\phi_j(t) - 2\int_0^t \sum_{j=0}^{k} \phi_j\left(\frac{s}{2}\right)\phi_{k-j}\left(\frac{s}{2}\right)\right\}\right] \tag{9.31}$$

By solving the relation equation (9.31), the first two iterations are obtained as:

$$\phi_0(t) = \frac{t^\alpha}{\alpha};$$

$$\phi_1(t) = \frac{t^{2\alpha}\left(2\alpha^2 + \frac{4^{-\alpha}\left(\frac{1}{\alpha}\right)^{-1/\alpha}\Gamma\left(2+\frac{1}{\alpha}\right)\left(\frac{t^\alpha}{\alpha}\right)^{1/\alpha}\left(4\Gamma\left(3+\frac{1}{\alpha}\right)t^\alpha - 4^\alpha\alpha^2\Gamma\left(4+\frac{1}{\alpha}\right)\right)}{\Gamma\left(3+\frac{1}{\alpha}\right)\Gamma\left(4+\frac{1}{\alpha}\right)}\right)}{2\alpha^4};$$

In this consequence, the kth iterative solutions $\phi_k(t)$ for $k \geq 2$ can be evaluated from equation (9.31). Thus sixth-order approximate solution for $\alpha = 1$ is obtained as:

$$S_6(t) = \sum_{k=0}^{6} \phi_k(t) = t + t^2 + \frac{t^3}{2} + \frac{t^4}{6} + \frac{t^5}{24} + \frac{t^6}{120} + \frac{t^7}{720} - \frac{t^8}{336} + \frac{97t^9}{40320} - \frac{39t^{10}}{44800}$$
$$+ \frac{11311t^{11}}{63866880} - \frac{2861t^{12}}{132710400} + \frac{20265209t^{13}}{12752938598400} - \frac{2448137t^{14}}{35267385753600}$$
$$+ \frac{49241431t^{15}}{27700928446464000} - \frac{40692071863t^{16}}{140410466109436723200 0}$$
$$+ \frac{33521726461t^{17}}{152766587127067154841600} - \frac{2460828499t^{18}}{244426539403307447746560 0}$$
$$+ \frac{77094233t^{19}}{357620861929117206970368 00} \tag{9.32}$$

which is nearest form of exact solution $\phi(t) = te^t$ at $\alpha = 1$.

Case 2. ${}_0\mathbb{HANITM}$: On implementing $\mathcal{K}_\alpha$ operator on the system of equation (9.30) and using the properties of NIT, we get

$$\mathcal{K}_\alpha\{\phi(t)\} - \mu^2\phi(0) + \mu^{2\alpha}\mathcal{K}_\alpha\left\{\left(\frac{t}{2} - 2\right)\phi(t) - 2\int_0^t \phi^2\left(\frac{s}{2}\right)ds - 1\right\} = 0$$

Construct nonlinear operator as:

$$\mathcal{N}\varphi(t,\aleph) = \mathcal{K}_\alpha\{\varphi(t;\aleph)\} + \mu^{2\alpha}\mathcal{K}_\alpha\left\{\left(\frac{t}{2} - 2\right)\varphi(t;\aleph) - 2\int_0^t \varphi^2\left(\frac{s}{2};\aleph\right)ds - 1\right\}. \tag{9.33}$$

Utilizing equation (9.33) in equation (9.19), to evaluate the term $\phi_k(t)$ for each $k \geq 1$:

$$\phi_k(t) = \chi_k\phi_{k-1}(t) + \hbar\mathcal{K}_\alpha^{-1}\left[\mathcal{P}_k(\vec{\phi}_{k-1}(t))\right] \tag{9.34}$$

where

$$\mathcal{P}_k(\vec{\phi}_{k-1}(t)) = \mathcal{K}_\alpha\{\phi_{k-1}(t)\} - (1-\chi_k)\mu^2$$
$$+\mu^{2\alpha}\mathcal{K}_\alpha\left[\left\{\left(\frac{t}{2} - 2\right)\phi(t) - 2\sum_{j=0}^{k-1}\left[\int_0^t \phi_j\left(\frac{s}{2}\right)\phi_{k-1-j}\left(\frac{s}{2}\right)ds\right]\right\}\right]$$

With the aid of Mathematica software, solve recurrence relation equation (34) for $\phi_k(t)$ for $k \geq 1$. On solving recurrence equation (9.34), the first three iterations are obtained as follows:

$$\phi_1(t) = -\frac{\hbar t^\alpha}{\alpha}; \quad \phi_2(t) = \frac{\hbar^2 t^\alpha}{2\alpha^2}\left(-2\alpha + 2t^\alpha - \frac{\left(\frac{1}{\alpha}\right)^{-1/\alpha}\Gamma\left(2+\frac{1}{\alpha}\right)t^\alpha\left(\frac{t^\alpha}{\alpha}\right)^{1/\alpha}}{\Gamma\left(3+\frac{1}{\alpha}\right)}\right) - \frac{\hbar t^\alpha}{\alpha};$$

$$\phi_3(t)=\frac{\hbar^2t^\alpha}{2\alpha^2}\left(-2\alpha+2t^\alpha-\frac{\left(\frac{1}{\alpha}\right)^{-1/\alpha}\Gamma\left(2+\frac{1}{\alpha}\right)t^\alpha\left(\frac{t^\alpha}{\alpha}\right)^{1/\alpha}}{\Gamma\left(3+\frac{1}{\alpha}\right)}\right)-\frac{\hbar t^\alpha}{\alpha}+\frac{\hbar^2t^{2\alpha}}{\alpha^2}-\frac{\hbar^2t^\alpha}{\alpha}-\frac{\hbar^3t^\alpha}{\alpha}$$

$$-\frac{2\hbar^3t^{3\alpha}}{3\alpha^3}+\frac{2\hbar^3t^{2\alpha}}{\alpha^2}-\frac{\left(\frac{1}{\alpha}\right)^{-1/\alpha}(2\hbar+1)\hbar^2\Gamma\left(2+\frac{1}{\alpha}\right)\left(\frac{t^\alpha}{\alpha}\right)^{\frac{1}{\alpha}+2}}{2\Gamma\left(3+\frac{1}{\alpha}\right)}$$

$$+\frac{(4\alpha+1)\left(\frac{1}{\alpha}\right)^{-1/\alpha}\hbar^3\Gamma\left(2+\frac{1}{\alpha}\right)\left(\frac{t^\alpha}{\alpha}\right)^{\frac{1}{\alpha}+3}}{2\alpha\Gamma\left(4+\frac{1}{\alpha}\right)}-\frac{\left(\frac{1}{\alpha}\right)^{-2/\alpha}\alpha\hbar^3\Gamma\left(3+\frac{2}{\alpha}\right)\left(\frac{t^\alpha}{\alpha}\right)^{\frac{2}{\alpha}+3}}{4(2\alpha+1)\Gamma\left(4+\frac{2}{\alpha}\right)}$$

$$+\frac{2\left(\frac{1}{\alpha}\right)^{-1/\alpha}\hbar^3\Gamma\left(2+\frac{1}{\alpha}\right)\left(\frac{t^\alpha}{\alpha}\right)^{\frac{1}{\alpha}+3}(\sinh(\alpha\log 4)-\cosh(\alpha\log 4))}{\alpha\Gamma\left(4+\frac{1}{\alpha}\right)}$$

In this consequence, the kth iterative solutions $\phi_k(t)$ for $k\geq 4$ can be evaluated from equation (9.34). Thus sixth-order approximate solution at $\alpha=1$ and $\hbar=-1$ is obtained as:

$$S_6(t)=\sum_{k=0}^{6}\phi_k(t)=t+t^2+\frac{t^3}{2}+\frac{t^4}{6}+\frac{t^5}{24}+\frac{t^6}{120}-\frac{19t^7}{1680}+\frac{61t^8}{13440}$$
$$-\frac{10189t^9}{11612160}+\frac{4769t^{10}}{58060800}-\frac{t^{11}}{332640} \tag{9.35}$$

which is nearest form of exact solution $\phi(t)=te^t$ at $\alpha=1$.

Example 9.3

Take the following initial valued problem of $_{\mathbb{CT}}\mathbb{GBE}^{\mathbb{PD}}$:

$$_tT^\alpha\phi(x,t)=\frac{\partial^2}{\partial x^2}[\phi(x,t)]+\frac{\partial}{\partial x}\left[\phi\left(x,\frac{t}{2}\right)\right]\phi\left(\frac{x}{2},\frac{t}{2}\right)+\frac{1}{2}\phi(x,t), \tag{9.36}$$
$$\phi(x,0)=x,\quad x,t\in[0,1],\quad 0<\alpha\leq 1.$$

Case 1. $\mathbb{NIDTM}$: On implementing $\mathbb{NIDTM}$ as mentioned in equation (9.12) on the system of equation (9.36), we obtain the following recursive equation:

$$\sum_{k=0}^{\infty}\phi_k(x,t)$$

$$=\phi(x,0)-\mathcal{K}_\alpha^{-1}\left[\mu^{2\alpha}\mathcal{K}_\alpha\left\{\frac{\partial^2}{\partial x^2}[\phi_{k-1}(x,t)]+\sum_{j=0}^{k-1}\left[\frac{\partial}{\partial x}\left[\phi_{k-1-j}\left(x,\frac{t}{2}\right)\right]\phi_j\left(\frac{x}{2},\frac{t}{2}\right)\right]+\frac{1}{2}\phi_{k-1}(x,t)\right\}\right] \tag{9.37}$$

By solving the recurrence relation equation (9.37), the first three iterative terms are obtained as:

$$\phi_0(x,t)=x;\quad \phi_1(x,t)=\frac{xt^{\alpha}}{\alpha};\quad \phi_2(x,t)=\frac{\left(2^{-\alpha}+\frac{1}{2}\right)xt^{2\alpha}}{2\alpha^2};$$

$$\phi_3(x,t)=\frac{2^{-3(\alpha+1)}\left(3\,2^{\alpha+1}+2^{2\alpha+1}+8^{\alpha}+4\right)xt^{3\alpha}}{3\alpha^3}$$

In this consequence, the kth iterative solutions $\phi_k(t)$ for $k\geq 4$ can be evaluated from equation (9.37). Thus sixth-order approximate solution for $\alpha=1$ is obtained as:

$$S_6(x,t)=\sum_{k=0}^{6}\phi_k(x,t)=x\left(1+t+\frac{t^2}{2!}+\frac{t^3}{3!}+\frac{t^4}{4!}+\frac{t^5}{5!}+\frac{t^6}{6!}\right) \tag{9.38}$$

which is nearest form of exact solution $\phi(x,t)=xe^{t}$ at $\alpha=1$.

Case 2. $_0\mathbb{HANITM}$: On implementing $\mathcal{K}_\alpha$ operator on the system of equation (9.36) and using the properties of NIT, we get

$$\mathcal{K}_\alpha\{\phi(x,t)\}-\mu\phi(x,0)-\mu^{2\alpha}\mathcal{K}_\alpha\left\{\frac{\partial^2}{\partial x^2}[\phi(x,t)]+\frac{\partial}{\partial x}\left[\phi\left(x,\frac{t}{2}\right)\right]\phi\left(\frac{x}{2},\frac{t}{2}\right)+\frac{1}{2}\phi(x,t)\right\}=0$$

Construct nonlinear operator as:

$$\mathcal{N}\varphi(t,\aleph)=\mathcal{K}_\alpha\{\varphi(x,t;\aleph)\}-x\mu-\mu^{2\alpha}$$
$$\times\mathcal{K}_\alpha\left\{\frac{\partial^2}{\partial x^2}[\varphi(x,t;\aleph)]+\frac{\partial}{\partial x}\left[\varphi\left(x,\frac{t}{2};\aleph\right)\right]\varphi\left(\frac{x}{2},\frac{t}{2};\aleph\right)+\frac{1}{2}\varphi(x,t;\aleph)\right\}. \tag{9.39}$$

Utilizing equation (9.39) in equation (9.19), we get the recursive formula to evaluate the term $\phi_k(x,t)$ for each $k\geq 1$:

$$\phi_k(x,t)=\chi_k\phi_{k-1}(x,t)+\hbar\mathcal{K}_\alpha^{-1}\left[\mathcal{P}_k(\vec{\phi}_{k-1}(x,t))\right] \tag{9.40}$$

where

$$\mathcal{P}_k(\vec{\phi}_{k-1}(t))=\mathcal{K}_\alpha\{\phi_{k-1}(t)\}-(1-\chi_k)\mu\phi_{k-1}(x,0)$$
$$+\mu^{2\alpha}\mathcal{K}_\alpha\left[\left[\left\{\frac{\partial^2}{\partial x^2}[\phi_{k-1}(x,t)]+\sum_{j=0}^{k-1}\left[\frac{\partial}{\partial x}\left[\phi_{k-1-j}\left(x,\frac{t}{2}\right)\right]\phi_j\left(\frac{x}{2},\frac{t}{2}\right)\right]+\frac{1}{2}\phi_{k-1}(x,t)\right\}\right]\right]$$

With the aid of Mathematica software, solve recurrence relation equation (9.40) for $\phi_k(x,t)$ for $k\geq 1$. On solving the recurrence equation (9.40), the first three iterations are obtained as follows:

$$\phi_1(x,t) = -\frac{\hbar x t^{\alpha}}{\alpha}; \quad \phi_2(x,t) = \hbar\left(\frac{2^{-\alpha-2}\left(2^{\alpha}+2\right)\hbar x t^{2\alpha}}{\alpha^2} - \frac{\hbar x t^{\alpha}}{\alpha}\right) - \frac{\hbar x t^{\alpha}}{\alpha};$$

$$\phi_3(x,t) = \frac{2^{-3\alpha-3}\hbar^3 x t^{3\alpha}}{3\alpha^3}\begin{bmatrix} 4+6\sinh(\alpha\log 2)+2\sinh(\alpha\log 4)+\sinh(\alpha\log 8)+6\cosh(\alpha\log 2) \\ +2\cosh(\alpha\log 4)\cosh(\alpha\log 8) \end{bmatrix}$$

$$+\frac{2^{-3\alpha-2}\hbar^2 x t^{2\alpha}}{\alpha^2}\big[2^{3\alpha}\hbar + 2^{2\alpha+1}\hbar + 2(\hbar+1)\sinh(\alpha\log 4) + 2(\hbar+1)\cosh(\alpha\log 4)$$

$$+(\hbar+1)\sinh(\alpha\log 4)\big(\sinh(\alpha\log 2)\big) + \cosh(\alpha\log 2)$$

$$+(\hbar+1)\cosh(\alpha\log 4)\big(\cosh(\alpha\log 2)+\sinh(\alpha\log 2)\big)\big]$$

$$+\frac{(\hbar^3(-x)-\hbar^2 x)t^{\alpha}}{\alpha} - \frac{\hbar x t^{\alpha}}{\alpha} + \hbar\left(\frac{2^{-\alpha-2}\left(2^{\alpha}+2\right)\hbar x t^{2\alpha}}{\alpha^2} - \frac{\hbar x t^{\alpha}}{\alpha}\right)$$

In this consequence, the kth iterative solutions $\phi_k(t)$ for $k \geq 4$ can be evaluated from equation (9.40). Thus sixth-order approximate solution at $\hbar = -1$ is obtained as:

$$S_6(x,t) = \frac{2^{-2\alpha-5}xt^{4\alpha}}{\alpha^4} + \frac{3\,2^{-3\alpha-5}xt^{4\alpha}}{\alpha^4} + \frac{2^{-5\alpha-4}xt^{4\alpha}}{\alpha^4} + \frac{2^{-\alpha-5}xt^{4\alpha}}{3\alpha^4}$$
$$+\frac{7\,2^{-4\alpha-4}xt^{4\alpha}}{3\alpha^4} + \frac{2^{-6\alpha-3}xt^{4\alpha}}{3\alpha^4} + \frac{xt^{4\alpha}}{192\alpha^4} + \frac{2^{-2\alpha-2}xt^{3\alpha}}{\alpha^3} + \frac{2^{-\alpha-2}xt^{3\alpha}}{3\alpha^3}$$
$$+\frac{2^{-3\alpha-1}xt^{3\alpha}}{3\alpha^3} + \frac{xt^{3\alpha}}{24\alpha^3} + \frac{2^{-\alpha-1}xt^{2\alpha}}{\alpha^2} + \frac{xt^{2\alpha}}{4\alpha^2} + \frac{xt^{\alpha}}{\alpha} + x + \ldots$$

which is nearest form of exact solution.

Example 9.4

In the last example consider the CTFPDEs as follows:

$$\begin{aligned} {}_t\mathcal{T}^{\alpha}\phi(x,t) &= \frac{\partial^2}{\partial x^2}\phi\left(\frac{x}{2},\frac{t}{2}\right)\frac{\partial}{\partial x}\phi\left(\frac{x}{2},\frac{t}{2}\right) - \frac{1}{8}\frac{\partial}{\partial x}\phi(x,t) - \phi(x,t) \\ \phi(x,0) &= x^2, \quad x,t \in [0,1], \quad 0 < \alpha \leq 1. \end{aligned} \tag{9.41}$$

Case 1. NIDTM : On implementing NIDTM as mentioned in equation (9.12) on the system of equation (9.41), we obtain the following recursive equation

$$\sum_{k=0}^{\infty}\phi_k(x,t) = \phi(x,0) - \mathcal{K}_{\alpha}^{-1}\left[\mu^{2\alpha}\mathcal{K}_{\alpha}\left\{\sum_{j=0}^{k}\frac{\partial^2}{\partial x^2}\phi_j\left(\frac{x}{2},\frac{t}{2}\right)\frac{\partial}{\partial x}\phi_{k-j}\left(\frac{x}{2},\frac{t}{2}\right) - \frac{1}{8}\frac{\partial}{\partial x}\phi_k(x,t) - \phi_k(x,t)\right\}\right] \tag{9.42}$$

By solving the relation equations (9.42), the first three iterations are obtained as

$$\phi_0(x,t)=x^2;\quad \phi_1(x,t)=-\frac{x^2t^\alpha}{\alpha};\quad \phi_2(x,t)=\frac{2^{-\alpha-3}xt^{2\alpha}\left(2^\alpha+2^{\alpha+2}x-2\right)}{\alpha^2};$$

$$\phi_3(x,t)=-\frac{8^{-\alpha-2}t^{3\alpha}\left(8^\alpha-2^{\alpha+1}-2^{2\alpha+1}+2^{3\alpha+5}x^2+2^{\alpha+4}\left(-2^\alpha+4^\alpha-2\right)x+4\right)}{3\alpha^3}$$

In this consequence, the kth iterative solutions $\phi_k(t)$ for $k\geq 4$ can be evaluated from equation (9.42). Thus sixth-order approximate solution for $\alpha=1$ is obtained as

$$S_6(x,t)=\sum_{k=0}^{6}\phi_k(x,t)=x^2\left(1-t+\frac{t^2}{2!}-\frac{t^3}{3!}+\frac{t^4}{4!}-\frac{t^5}{5!}+\frac{t^6}{6!}\right)$$

Which is nearest form of exact solution $\phi(x,t)=x^2e^{-t}$ at $\alpha=1$.

Case 2. $_0$HANITM: On implementing $\mathcal{K}_\alpha$ operator on the system of equation (9.41) and using the properties of NIT, we get

$$\mathcal{K}_\alpha\{\phi(x,t)\}-\mu\phi(x,0)-\mu^{2\alpha}\mathcal{K}_\alpha\left\{\frac{\partial^2}{\partial x^2}\phi\left(\frac{x}{2},\frac{t}{2}\right)\frac{\partial}{\partial x}\phi\left(\frac{x}{2},\frac{t}{2}\right)-\frac{1}{8}\frac{\partial}{\partial x}\phi(x,t)-\phi(x,t)\right\}=0$$

Construct nonlinear operator as:

$$\mathcal{N}\varphi(t,\aleph)=\mathcal{K}_\alpha\{\varphi(x,t;\aleph)\}-x^2\mu-\mu^{2\alpha}\mathcal{K}_\alpha$$
$$\left\{\frac{\partial^2}{\partial x^2}\varphi\left(\frac{x}{2},\frac{t}{2};\aleph\right)\frac{\partial}{\partial x}\varphi\left(\frac{x}{2},\frac{t}{2};\aleph\right)-\frac{1}{8}\frac{\partial}{\partial x}\varphi(x,t;\aleph)-\varphi(x,t;\aleph)\right\}. \tag{9.43}$$

Utilizing equation (9.43) in equation (9.19), we get the reclusive formula to evaluate the term $\phi_k(x,t)$ for each $k\geq 1$:

$$\phi_k(x,t)=\chi_k\phi_{k-1}(x,t)+\hbar\mathcal{K}_\alpha^{-1}\left[\mathcal{P}_k(\vec{\phi}_{k-1}(x,t))\right] \tag{9.44}$$

where

$$\mathcal{P}_k(\vec{\phi}_{k-1}(t))=\mathcal{K}_\alpha\{\phi_{k-1}(t)\}-(1-\chi_k)x^2\mu+\mu^{2\alpha}$$
$$\times\mathcal{K}_\alpha\left[\left\{\sum_{j=0}^{k-1}\left[\frac{\partial^2}{\partial x^2}\phi_{k-1-j}\left(\frac{x}{2},\frac{t}{2}\right)\frac{\partial}{\partial x}\phi_j\left(\frac{x}{2},\frac{t}{2}\right)\right]-\frac{1}{8}\frac{\partial}{\partial x}\phi_{k-1}(x,t)-\phi_{k-1}(x,t)\right\}\right]$$

With the aid of Mathematica software, solve recurrence relation equation (9.44) for $\phi_k(x,t)$ for $k\geq 1$. On solving recurrence equation (9.44), the first three iterations are obtained as follows:

$$\phi_1(x,t)=\frac{hx^2t^\alpha}{\alpha};\phi_2(x,t)=\frac{h^2x^2t^{2\alpha}}{2\alpha^2}-\frac{2^{-\alpha-2}h^2xt^{2\alpha}}{\alpha^2}+\frac{h^2xt^{2\alpha}}{8\alpha^2}+\frac{h^2x^2t^\alpha}{\alpha}+\frac{hx^2t^\alpha}{\alpha};$$

$$\phi_3(x,t)=\frac{\hbar^3x^2t^\alpha}{\alpha}+\frac{2\hbar^2x^2t^\alpha}{\alpha}+\frac{\hbar x^2t^\alpha}{\alpha}$$

$$+\frac{2^{-3\alpha-3}\hbar^2(\hbar+1)x\cosh(\alpha\log 4)\left(\cosh(\alpha\log 2)+\sinh(\alpha\log 2)\right)t^{2\alpha}}{\alpha^2}$$

$$+\frac{2^{-3\alpha-1}\hbar^2(\hbar+1)x^2\left(\cosh(\alpha\log 8)+\sinh(\alpha\log 8)\right)t^{2\alpha}}{\alpha^2}$$

$$+\frac{2^{-3\alpha-3}\hbar^2(\hbar+1)x\sinh(\alpha\log 4)\left(\cosh(\alpha\log 2)+\sinh(\alpha\log 2)\right)t^{2\alpha}}{\alpha^2}$$

$$+\frac{2^{-3\alpha-4}h^3t^{3\alpha}}{3\alpha^3}-\frac{2^{-3\alpha-5}h^3\left(\cosh(\alpha\log 2)+\sinh(\alpha\log 2)\right)t^{3\alpha}}{3\alpha^3}$$

$$-\frac{2^{-\alpha-2}\hbar^2(\hbar+1)xt^{2\alpha}}{\alpha^2}-\frac{\hbar^2(\hbar+1)x^2t^{2\alpha}}{2\alpha^2}+\frac{\hbar^2(\hbar+1)xt^{2\alpha}}{8\alpha^2}$$

$$+\frac{2^{-3\alpha-2}\hbar^2(\hbar+1)x\left(\cosh(\alpha\log 4)+\sinh(\alpha\log 4)\right)t^{2\alpha}}{\alpha^2}$$

$$+\frac{2^{-3\alpha-2}h^3x\left(\cosh^3(\alpha\log 2)+\sinh^3(\alpha\log 2)\right)t^{3\alpha}}{3\alpha^3}$$

$$+\frac{2^{-3\alpha-2}h^3x\left(-1/3+\cosh(\alpha\log 2)\right)\sinh^2(\alpha\log 2)t^{3\alpha}}{\alpha^3}$$

$$+\frac{2^{-3\alpha-6}h^3\left(\cosh(\alpha\log 8)+\sinh(\alpha\log 8)\right)t^{3\alpha}}{3\alpha^3}$$

$$+\frac{2^{-3\alpha-1}h^3x^2\left(\cosh(\alpha\log 8)+\sinh(\alpha\log 8)\right)t^{3\alpha}}{3\alpha^3}$$

$$+\frac{2^{-3\alpha-2}h^3x\cosh^2(\alpha\log 2)\left(-1/3+\sinh(\alpha\log 2)\right)t^{3\alpha}}{\alpha^3}$$

$$-\frac{2^{-3\alpha-1}h^3x\left(\cosh(\alpha\log 2)-\sinh(\alpha\log 2)-\cosh(\alpha\log 2)\sinh(\alpha\log 2)\right)t^{3\alpha}}{3\alpha^3}$$

$$-\frac{2^{-3\alpha-5}h^3\left(\cosh(\alpha\log 4)+\sinh(\alpha\log 4)\right)t^{3\alpha}}{3\alpha^3}$$

In this consequence, the kth iterative solutions $\phi_k(t)$ for $k\geq 4$ can be evaluated from equation (9.44). Thus sixth-order approximate solution at and $\alpha=1$ reduces to

$$S_6(x,t)=\sum_{k=0}^{6}\phi_k(x,t)=x^2\left(1-t+\frac{t^2}{2!}-\frac{t^3}{3!}+\frac{t^4}{4!}-\frac{t^5}{5!}+\frac{t^6}{6!}\right)$$

Which is nearest form of exact solution.

9.6 Conclusion

In this chapter, a comparative study for CTFPDEs with proportional delay has been made by adopting two efficient techniques: New integral decomposition transform method (NIDTM) and optimal homotopy analysis NIT method (${}_{O}$HANITM). Specially, conformable time-fractional nonlinear nonhomogeneous pantograph-type integro-differential equation and Burger equation with proportional delay along with two more problems are effectively solved and analyzed via these two techniques. The findings from the graphics and tables represent that the approximate solutions from both techniques agreed well with the exact solution and the error decreases with increasing the order of iterations. Moreover, (NIDTM) converges faster than (${}_{O}$HANITM) but computational cost of (NIDTM) is higher than (${}_{O}$HANITM) for Examples 9.1 and 9.2. But for Examples 9.3 and 9.4 the solutions obtained from NIDTM and (${}_{O}$HANITM) for $\hbar = -1$ coincides. Although (${}_{O}$HANITM) produces more accurate results in comparison to NIDTM and existing results [19] at optimal value of $\hbar$ with higher computational cost.

9.7 Result and Discussion

Tables 9.1 and 9.2 report approximate solutions obtained from ${}_{O}$HANITM at optimal value $\hbar = -1$ for different values of $\alpha = 0.8, 0.9, 1$ and exact solution at $\alpha = 1$.

In addition, absolute error in sixth-order results computed from NIDTM and ${}_{O}$HANITM at different time levels are also reported in the table. Figures 9.1a and 9.2a depicts the behavior of sixth-order results for different $\alpha = 0.8, 0.9, 1$; Figures 9.1b and 9.2b show the two-dimensional plots of the absolute error for $\alpha = 1$ in k^{th} order results (k=4, 5, 6); Figures 9.1c

TABLE 9.1

Approximate Solutions in k=6th Order Results of Example 9.1

					Absolute Error	
t	$\alpha = 0.8$	$\alpha = 0.9$	$\alpha = 1$	Exact	$\hbar = -1$	NIDTM
0.1	0.687114809	0.338579687	0.099833417	0.099833417	1.98387E-11	1.38778E-17
0.2	1.060792659	0.561720142	0.198669333	0.198669331	2.53827E-09	2.77556E-17
0.3	1.338783663	0.743043822	0.295520250	0.295520207	4.33387E-08	0.00000E+00
0.4	1.555763829	0.895494803	0.389418667	0.389418342	3.24358E-07	0.00000E+00
0.5	1.726874559	1.024671393	0.479427083	0.479425539	1.54473E-06	0.00000E+00
0.6	1.860708146	1.133664777	0.564648000	0.564642473	5.52660E-06	3.33067E-16
0.7	1.962954201	1.224464662	0.644233917	0.644217687	1.62294E-05	3.55271E-15
0.8	2.037818345	1.298519863	0.717397333	0.717356091	4.12424E-05	2.68674E-14
0.9	2.088683584	1.357005549	0.783420750	0.783326910	9.38404E-05	1.56986E-13
1	2.118457180	1.400966770	0.841666667	0.841470985	1.95682E-04	7.61946E-13
CPU	Time (In sec)				144.187	79.8

For $\alpha = 0.8, 0.9$ & 1 compared with exact result and absolute error in 6th order results obtained from ${}_{O}$HANITM for optimal value of $\hbar$ ($\hbar = -1$) and NIDTM at distinct time levels $0 < t \leq 1$.

TABLE 9.2

Approximate Solutions in k=6th Order Results of Example 9.2

					Absolute Error		
t	$\alpha=0.8$	$\alpha=0.9$	$\alpha=1$	Exact	$\hbar=-1$	$\hbar=-1.1240$	NIDTM
0.1	0.4836016	0.2316341	0.1105170	0.1105170	1.2273E-09	1.2908E-09	2.9451E-11
0.2	1.0779989	0.5375138	0.2442804	0.2442805	1.5188E-07	9.6011E-08	6.9938E-09
0.3	1.7403048	0.9009098	0.4049551	0.4049576	2.5097E-06	6.6505E-07	1.6628E-07
0.4	2.4548178	1.3149513	0.5967116	0.5967298	1.8189E-05	1.4747E-08	1.5409E-06
0.5	3.2151700	1.7772602	0.8242766	0.8243606	8.3938E-05	4.3351E-06	8.5221E-06
0.6	4.0193992	2.2879798	1.0929800	1.0932712	2.9120E-04	7.3893E-06	3.4011E-05
0.7	4.8681144	2.8489681	1.4087971	1.4096268	8.2979E-04	6.2423E-06	1.0839E-04
0.8	5.7636440	3.4634006	1.7783850	1.7804327	2.0477E-03	5.5071E-05	2.9302E-04
0.9	6.7095869	4.1355520	2.2091146	2.2136428	4.5282E-03	1.3852E-04	6.9885E-04
1	7.7105552	4.8706663	2.7090975	2.7182818	9.1843E-03	1.9646E-04	1.5102E-03
CPU	(In sec)					95.657	154.031

For $\alpha = 0.8, 0.9$ &1 compared with exact result and absolute error in 6th order results obtained from $_0$HANITM for $\hbar = -1$ and optimal value of $\hbar$) and NIDTM at distinct time levels $0 < t \leq 1$.

and 9.2c display the comparison of the absolute error in k=6th order solution obtained from NIDTM and $_0$HANITM. The findings show that the computed result obtained from both techniques agreed well with the exact solution and the error decreases with increasing the iterations. It can also be depicted from table that NIDTM converges faster than $_0$HANITM but the computational cost of NIDTM is much higher than $_0$HANITM for Examples 9.1 and 9.2.

Tables 9.3 and 9.4 report approximate solutions obtained from $_0$HANITM at $\hbar = -1$ and at the optimal value $\hbar = -1.1240$ for different values of $\alpha = 0.8, 0.9, 1$ and exact solution at $\alpha = 1$. In addition, absolute error in sixth-order results computed from NIDTM and $_0$HANITM & comparison with [19] at different time levels also reported in table.

Figures 9.3a and 9.4a depict the behavior of sixth-order results for different $\alpha = 0.8, 0.9, 1$; Figures 9.3b and 9.4b show 3D plots of the absolute error for $\alpha = 1$ in kth order results (k=4, 5, 6); Figures 9.3c and 9.4c show the two-dimensional plots of the absolute error for $\alpha = 1$ in kth order results (k=4, 5, 6); Figures 9.3d and 9.4d display the comparison of absolute error in k=6th order solution obtained from NIDTM and $_0$HANITM.The finding shows that the computed result obtained from both techniques agreed well with the exact solution and the error decreases with increasing the iterations. It can also be depicted from table that the solutions obtained from NIDTM and $_0$HANITM for -coincides but at optimal value of $\hbar$ absolute error in k=6th order solution decrease more rapidly than NIDTM and ERCDTM [19]. Also, computational cost of $_0$HANITM is higher than NIDTM for Examples 9.3 and 9.4.

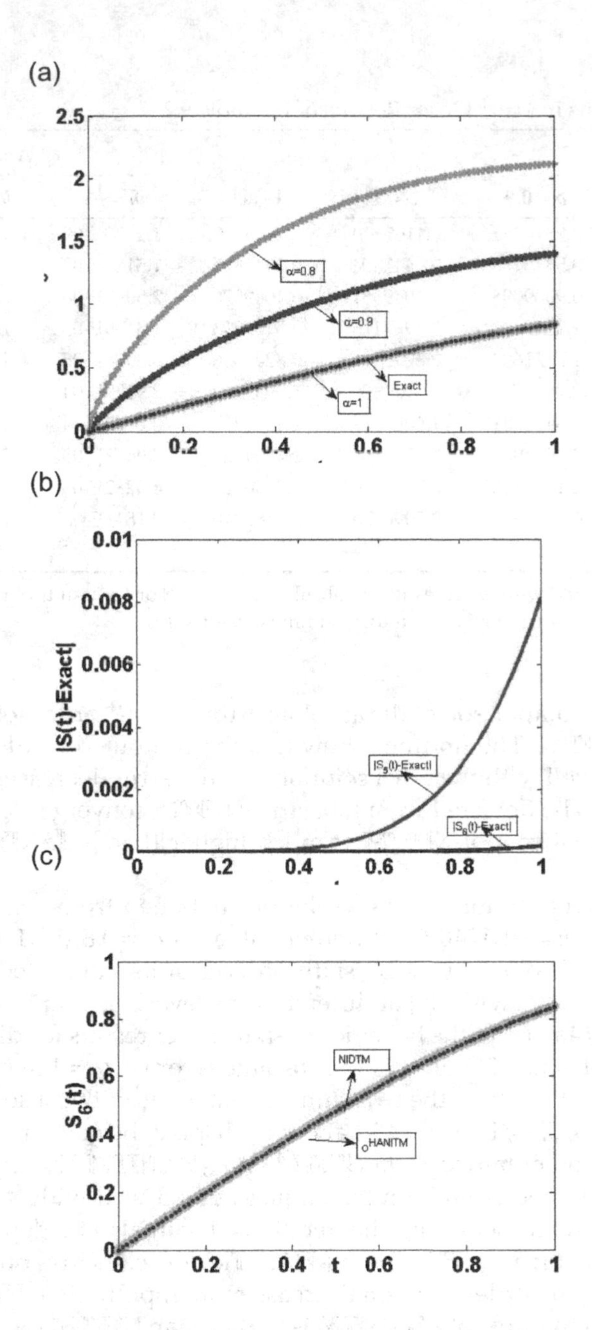

FIGURE 9.1
(a) 2D plots of the behavior k=6th order results for $\alpha = 0.8, 0.9, 1$ and exact result in $t \in (0,1)$ for Example 9.1. (b) Absolute error plot for kth iterative (k=4, 5, 6) results for Example 9.1. (c) Comparison in absolute error for k=6th order iterative results obtained from both techniques in Example 9.1.

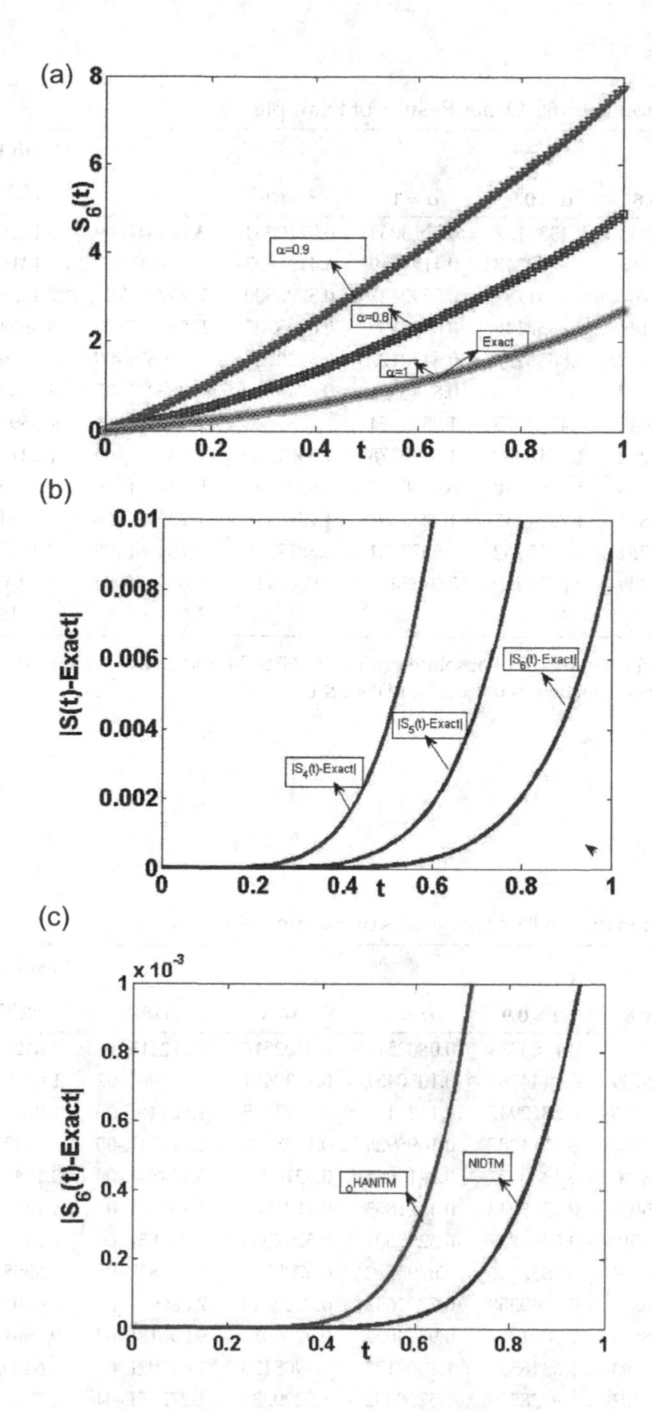

FIGURE 9.2
(a) 2D plots of the behavior k=6th order results for $\alpha = 0.8, 0.9, 1$ and exact result in $t \in (0,1)$ for Example 9.2. (b) Absolute error plot for k^{th} iterative (k=4, 5, 6) results for Example 9.2. (c) Comparison in absolute error for k=6th order iterative results obtained from both techniques in Example 9.2.

TABLE 9.3

Approximate Solution in k=6th Order Results of Example 9.3

					Absolute Error		
(x, t)	**$\alpha = 0.8$**	**$\alpha = 0.9$**	**$\alpha = 1$**	**Exact**	**[19]**	NIDTM	**$\hbar = -1$**
(0.25,0.25)	0.5044843	0.4333137	0.3210064	0.3210064	3.12484E-09	3.12484E-09	3.58611E-10
(0.25,0.5)	0.6419288	0.5778248	0.4121799	0.4121803	4.13161E-07	4.13161E-07	1.44368E-08
(0.25,0.75)	0.7627260	0.7251787	0.5292427	0.5292500	7.29785E-06	7.29785E-06	7.95825E-08
(0.25,1)	0.8857146	0.8857146	0.6795139	0.6795705	5.65682E-05	5.65682E-05	8.30344E-07
(0.5,0.25)	1.0089687	0.8666274	0.6420127	0.6420127	6.24969E-09	6.24969E-09	7.17221E-10
(0.5,0.5)	1.2838577	1.1556495	0.8243598	0.8243606	8.26322E-07	8.26322E-07	2.88735E-08
(0.5,0.75)	1.5254521	1.4503575	1.0584854	1.0585000	1.45957E-06	1.45957E-06	1.59165E-07
(0.5,1)	1.7714292	1.7714292	1.3590278	1.3591409	1.13136E-05	1.13136E-05	1.66069E-06
(0.75,0.25)	1.5134530	1.2999410	0.9630191	0.9630191	9.37453E-09	9.37453E-09	1.07583E-09
(0.75,0.5)	1.9257865	1.7334743	1.2365397	1.2365410	1.23948E-06	1.23948E-06	4.33103E-08
(0.75,0.75)	2.2881781	2.1755362	1.5877281	1.5877500	2.18936E-05	2.18936E-05	2.38748E-07
(0.75,1)	2.6571439	2.6571439	2.0385417	2.0387114	1.69705E-04	1.69705E-04	2.49103E-06
CPU (In sec)					1.280	75.155	230.531

For $\alpha = 0.8, 0.9$ & 1 and comparison of absolute error from NIDTM and $_0$HANITM in M=6th order results with ERC-DTM [19] at distinct time levels $0 < x \leq 1$ and $0 < t \leq 1$.

TABLE 9.4

Approximate Solution in k=6th order Results of Example 9.4

					Absolute Error		
(x, t)	**$\alpha = 0.8$**	**$\alpha = 0.9$**	**$\alpha = 1$**	**Exact**	**[19]**	NIDTM	**$\hbar = -1.06462$**
(0.25,0.25)	0.1459650	0.1083284	0.0802516	0.0802516	7.81211E-10	7.81211E-10	1.09656E-10
(0.25,0.5)	0.1968759	0.1444562	0.1030451	0.1030451	1.03290E-07	1.03290E-07	3.66342E-09
(0.25,0.75)	0.2416755	0.1812947	0.1323125	0.1323125	1.82446E-06	1.82446E-06	2.25854E-08
(0.25,1)	0.2844234	0.2214287	0.1698928	0.1698926	1.41421E-05	1.41421E-05	1.84690E-07
(0.5,0.25)	0.5838600	0.4333137	0.3210064	0.3210064	3.12484E-09	3.12484E-09	4.38623E-10
(0.5,0.5)	0.7875038	0.5778248	0.4121803	0.4121803	4.13161E-07	4.13161E-07	1.46537E-08
(0.5,0.75)	0.9667019	0.7251787	0.5292501	0.5292500	7.29785E-06	7.29785E-06	9.03416E-08
(0.5,1)	1.1376934	0.8857146	0.6795712	0.6795705	5.65682E-05	5.65682E-05	7.38759E-07
(0.75,0.25)	1.3136851	0.9749558	0.7222643	0.7222643	7.03090E-09	7.03090E-09	9.86902E-10
(0.75,0.5)	1.7718835	1.3001057	0.9274057	0.9274057	9.29613E-07	9.29613E-07	3.29708E-08
(0.75,0.75)	2.1750793	1.6316522	1.1908127	1.1908125	1.64202E-05	1.64202E-05	2.03268E-07
(0.75,1)	2.5598102	1.9928579	1.5290352	1.5290335	1.27279E-04	1.27279E-04	1.66221E-06
CPU (In sec)					2.406	80.094	379.531

For $\alpha = 0.8, 0.9$ & 1 and comparison of absolute error obtained from NIDTM and $_0$HANITM in k=6th order results with [19] at distinct time levels $0 < x \leq 1$ and $0 < t \leq 1$.

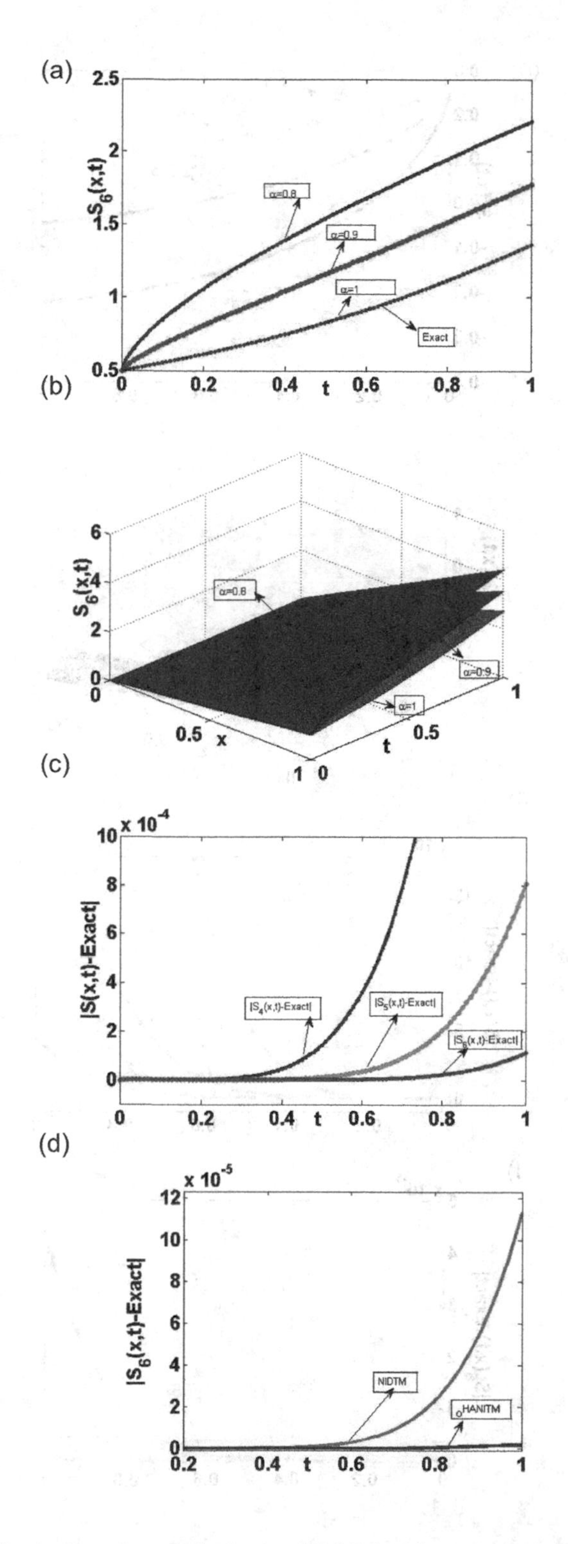

FIGURE 9.3
(a) 2D plots of the behavior k=6th order results for $\alpha = 0.8, 0.9, 1$ and exact result in $t \in (0,1)$ for Example 9.3. (b) 3D plots of the behavior k=6th order results for $\alpha = 0.8, 0.9, 1$ and exact result in $x, t \in (0,1)$ for Example 9.3. (c) Absolute error plot for kth iterative (k=4, 5, 6) results for Example 9.3. (d) Comparison in absolute error for k=6th order iterative results obtained from both techniques for $x = 0.5$ and $t \in (0,1)$ in Example 9.3.

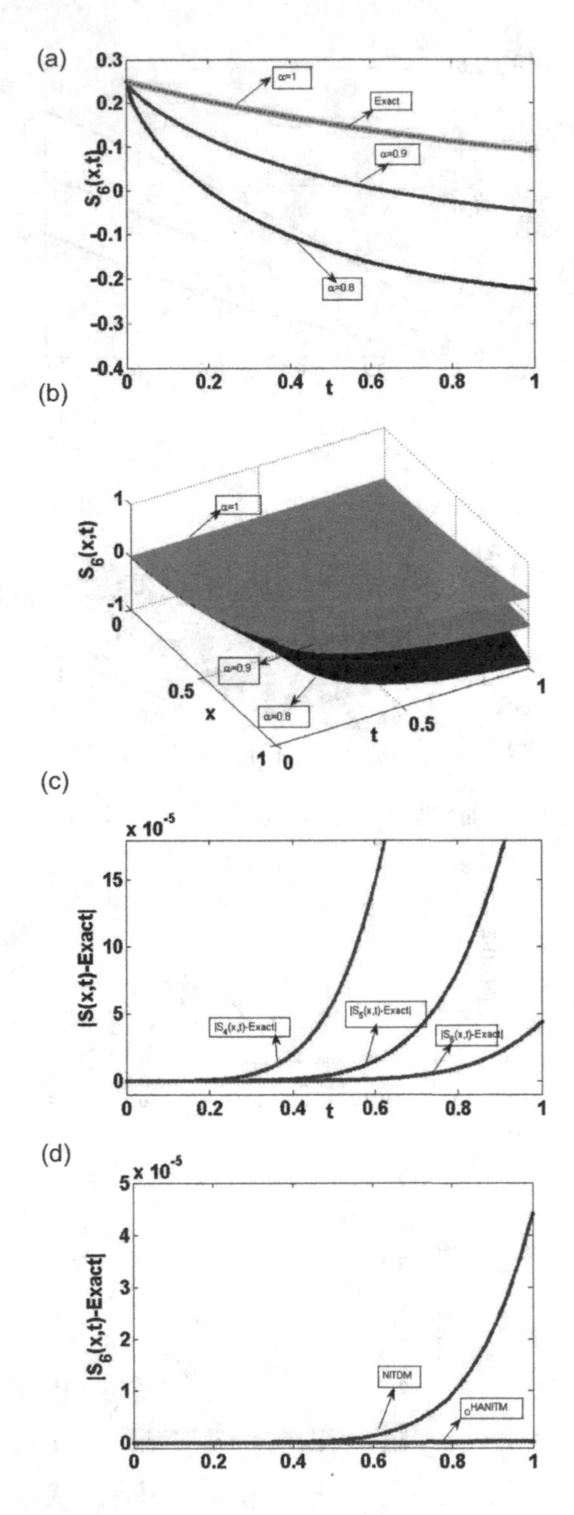

FIGURE 9.4
(a) 2D plots of the behavior k=6th order results for $\alpha = 0.8, 0.9, 1$ and exact result in $t \in (0,1)$ for Example 9.4. (b) 3D plots of the behavior k=6th order results for $\alpha = 0.8, 0.9, 1$ and exact result in $x, t \in (0,1)$ for Example 9.4. (c) Absolute error plot for k^{th} iterative (M=4, 5, 6) results for Example 9.4. (d) Comparison in absolute error for k=6th order iterative results obtained from both techniques for $x = 0.5$ and $t \in (0,1)$ in Example 9.4.

Acknowledgments

The authors are obliged to the editor and reviewers for their advice & instructions. S. Agrawal thanks to BBA University, Lucknow, India for their economic support to perform her research article.

References

1. Miller, K. S., Ross, B., *An Introduction to the Fractional Calculus and Fractional Differential Equations,* Wiley, New York (1993).
2. Podlubny, I., *Fractional Differential Equations,* Academic Press, San Diego (1999).
3. Caputo, M., Mainardi, F., Linear models of dissipation in anelastic solids, *Rivista del Nuovo Cimento,* 1 (1971) 161–98.
4. He, J.H., Nonlinear oscillation with fractional derivative and its applications. In *International Conference on Vibrating Engineering* (1998) 288–291.
5. Burgers, J.M., A mathematical model illustrating the theory of turbulence, *Adv. Appl. Mech.* 1 (1948) 171–199.
6. Duan, Y., Tian, P., Zhang, S., Oscillation and stability of nonlinear neutral impulsive delay differential equations, *J. Appl. Math. Comput.* 11 (2003) 243–253.
7. Jordan, P.M., A note on Burgers' equation with time delay: Instability via finite-time blow-up, *Phys. Lett. A* 372 (2008) 6363–6367.
8. Kolmanovskii, V., Myshkis, A., Introduction to the theory and applications of functional differential equations, 1999.
9. Ockendon, J.R., Tayler, A.B., The dynamics of a current collection system for an electric locomotive, *Proc. R. Soc. Lond. Ser. A.* 322(1971) 447–468.
10. Khalil, R., Al Horani, M., Yousef, A., Sababheh, M., A new definition of fractional derivative, *J. Comput Appl Math,* 264 (2014) 65–70. doi: 10.1016/j.cam.2014.01.002
11. Lio, S., The proposed homotopy analysis techniques for the solution of nonlinear problems (Ph.D. Thesis), Shanghai Jiao Tong University, Shanghai (1992).
12. Liao, S., *Homotopy Analysis Method in Nonlinear Differential Equations,* Springer, Heidelberg, Dordrecht, London (2012).
13. Iyiola, O. S., Tasbozan, O., Kurt, A., Çenesiz, Y., On the analytical solutions of the system of conformable time-fractional Robertson equations with 1-D diffusion, *Chaos, Solitons and Fractals* 94 (2017), 1–7. doi: 10.1016/j.chaos.2016.11.003.
14. Kumar, S., Kumar, A., Baleanu, D., Two analytical methods for time fractional nonlinear coupled Boussinesq Burger's equations arise in propagation of shallow water waves, *Nonlinear Dyn.* 85 (2) (2016) 699–715.
15. Zurigat, M., Momani, S., Odibat, Z., Alawneh, A., The homotopy analysis method for handling systems of fractional differential equations, *Appl. Math. Model.* 34 (2010) 24–35.
16. Ray, S.S., Patra, A., Application of homotopy analysis method and adomian decomposition method for the solution of neutron diffusion equation in the hemisphere and cylindrical reactors, *J. Nucl. Eng. Technol.* 1 (2011) 1–12.
17. Cenesiz, Y., Kurt, A., Tasbozan, O., On the new solutions of the conformable time fractional generalized hirota-satsuma coupled KdV system, *Ann. West Univ. Timisoara-Math. Comput. Sci.* 1 (2017) 37–49. doi: 10.1515/awutm–2017–0003
18. Saad, K.M., Al-Shareef, E.H., Mohamed, M.S., Yang, X.-J., Optimal q-homotopy analysis method for time-space fractional gas dynamics equation, *Eur. Phys. J. Plus* 132 (2017) 23.

19. Singh, B.K., Agrawal, S., A new approximation of conformable time fractional partial differential equations with proportional delay, *App. Numer. Math.* 157(2020) 419–433. doi: 10.1016/j.apnum.2020.07.001
20. Thabet, H., Kendre, S., Analytical solutions for conformable space-time fractional partial differential equations via fractional differential transform, *Chaos Solitons Fractals* 109 (2018) 238–245. doi: 10.1016/j.chaos.2018.03.001
21. Abdeljawad, T., On conformable fractional calculus, *J. Comput. Appl. Math.* 279 (2015) 57–66. doi: 10.1016/j.cam.2014.10.016.
22. Kashuri, A., Fundo, A., A new integral transform, *Adv. Theor. Appl. Math.* 8(1) (2013) 27–43.
23. Singh, B.K., Homotopy perturbation new integral transform method for numeric study of spaceand time- fractional (n+1)-dimensional heatand wave-like equations, *Waves Wavelets Fractals* 4 (2018) 19–36. doi: 10.1515/wwfaa-2018-0003
24. Shah, K., Twinkle, S., Kilicman, A., Combination of integral and projected differential transform methods for time-fractional gas dynamics equations, *Ain Shams Eng. J.* 9(4) (2017) 1683–1688. doi: 10.1016/j.asej.2016.09.012
25. Khan, H., Khan, A., Chen, W., Shah, K., Stability analysis and a numerical scheme for fractional Klein-Gordon equations, *Math Methods Appl. Sci.* 42 (2019) 723–732. doi: 10.1002/mma.5375
26. Smardal, Z., Diblikl, J., Khan, Y., Extension of the differential transformation method to nonlinear differential and integro-differential equations with proportional delays, *Adv. Differ. Equ.* 2013 (2013) 1–12.

10

Numerical Treatment on the Convective Instability in a Jeffrey Fluid Soaked Permeable Layer with Through-Flow

Dhananjay Yadav
University of Nizwa

Mukesh Kumar Awasthi
Babasaheb Bhimrao Ambedkar University

U. S. Mahabaleshwar
Davangere University

Krishnendu Bhattacharyya
Banaras Hindu University

CONTENTS

10.1 Introduction 159
10.2 Mathematical Model 160
10.2.1 Basic State 162
10.3 Stability Analysis 162
10.4 Procedure of Solution 163
10.5 Outcomes and Discussion 163
10.6 Conclusions 166
Acknowledgments 166
References 167

10.1 Introduction

Convective instability is a phenomenon that frequently occurs in a horizontal layer of fluid warmed from the bottom. Due to heat from the bottom, the density of the bottom fluid decreases, so buoyancy force drives the less-dense fluid upward in the direction of the colder end of the plate [1–4]. In the meantime, the colder fluid at the top is denser, so it goes down and substitutes the hot fluid. This convective instability problem in a layer of liquid-soaked permeable media has been extensively explored by numerous investigators in the past as current because of its significant applications in various areas, like geosciences (geothermal basins, geological carbon repository), astrophysics (pore-liquid convective

DOI: 10.1201/9781003291916-10

motion within carbonaceous chondrite parent bodies), manufacturing and industrialized procedure (water management practice, nuclear dissipate removal) [5–10].

In many such applications, the control of the instability of fluid flow is a very important and fascinating task. Throughflow is one of the effectual ways that manages the convective fluid movement in permeable media and it is studied by many researchers. Throughflow amends the basic shape for the temperature field from linear to nonlinear with a depth of the layer, which influences the steadiness of the scheme considerably. Higgins [11] and Sherwood and Homsy [12] analyzed the impact of throughflow on convective instability for the situ dealing of power resources in coal gasification. The limit for the occurrence of convective progress in permeable media with throughflow was studied by Nield [13]. Chen [14] inspected the weight of upright throughflow on the convective activity in permeable layers and verified the limit presented by Nield [13] numerically. An analytical examination of the power of perpendicular throughflow on the start of convection in a layered permeable matrix was made by Nield and Kuznetsov [15]. They established that the throughflow has a stabilizing influence whose amount may be augmented or diminished with the heterogeneity. Later, Hill [16], Hill et al. [17], Yadav [18–21], Shankar and Shivakumara [22–24], Brevdo and Ruderman [25], and Kiran [26] inspected the power of throughflow on the convective instability under various conditions.

The convective flow for non-Newtonian fluids involving the Jeffrey model is an area of research undergoing rapid growth in the past few years due to the best fit model for biological fluid (spit, synovia, blood, and chyme) and thin polymer solution [27–39]. Nazeer et al. [40] investigated the multiphase motion of Jeffrey liquid with warmth transport throughout a flat channel. They observed that the drive of multiphase movement improves due to shear thinning influences sourced by Jeffrey parameter. Impacts of radiation and magnetic force on compressible Jeffrey fluid were explored by Khan and Rafaqat [41]. Mehboob et al. [42] deliberated the transportation of a Jeffrey liquid throughout a permeable slit of microchannel. They found that the flood rate at the middle line of slit and shear stress on the boundaries of slit decompose due to the occurrence of permeable medium and Jeffrey parameter. Ali et al. [43] investigated the magnetic force impact on the nonisothermal movement of an incompressible Jeffrey liquid. They detected that the Jeffrey parameter plays an important role to manage the velocity, flood rate, and web coating width. The force of electric power on the initiate of stationary Jeffrey-nanofluid convection in a permeable matrix was scrutinized by Gautam et al. [44]. They created that the Jeffrey parameter progress in the launch of marginal convection. Yadav [45–47] studied the effects of warmth nonequilibrium, anisotropy, and inner heating on the arrival of Jeffrey fluid convection and found that the steadiness of the arrangement reduced with the expansion of the Jeffrey parameter. The magnetic power influence on the convective development in a permeable medium layer soaked by Jeffrey liquid was inspected by Yadav et al. [48]. They found that the extent of convective cells reduces with Jeffrey and magnetic force parameters.

The literature evaluation confirms that till now no exertion has been taken to explore the impact of throughflow on the Jeffrey fluid convection, while it is important in many applications as cited above. Therefore, the chief aspiration of the present work is to study the impact of throughflow on the convective instability in a Jeffrey fluid-soaked porous layer.

10.2 Mathematical Model

The physical setup engages a horizontal Jeffrey fluid flooded permeable layer bounded by restrictions at $z = 0$ and $z = d$ as exposed in Figure 10.1. The base boundary is mentioned at an identical temperature $T_0 + \Delta T$, while the top boundary is kept at a temperature T_0.

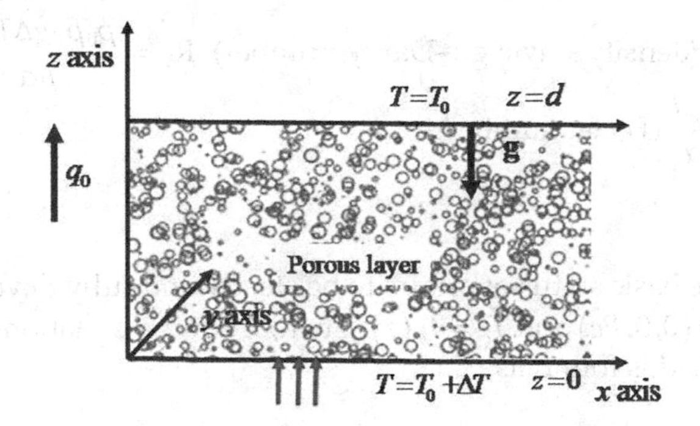

FIGURE 10.1
Graphic representation of the considered problem.

The layer is activated by a uniform upward throughflow q_0 and the gravity which proceeds in the reverse z-pathway. The Darcy law is expanded to comprise the viscoelasticity of the Jeffrey fluid and the Boussinesq finding is employed for density digressions. The pertinent governing equations are [5,44,46]:

$$\nabla \cdot \mathbf{q} = 0, \tag{10.1}$$

$$\nabla P + \frac{\mu K}{(1+\eta)}\mathbf{q} = \rho_0\left[1-\beta_T(T-T_0)\right]\mathbf{g}, \tag{10.2}$$

$$\gamma\frac{\partial T}{\partial t} + (\mathbf{q}\cdot\nabla)T = \alpha\nabla^2 T, \tag{10.3}$$

$$w = q_0,\ T = T_0 + \Delta T,\ \text{at } z = 0 \text{ and } w = q_0,\ T = T_0,\ \text{at } z = d. \tag{10.4}$$

where $\mathbf{q}(u,v,w)$ means the velocity vector, t corresponds to the time, ρ_0 represents the reference density at $T = T_0$, K and α are the permeability and the thermal diffusivity of the permeable medium, correspondingly, β_T illustrates the heat expansion coefficient, P designates the pressure, γ characterizes the heat capacity fraction, η symbolizes the Jeffrey parameter and described by the fraction of relaxation to retardation instants, and μ denotes the viscosity of the Jeffrey liquid.

The following scaling is used to get the nondimensional formulation of the problem:

$$\tilde{\mathbf{x}} \to \frac{\mathbf{x}}{d},\ \tilde{t} \to \frac{t\alpha}{\gamma d^2},\ \tilde{\mathbf{q}} = \frac{\mathbf{q}d}{\alpha},\ \tilde{P} = \frac{PK}{\mu\alpha},\ \tilde{T} = \frac{T-T_0}{\Delta T}, \tag{10.5}$$

By using equation (10.5) in to equations (10.1)–(10.4), we have

$$\tilde{\nabla}\cdot\tilde{\mathbf{q}} = 0, \tag{10.6}$$

$$\frac{\tilde{\mathbf{q}}}{(1+\eta)} = -\tilde{\nabla}\left(\tilde{P} + \tilde{z}\mathrm{R}_{\rho 0}\hat{\mathbf{e}}_z\right) + \mathrm{R}_T\tilde{T}\hat{\mathbf{e}}_z, \tag{10.7}$$

$$\frac{\partial\tilde{T}}{\partial\tilde{t}} + \left(\tilde{\mathbf{q}}\cdot\tilde{\nabla}\right)\tilde{T} = \tilde{\nabla}^2\tilde{T}, \tag{10.8}$$

$$\tilde{w} = Pe \text{ at } z = 0,1,\ \tilde{T} = 1 \text{ at } z = 0 \text{ and } \tilde{T} = 0 \text{ at } z = 1. \tag{10.9}$$

Here, $R_{\rho_0} = \dfrac{\rho_0 g K d}{\mu\alpha}$ (density Rayleigh–Darcy number), $R_T = \dfrac{\rho_0 \beta_T g \Delta T K d}{\mu\alpha}$ (Rayleigh–Darcy number), and $Pe = \dfrac{q_0 d}{\alpha}$ (Péclet number).

10.2.1 Basic State

It is believed that the basic status of affairs to be steady and fully developed in the upright z—direction as $\tilde{\mathbf{q}}_b = (0,0,Pe)$ and $\tilde{T}_b = \tilde{T}_b(\tilde{z})$. Then, from the equations (10.8) and (10.9), the basic temperature field is found as:

$$\tilde{T}_b = \frac{e^{\mathrm{Pe}} - e^{\mathrm{Pe}\tilde{z}}}{e^{\mathrm{Pe}} - 1}. \tag{10.10}$$

10.3 Stability Analysis

For stability analysis, we disturb the basic state by very small amplitude disturbances as

$$\tilde{P} = \tilde{P}_b + \phi \tilde{P}', \ \tilde{\mathbf{q}} = \tilde{\mathbf{q}}_b + \phi \tilde{\mathbf{q}}', \ \tilde{T} = \tilde{T}_b + \phi \tilde{T}' \tag{10.11}$$

Here prime denotes the perturbed nondimensional variables and $\phi << 1$. Replacing equation (10.11) into equations.(10.6)–(10.9) with the basic state results and eliminating the pressure term from equation (10.7), we have the stability equations as:

$$\frac{\tilde{\nabla}^2 \tilde{w}'}{(1+\eta)} = R_T \tilde{\nabla}_H^2 \tilde{T}', \tag{10.12}$$

$$\frac{\partial \tilde{T}'}{\partial \tilde{t}} + \left(\tilde{\mathbf{q}}' \cdot \tilde{\nabla}\right)\tilde{T}_b + \left(\tilde{\mathbf{q}}_b \cdot \tilde{\nabla}\right)\tilde{T}' = \tilde{\nabla}^2 \tilde{T}'. \tag{10.13}$$

$$\tilde{w}' = 0, \ \tilde{T}' = 0, \ \text{at } z = 0,1. \tag{10.14}$$

Here $\tilde{\nabla}_H^2 = \dfrac{\partial^2}{\partial \tilde{x}^2} + \dfrac{\partial^2}{\partial \tilde{y}^2}$. By means of the normal modes [1, 49–58], the perturbed variables can be taken as:

$$\begin{bmatrix} \tilde{w}' \\ \tilde{T}' \end{bmatrix} = \begin{bmatrix} W(\tilde{z}) \\ \Theta(\tilde{z}) \end{bmatrix} \exp\left[i\lambda_x x + i\lambda_y y + \omega t\right]. \tag{10.15}$$

where λ_x and λ_y are the wave numbers in the x and y directions, respectively and ω is the growth rate of instability. On using equation (10.15) into equations (10.12)–(10.14), we have:

$$(D^2 - \lambda^2)W + (1+\eta)\lambda^2 R_T \Theta = 0, \tag{10.16}$$

$$-\frac{d\tilde{T}_b}{d\tilde{z}} W + \left[D^2 - \lambda^2 - \omega - PeD\right]\Theta = 0, \tag{10.17}$$

$$W = \Theta = 0, \text{ at } z = 0,1. \tag{10.18}$$

where $\frac{d}{d\tilde{z}} \equiv D$ and $\lambda = \sqrt{\lambda_x^2 + \lambda_y^2}$ is the resulting dimensionless wave number.

10.4 Procedure of Solution

The resulting system of equations (10.16)–(10.18) are solved with high numerical correctness by utilizing the higher-order Galerkin process. Thus, we assumed as

$$W = \sum_{p=1}^{n} E_p \sin p\pi z \quad \text{and} \quad \Theta = \sum_{p=1}^{N} F_p \sin p\pi z \tag{10.19}$$

where E_p and F_p are undisclosed coefficients. Exploiting equation (10.19) into equations (10.16) and (10.17) and using the orthogonal properties, we have

$$\begin{gathered} G_{jp}E_p + H_{jp}F_p = 0, \\ I_{jp}E_p + J_{jp}F_p = \omega K_{jp}F_p. \end{gathered} \tag{10.20}$$

Here G_{jp} to K_{jp} are $G_{jp} = \langle DW_jDW_p - \lambda^2 W_jW_p \rangle$, $H_{js} = \langle \lambda^2 R_T W_j \Theta_p (1+\eta) \rangle$, $I_{jp} = \langle -\Theta_j W_p D\tilde{T}_b \rangle$, $J_{jp} = \langle D\Theta_j D\Theta_p - \lambda^2 \Theta_j \Theta_p - Pe\Theta_j D\Theta_p \rangle$, $K_{jp} = \langle \Theta_j \Theta_p \rangle$, where $\langle AB \rangle = \int_0^1 AB\,dz$.

The above arrangement of equation (10.20) is a generalized eigenvalue condition and resolved in MATLAB® exploiting QZ algorithm, EIG function and Newton's routine. The pattern of the convective activity is stationary for an investigated problem.

10.5 Outcomes and Discussion

The criterion for the start of convective progress in a flat Jeffrey fluid-soaked permeable layer with a continuous upward throughflow is analyzed numerically applying the higher-order Galerkin scheme.

The essential dimensionless parameters that direct the movement are the Rayleigh–Darcy number R_T, the Péclet number Pe, and the Jeffrey parameter η. The outcomes are given in terms of computed critical estimates of R_T with respect to the wavenumber λ for diverse estimates of Péclet number Pe and the Jeffrey parameter η.

The rightness of our numerical results has been confirmed by contrasting the results obtained under the subcase of this analysis with the results obtained by Yadav [20]. The comparisons are also listed in Table 10.1. From this comparison, it is found that the outcomes calculated from the current code are in outstanding accord with the available ones.

TABLE 10.1
Contrast of $R_{T,c}$ and λ_c with Pe in the Nonexistence of η

	Present Study		Yadav [20]	
Pe	$R_{T,c}$	λ_c	$R_{T,c}$	λ_c
0	39.478	3.14	39.478	3.142
0.5	39.826	3.15	39.827	3.151
1.0	40.873	3.18	40.873	3.179
1.5	42.621	3.23	42.621	3.225
2.0	45.071	3.29	45.071	3.292

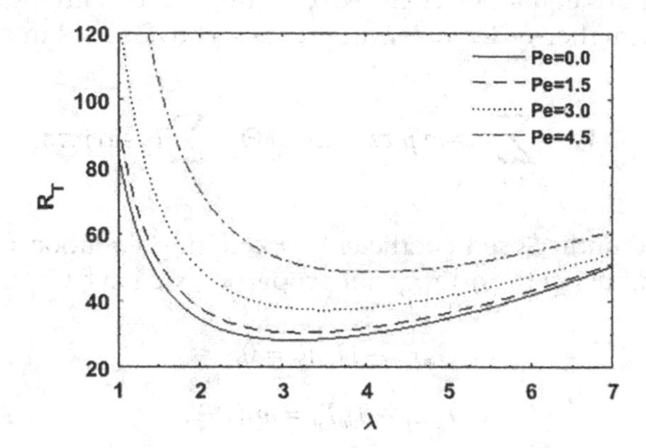

FIGURE 10.2
The impact of Péclet number Pe on neutral stability curves at $\eta = 0.4$.

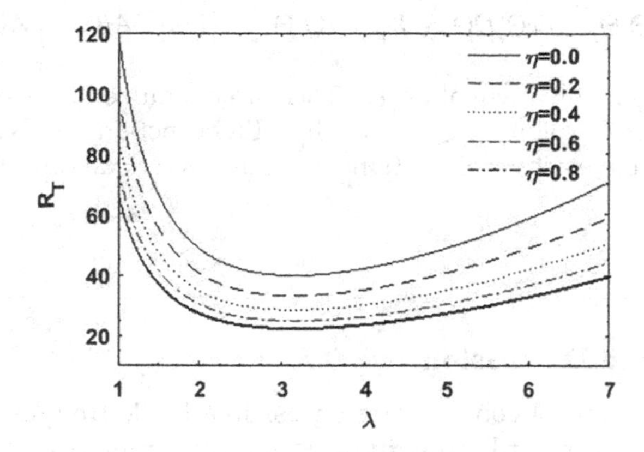

FIGURE 10.3
The effect of Jeffrey parameter η on neutral stability diagrams at $Pe = 0.5$.

Figures 10.2 and 10.3 show the neutral stability diagrams in the (R_T, λ)-plane for diverse estimates of Péclet number Pe and the Jeffrey parameter η, respectively, using two-term Galerkin method. The neutral stability graphs provide the threshold among the stability and instability. The smallest of the neutral stability diagrams gives the critical assessments of R_T and λ which are indicated by $R_{T,c}$ and λ_c, correspondingly. The physical significance of the critical assessments is that the system will be stable for $R_T < R_{T,c}$.

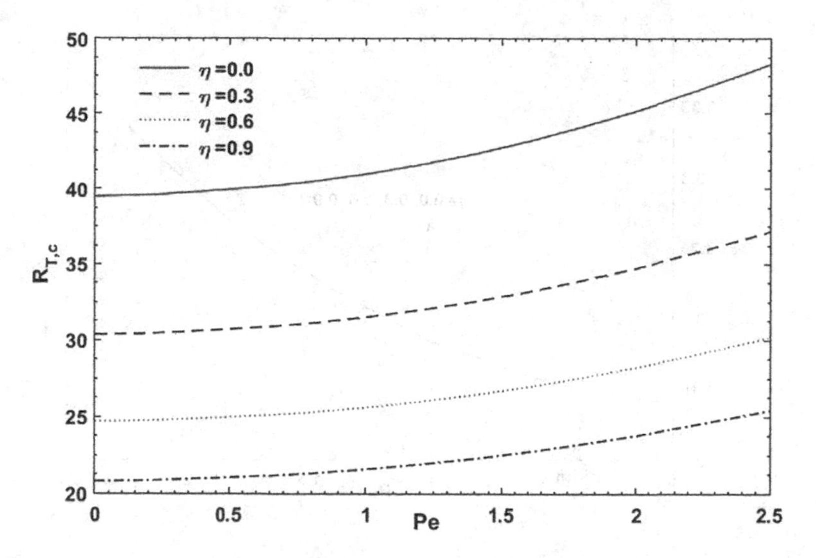

FIGURE 10.4
Digression of $R_{T,c}$ with *Pe* for diverse estimates of η.

TABLE 10.2
Estimation of the $R_{T,c}$ and λ_c for Diverse Estimates of *Pe* and η

Pe	η	$R_{D,c}$	a_c	*Pe*	η	$R_{D,c}$	a_c
0	0	39.478	3.14	1.5	0	42.621	3.23
	0.2	32.899	3.14		0.2	35.518	3.23
	0.4	28.199	3.14		0.4	30.444	3.23
	0.6	24.674	3.14		0.6	26.638	3.23
	0.8	21.932	3.14		0.8	23.678	3.23
	1	19.739	3.14		1	21.311	3.23
0.5	0	39.827	3.15	2.0	0	45.071	3.29
	0.2	33.189	3.15		0.2	37.559	3.29
	0.4	28.448	3.15		0.4	32.194	3.29
	0.6	24.892	3.15		0.6	28.169	3.29
	0.8	22.126	3.15		0.8	25.039	3.29
	1	19.913	3.15		1	22.536	3.29
1.0	0	40.873	3.18	2.5	0	48.220	3.38
	0.2	34.061	3.18		0.2	40.183	3.38
	0.4	29.195	3.18		0.4	34.443	3.38
	0.6	25.546	3.18		0.6	30.137	3.38
	0.8	22.707	3.18		0.8	26.789	3.38
	1	20.437	3.18		1	24.110	3.38

Figure 10.4 shows the disparity of the critical Rayleigh–Darcy number $R_{T,c}$ with a variation of Péclet number *Pe* for assorted estimates of the Jeffrey parameter η. The outcomes are also listed in Table 10.2. The corresponding critical wave number λ_c is presented in Figure 10.5. From Figure 10.4, it is established that on increasing the worth of *Pe*, the assessment of $R_{T,c}$ boosts. Thus, the effect of *Pe* is to postponement the start of convective movement. This came about because the throughflow transports the imperative warmth

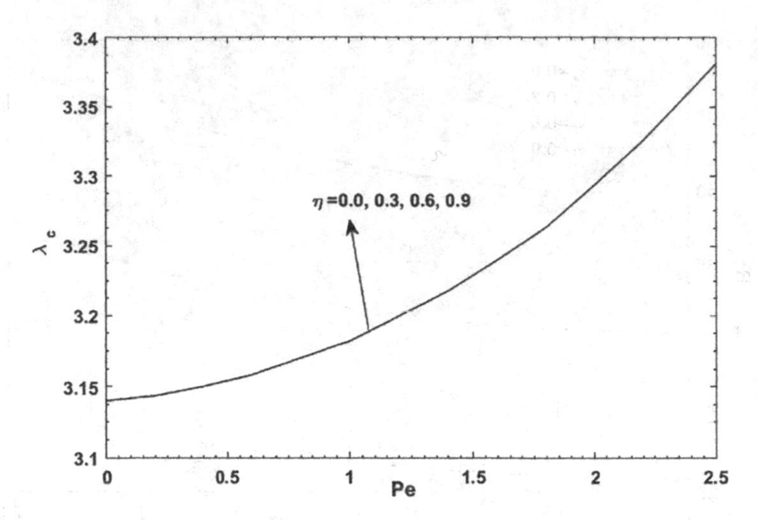

FIGURE 10.5
Digression of λ_c with *Pe* for diverse values of η.

gradient into a thermal boundary layer at the boundary toward which it is implemented. From Figure 10.4, we can also find that the stability of the system decreases with raising the Jeffery parameter λ. This is because increasing value of Jeffery parameter η reduces the retardation time of Jeffery fluid, thus making the system more unstable.

From Figure 10.5, it is found that the worth of λ_c enlarges with an increase in the values of *Pe*. Thus, *Pe* reduces the magnitude of convective cells, whereas the Jeffery parameter η has no major control on λ_c.

10.6 Conclusions

The influence of throughflow is analyzed on the convective movement in a Jeffrey fluid-saturated permeable layer numerically. The following conclusions are made:

Increasing Péclet number *Pe* increases the critical Rayleigh–Darcy number $R_{T,c}$. Thus, it postpones the onset of convective instability by increasing $R_{T,c}$.

The arrangement becomes more unstable with escalating Jeffrey parameter η.

The size of convective cell shrinks with *Pe*, while the Jeffrey parameter η has no effect on it.

Acknowledgments

This study was supported by the University of Nizwa Research Grant (Grant No.: A/2021-2022-UoN/3/CAS/IF), the Sultanate of Oman.

References

1. Chandrasekhar S., *Hydrodynamic and hydromagnetic stability*. Dover Publication, New York, (2013).
2. Banzhaf, W., Self-organizing systems, in: R.A. Meyers (Ed.) *Encyclopedia of physical science and technology* (Third Edition), Academic Press, New York (2003) 589–598.
3. Chandrasekhar, S., Thermal convection, *Daedalus*, 86(4) (1957) 323–339.
4. Kadanoff, L. P., Turbulent heat flow: structures and scaling, *Physics Today*, 54(8) (2001) 34–39.
5. Nield, D.A., Bejan, A., *Convection in porous media*, Springer (2006).
6. Capone, F., Gentile, M., Massa, G., The onset of thermal convection in anisotropic and rotating bidisperse porous media, *Zeitschrift für angewandte Mathematik und Physik*, New York, 72(4) (2021) 1–16.
7. Xu, X., Chen, S., Zhang, D., Convective stability analysis of the long-term storage of carbon dioxide in deep saline aquifers, *Advances in Water Resources*, 29(3) (2006) 397–407.
8. Pozzoni, P., Di Filippo, S., Manzoni, C., Locatelli, F., The relevance of convection in clinical practice: a critical review of the literature, *Hemodialysis International*, 10 (2006) S33–S38.
9. Zhang, K., Liao, X., Schubert, G., Pore water convection within carbonaceous chondrite parent bodies: Temperature-dependent viscosity and flow structure, *Physics of Fluids*, 17(8) (2005) 086602.
10. Kim, M. C., Yadav, D., Linear and nonlinear analyses of the onset of buoyancy-induced instability in an unbounded porous medium saturated by miscible fluids, *Transport in Porous Media*, 104(2) (2014) 407–433.
11. G. Higgins, A new concept for in situ coal gasification: Livermore, Ca, Univ. Ca., Lawrence Livermore Lab., Rept. UCRL-51217, (1972).
12. Sherwood, A., Homsy, G. M., *Convective instability during in situ coal gasification*, Lawrence Livermore Laboratory (1975).
13. Nield, D., Convective instability in porous media with throughflow, *AIChE Journal*, California, 33(7) (1987) 1222–1224.
14. Chen, F., Throughflow effects on convective instability in superposed fluid and porous layers, *Journal of Fluid Mechanics*, 231 (1991) 113–133.
15. Nield, D., Kuznetsov, A., The onset of convection in a layered porous medium with vertical throughflow, *Transport in Porous Media*, 98(2) (2013) 363–376.
16. Hill, A. A., Unconditional nonlinear stability for convection in a porous medium with vertical throughflow, *Acta Mechanica*, 193(3) (2007) 197–206.
17. Hill, A. A., Rionero, S., Straughan, B., Global stability for penetrative convection with throughflow in a porous material, *IMA Journal of Applied Mathematics*, 72(5) (2007) 635–643.
18. Yadav, D. The influence of pulsating throughflow on the onset of electro-thermo-convection in a horizontal porous medium saturated by a dielectric nanofluid, *Journal of Applied Fluid Mechanics*, 11(6) (2018) 1679–1689.
19. Yadav, D., Throughflow and magnetic field effects on the onset of convection in a Hele Shaw cell, *Revista Cubana de Física*, 35(2) (2018) 108–114.
20. Yadav, D., Numerical investigation of the combined impact of variable gravity field and throughflow on the onset of convective motion in a porous medium layer, *International Communications in Heat and Mass Transfer*, 108 (2019) 104274.
21. Yadav, D., The effect of pulsating throughflow on the onset of magneto convection in a layer of nanofluid confined within a Hele-Shaw cell, Proceedings of the Institution of Mechanical Engineers, Part E: *Journal of Process Mechanical Engineering*, 233(5) (2019) 1–12.
22. Shankar, B., Shivakumara, I., Stability of porous-Poiseuille flow with uniform vertical throughflow: High accurate solution, *Physics of Fluids*, 32(4) (2020) 044101.
23. Shankar, B., Shivakumara, I., Changes in the hydrodynamic stability of plane porous-Couette flow due to vertical throughflow, *Physics of Fluids*, 33(7) (2021) 074103.

24. Shankar, B., Shivakumara, I., Instability of natural convection in a vertical fluid layer with net horizontal throughflow, *Zeitschrift für angewandte Mathematik und Physik*, 72(3) (2021) 1–17.
25. Brevdo, L., M.S. Ruderman, On the convection in a porous medium with inclined temperature gradient and vertical throughflow. Part I. Normal modes, *Transport in Porous Media*, 80(1) (2009) 137–151.
26. Kiran, P., Nonlinear throughflow and internal heating effects on vibrating porous medium, *Alexandria Engineering Journal*, 55(2) (2016) 757–767.
27. Ahmad, K., Hanouf, Z., Ishak, A., Mixed convection Jeffrey fluid flow over an exponentially stretching sheet with magnetohydrodynamic effect, *AIP Advances*, 6(3) (2016) 035024.
28. Bhatti, M., Abbas, A., Simultaneous effects of slip and MHD on peristaltic blood flow of Jeffrey fluid model through a porous medium, *Alexandria Engineering Journal*, 55(2) (2016) 1017–1023.
29. Hayat, T., Qayyum, S., Imtiaz, M., Alsaedi, A., Impact of Cattaneo-Christov heat flux in Jeffrey fluid flow with homogeneous-heterogeneous reactions, *PLoS One*, 11(2) (2016) e0148662.
30. Javed, M., Farooq, M., Anjum, A., Ahmad, S., Insight of thermally stratified Jeffrey fluid flow inside porous medium subject to chemical species and melting heat transfer, *Advances in Mechanical Engineering*, 11(9) (2019) 1687814019876187.
31. Kahshan, M., Lu, D., Siddiqui, A., A Jeffrey fluid model for a porous-walled channel: Application to flat plate dialyzer, *Scientific Reports*, 9(1) (2019) 1–18.
32. Krishna, M. V., Bharathi, K., Chamkha, A. J., Hall effects on MHD peristaltic flow of Jeffrey fluid through porous medium in a vertical stratum, *Interfacial Phenomena and Heat Transfer*, 6(3) (2018) 253–268.
33. Naganthran, K., Nazar, R., Pop, I., Effects of heat generation/absorption in the Jeffrey fluid past a permeable stretching/shrinking disc, *Journal of the Brazilian Society of Mechanical Sciences and Engineering*, 41(10) (2019) 1–12.
34. Nallapu, S., Radhakrishnamacharya, G., Jeffrey fluid flow through porous medium in the presence of magnetic field in narrow tubes, *International Journal of Engineering Mathematics*, 2014 (2014) 713831.
35. Narayana, P. S., Babu, D. H., Numerical study of MHD heat and mass transfer of a Jeffrey fluid over a stretching sheet with chemical reaction and thermal radiation, *Journal of the Taiwan Institute of Chemical Engineers*, 59 (2016) 18–25.
36. Noor, N A., Shafie, M. S., Admon, M. A., Unsteady MHD squeezing flow of Jeffrey fluid in a porous medium with thermal radiation, heat generation/absorption and chemical reaction, *Physica Scripta*, 95(10) (2020) 105213.
37. Ojjela, O., Raju, A., Kumar, N. N., Influence of induced magnetic field and radiation on free convective Jeffrey fluid flow between two parallel porous plates with Soret and Dufour effects, *Journal of Mechanics*, 35(5) (2019) 657–675.
38. Santhosh, N., Radhakrishnamacharya, G., Chamkha, A. J., Flow of a Jeffrey fluid through a porous medium in narrow tubes, *J. Porous Media*, 18(1) (2015) 71–78.
39. Vajravelu, K., Sreenadh, S., Lakshminarayana, P., The influence of heat transfer on peristaltic transport of a Jeffrey fluid in a vertical porous stratum, *Communications in Nonlinear Science and Numerical Simulation*, 16(8) (2011) 3107–3125.
40. Nazeer, M., Hussain, F., Ahmad, M. O., Saeed, S., Khan, M. I., Kadry, S., Chu, Y.-M., Multi-phase flow of Jeffrey Fluid bounded within magnetized horizontal surface, *Surfaces and Interfaces*, 22 (2021) 100846.
41. Khan, A., Rafaqat, A. R., Effects of radiation and MHD on compressible Jeffrey fluid with peristalsis, *Journal of Thermal Analysis and Calorimetry*, 143(3) (2021) 2775–2787.
42. Mehboob, H., Maqbool, K., Ellahi, R., Sait, S. M., Transport of Jeffrey fluid in a rectangular slit of the microchannel under the effect of uniform reabsorption and a porous medium, *Communications in Theoretical Physics*, 73(11) (2021) 115003.
43. Ali, F., Hou, Y., Zahid, M., Rana, M. A., Usman, M., Influence of magnetohydrodynamics and heat transfer on the reverse roll coating of a Jeffrey fluid: A theoretical study, *Journal of Plastic Film & Sheeting*, 38(1) (2022) 72–104.

44. Gautam, P. K., Rana, G. C., Saxena, H., Stationary convection in the electrohydrodynamic thermal instability of Jeffrey nanofluid layer saturating a porous medium: free-free, rigid-free, and rigid-rigid boundary conditions, *Journal of Porous Media*, 23(11) (2020) 1043–1063.
45. Yadav, D., Thermal non-equilibrium effects on the instability mechanism in a non-Newtonian Jeffery fluid saturated porous layer, *Journal of Porous Media*, 25(2) (2022) 1–12.
46. Yadav, D., Influence of anisotropy on the Jeffrey fluid convection in a horizontal rotary porous layer, *Heat Transfer*, 50(5) (2021) 4595–4606.
47. Yadav, D., Electric field effect on the onset of Jeffery fluid convection in a heat generating porous medium layer, *Pramana - Journal of Physics*, 96, 19, (2022).
48. Yadav, D., Mohamad, A. A., Awasthi, M. K., The Horton–Rogers–Lapwood problem in a Jeffrey fluid influenced by a vertical magnetic field, *Proceedings of the Institution of Mechanical Engineers, Part E: Journal of Process Mechanical Engineering*, 235(6) (2021) 2119–2128.
49. Akbarzadeh, P., Mahian, O., The onset of nanofluid natural convection inside a porous layer with rough boundaries, *Journal of Molecular Liquids*, 272 (2018) 344–352.
50. Alex, S. M., Patil, P. R., Effect of variable gravity field on Soret driven thermosolutal convection in a porous medium, *International Communications in Heat and Mass Transfer*, 28(4) (2001) 509–518.
51. Awad, F., Sibanda, P., Motsa, S. S., On the linear stability analysis of a Maxwell fluid with double-diffusive convection, *Applied Mathematical Modelling*, 34(11) (2010) 3509–3517.
52. Banu, N., Rees, D., Onset of Darcy–Benard convection using a thermal non-equilibrium model, *International Journal of Heat and Mass Transfer*, 45(11) (2002) 2221–2228.
53. Capone, F., Gentile, M., Hill, A. A., Penetrative convection via internal heating in anisotropic porous media, *Mechanics Research Communications*, 37(5) (2010) 441–444.
54. Yadav, D., Mohamad, A., Rana, G., Effect of throughflow on the convective instabilities in an anisotropic porous medium layer with inconstant gravity, *Journal of Applied and Computational Mechanics*, 7(4) (2020) 1964–1972.
55. Yadav, D., Effects of rotation and varying gravity on the onset of convection in a porous medium layer: A numerical study, *World Journal of Engineering*, 17(6) (2020) 785–793.
56. Yadav, D., Maqhusi, M., Influence of temperature dependent viscosity and internal heating on the onset of convection in porous enclosures saturated with viscoelastic fluid, *Asia-Pacific Journal of Chemical Engineering*, 15(6) (2020) e2514.
57. Yadav, D., Numerical solution of the onset of Buoyancy-driven nanofluid convective motion in an anisotropic porous medium layer with variable gravity and internal heating, *Heat Transfer—Asian Research*, 49(3) (2020) 1170–1191.
58. Yadav, D., The density-driven nanofluid convection in an anisotropic porous medium layer with rotation and variable gravity field: A numerical investigation, *Journal of Applied and Computational Mechanics*, 6(3) (2020) 699–712.

11

Computational Modeling of Nonlinear Reaction-Diffusion Fisher–KPP Equation with Mixed Modal Discontinuous Galerkin Scheme

Satyvir Singh

Nanyang Technological University

CONTENTS

11.1 Introduction 171
11.2 Mixed Modal Discontinuous Galerkin Scheme 173
11.3 Numerical Results 176
11.4 Concluding Remarks 181
Acknowledgements 182
References 182

11.1 Introduction

It is constantly difficult for researchers to simulate nonlinear partial differential equations (PDEs) since these nonlinear PDEs are utilized to solve several challenging problems that occurred in science, nature, engineering, and other fields. For example, Cahn–Hilliard PDE in material sciences; nonlinear Schrödinger PDE sinquantum mechanics; nonlinear reaction–diffusion PDEs in biology; Euler and Navier–Stokes PDEs in fluid dynamics (Zheng, 2004), Schrödinger/Gross–Pitaevskii and Klein–Gordon–Zakharov PDEs (Dehghan and Abbaszadeh, 2018), generalized Zakharov and Gross–Pitaevskii PDEs (Dehghan and Abbaszadeh, 2017) are several illustrious examples of the nonlinear PDEs. Many scholars have spent decades trying to find effective numerical methods for solving nonlinear PDEs. The purpose of this chapter is to describe an efficient numerical technique based on modal discontinuous Galerkin (DG) method for simulating nonlinear reaction–diffusion PDEs in two dimensions.

A nonlinear reaction–diffusion model was introduced by Fisher (1937) and Kolmogorov (1937) autonomously to delineate the propagation of a mutant gene across a population. This model is commonly referred to as the Fisher or Fisher–Kolmogorov–Petrovsky–Piscounov (Fisher–KPP) equation in the literature. This Fisher–KPP equation has since been applied in a variety of fields of research, most notably biological applications. The Fisher–KPP equation, for example, can be used to represent the logistic growth of coral

DOI: 10.1201/9781003291916-11

reefs (Roessler and Hüssner, 1997) and the stirrings of neutrons in a nuclear reactor (José, 1969). The mathematical expression of the Fisher–KPP equation is given by

$$\frac{\partial w}{\partial T} = \nabla \cdot (\lambda \nabla w) + \mu f(w), \tag{11.1}$$

where $w \equiv w(X,Y,T)$ denotes the two-dimensional unknown population density, λ is the population diffusivity, μ is the population proliferation rate and $\nabla^2 \equiv \partial^2/\partial X^2 + \partial^2/\partial Y^2$ presents the two-dimensional gradient operator. The function $f(w) = w(1-w)$ is a nonlinear term that represents the effects of reaction or multiplication. Assuming λ_1 and λ_2 as the paramount values of λ with (X,Y) coordinates along the principal axes, the Fisher–KPP equation (11.1) is expressed as follows:

$$\frac{\partial w}{\partial T} = \lambda_1 w_{XX} + \lambda_2 w_{YY} + \mu w(1-w). \tag{11.2}$$

On rescaling X, Y and T as

$$t = \mu T, \quad x = \sqrt{\frac{\mu}{\lambda_1}} X, \text{ and } y = \sqrt{\frac{\mu}{\lambda_2}} Y, \tag{11.3}$$

Equation (11.2) becomes

$$\frac{\partial w}{\partial t} = \lambda_1 w_{xx} + \lambda_2 w_{yy} + \mu w(1-w), \tag{11.4}$$

which is a general form of two-dimensional Fisher–KPP equation.

In literature, several analytical and numerical approaches exist for solving the Fisher–KPP equation. In 1974, the Fisher–KPP equation was examined numerically for the first time using a pseudo-spectral scheme (Gazdag and Canosa, 1974) after which it was investigated by a variety of additional numerical schemes, including Petrov–Galerkin finite element method (Tang and Weber, 1991), tanh method (Wazwaz, 2004), Homotopy analysis method (Tan et al., 2007), alternating group explicit iterative method (Evans and Sahimi, 1989), hybrid approach of homotopic method and theory of traveling wave transform (Mo et al., 2007), centered finite-difference method (Hagstrom and Keller, 1986), implicit and explicit finite-differences methods(Parekh and Puri, 1990), wavelet Galerkin method (Mittal and Kumar, 2006), collocation of cubic B-splines basis (Mittal and Arora, 2010), differential quadrature method (Arora and Joshi, 2021).

In this study, a mixed modal DG method is proposed for simulating the two-dimensional Fisher–KPP equation. The DG approaches for solving nonlinear PDEs have been increasingly popular in recent decades. Initially, Reed and Hill (1973) proposed the DG method to simulate the unsteady neutron transport equation. Afterward, several DG algorithms were proposed and developed numerically for solving nonlinear hyperbolic systems by Cockburn and Shu (1989a, 1989b, 1998, 2001). The DG approaches integrate the advantages of modern CFD methods, including Finite Element (FE), and Finite Volume (FV) methods and have been effectively utilized to a broad spectrum of scientific problems, including computational fluid dynamics, plasma physics, quantum physics, biological sciences, and many others (Le et al. 2014; Raj et al., 2017; Singh and Myong, 2017; Singh, 2018; Chourushi et al., 2020; Singh, 2020; Singh and Battiato, 2020a; Singh and Battiato, 2020b; Singh and

Battiato, 2021a; Singh and Battiato, 2021b; Singh et al., 2021; Singh, 2021a; Singh, 2021b; Singh, 2021c). The DG approaches have several key characteristics that make them interesting for usage in applications. These characteristics include their capacity to easily address complex geometry and boundary conditions, their flexibility for easy hp-adaptivity, their ability for nonconforming elements having hanging nodes, and efficient parallel implementation along with time-stepping algorithms.

This study presents a two-dimensional mixed modal DG scheme different from previous studies for solving the nonlinear Fisher–KPP equations. The third-order scaled Legendre polynomials are adopted for DG spatial discretization, while a third-order Strong Stability Preserving Runge–Kutta (SSP-RK33) scheme is employed for a temporal marching algorithm is utilized to temporally discretize the emerging semi-discrete differential equation. The accuracy and efficiency of the proposed algorithm are evaluated by solving four well-known Fisher–KPP test problems.

11.2 Mixed Modal Discontinuous Galerkin Scheme

Here, a mixed modal DG scheme is described for solving the nonlinear Fisher–KPP equation. Thus, in order to employ this numerical scheme, equation (11.4) can be reformulated as

$$\frac{\partial w}{\partial t} + \nabla F(\nabla w) = S(w), \tag{11.5}$$

where $F(\nabla w) = \lambda_1 w_x + \lambda_2 w_y$, and $S(w) = \mu w(1 - w)$. For the spatial discretization of such type of nonlinear equation that contains second-order derivatives in diffusion part, a mixed DG formulation was developed previously in literature. An auxiliary variable Θ is provided in this formulation to address the higher-order derivatives which are considered as the derivative of solution variable. Thus, the Fisher–KPP equation (11.5) is rewritten as a coupled system of w and Θ for the mixed DG construction,

$$\begin{aligned} & w - \nabla \Theta = 0, \\ & \frac{\partial w}{\partial t} + \nabla F(\Theta) = S(w). \end{aligned} \tag{11.6}$$

In this work, the computational simulation of Fisher–KPP equation is considered in one and two-dimensional spaces. Recently, a mixed modal DG scheme for solving one-dimensional nonlinear Fisher–KPP equation has been developed (Singh, 2021c; Singh, 2021d). A two-dimensional Fisher–KPP equation is here computed by an explicit mixed modal DG approach based on rectangular elements. A rectangular element based on the tensor product of mesh points in x- and y-directions has been used to discretize the coupled system of equation (11.6). Let $[\Im]$ be a family of partitions of the physical domain $\Omega = \Omega_x \times \Omega_y$ which is divided into $N_x \times N_y$ regular rectangular elements as

$$\begin{aligned} & x_L = x_{1/2} < x_{3/2} < \cdots\cdots < x_{i-1/2} < x_{i+1/2} < \cdots\cdots < x_{N+1/2} = x_R, \\ & y_B = y_{1/2} < y_{3/2} < \cdots\cdots < y_{j-1/2} < y_{j+1/2} < \cdots\cdots < y_{N+1/2} = y_U. \end{aligned} \tag{11.7}$$

A two-dimensional Cartesian mesh $\Im$ is defined as

$$\Im := \{T_{ij} = I_i \times J_j, 1 \le i \le N_x, 1 \le j \le N_y\},$$
$$I_i := [x_{i-1/2}, x_{i+1/2}], \quad \forall i = 1 \cdots N_x; \tag{11.8}$$
$$J_j := [y_{j-1/2}, y_{j+1/2}], \quad \forall j = 1 \cdots N_y.$$

where $(x_i, y_j) = ((x_{i-1/2} + x_{i+1/2})/2, (y_{j-1/2} + y_{j+1/2})/2)$ denotes the cell-center points of the elements. In the case of Ω_x – domain, the piecewise polynomial space of the functions $\xi : \Omega_x \to \Re$ is defined as

$$Z_h^k = \{\xi : \xi|_{\Omega_x} \in \mathbb{S}^k(I_i), \ i = 1 \cdots N_x\}, \tag{11.9}$$

where $\mathbb{S}^k(I_i)$ is the space of polynomial functions of degree at most k on I_i. For the domain $\Omega = \Omega_x \times \Omega_y$, it is defined $\varphi : \Omega \to \Re$ as

$$V_h^k = \{\varphi : \varphi|_{\Omega} \in \mathbb{F}^k(T_{ij}), \ 1 \le i \le N_x, 1 \le j \le N_y\}, \tag{11.10}$$

where $\mathbb{F}^k(T_{ij}) = \mathbb{S}^k(I_i) \otimes \mathbb{S}^k(J_j)$ denotes the space of polynomials of degree at most k on T_{ij}. The number of degrees of freedom of $\mathbb{F}^k(T_{ij})$ can be calculated by $N_k = (k+1)(k+2)/2$. After then, the exact solutions of w and Θ can be approximated by the DG polynomial approximations of $w_h \in V_h^k$ and $\Theta_h \in V_h^k$ as

$$\Theta_h(x,y) = \sum_{i=0}^{N_k} \Theta_h^i(t) b_i(x,y),$$
$$w_h(x,y,t) = \sum_{i=0}^{N_k} w_h^i(t) b_i(x,y), \quad \forall (x,y) \in T_{ij}, \tag{11.11}$$

where w_h^i denotes the unknown coefficients for w to be corrected with time, and $b_i(x,y)$ represents the polynomial (basis) function. The two-dimensional scaled Legendre polynomial functions are formulated here as a tensor product of the so-called principal functions, which are defined as (Singh, 2018)

$$b_k(\xi,\eta) = \psi_i(\xi) \otimes \psi_j(\eta), \tag{11.12}$$

with

$$\psi_i(\xi) = \frac{2^i (i!)^2}{(2i)!} P^{0,0}(\xi), \ -1 \le \xi \le 1,$$

$$\psi_j(\eta) = \frac{2^j (j!)^2}{(2j)!} P^{0,0}(\eta), \ -1 \le \eta \le 1,$$

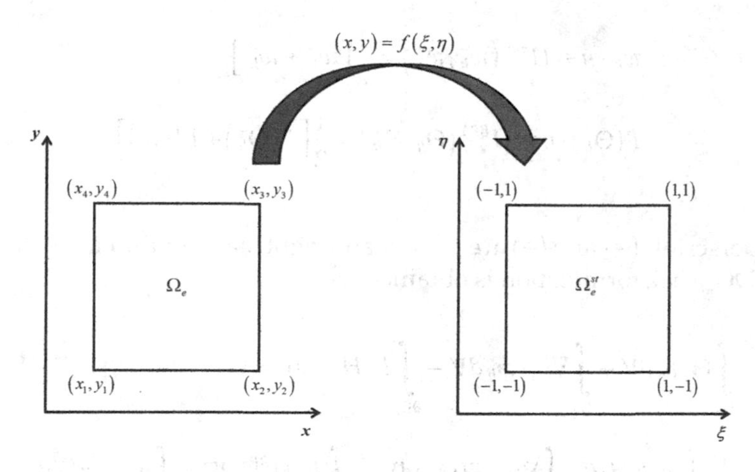

FIGURE 11.1
Elemental transformation from physical element to reference element in a 2D computational domain.

where $P^{0,0}(\xi)$ is the Legendre polynomial function. A standard rectangular element Ω_e^{st} is defined using a local Cartesian coordinate system $(\xi,\eta)\in[-1,1]$, which is illustrated in Figure 11.1. The standard element can be mapped from the computational space (ξ,η) to an arbitrary rectangular element in the physical space (x,y) under the linear transformation $T:\Omega_e^{st}\rightarrow\Omega_e$ defined by

$$\begin{aligned} x &= \frac{1}{2}\left[(1-\xi)x_1+(1+\xi)x_2\right], \\ y &= \frac{1}{2}\left[(1-\eta)y_1+(1+\eta)y_2\right], \end{aligned} \tag{11.13}$$

where $(x_i, y_i : i=1,4)$ denotes the physical coordinates of the vertices of Ω_e.

The DG discretization of the coupled system of Fisher–KPP equation (11.6) can be obtained by replacing the exact solutions with the corresponding approximations defined in equation (11.11), and multiplying by a polynomial function $b_h(x,y)$ and then integrated by parts over the element $T_{ij}\in\Im$. Taking $b_h\in V_h^k(\Im)$, $w_h\in V_h^k(\Im)$, and $\Theta_h\in V_h^k(\Im)$, we obtain

$$\begin{aligned} &\int_{\Omega_e}\Theta_h\,b_h\,dV+\int_{\Omega_e}\nabla b_h\cdot w_h\,dV-\int_{\partial\Omega_e}b_h\,w_h\cdot n\,d\Gamma=0, \\ &\frac{\partial}{\partial t}\int_{\Omega_e}w_h\,b_h\,dV-\int_{\Omega_e}\nabla b_h\cdot F(\Theta_h)dV+\int_{\partial\Omega_e}b_h\,F(\Theta_h)\cdot n\,d\Gamma=\int_{\Omega_e}b_h\,S(w_h)\,dV. \end{aligned} \tag{11.14}$$

where n denotes the outward unit normal vector; Γ and V are the boundary and volume of the element Ω_e, respectively. The interface fluxes are not uniquely defined because of the discontinuity in the solution w_h and Θ_h at the elemental interfaces. The functions $F(\Theta_h)\cdot n$ and $w_h\cdot n$ emerging in equation (11.14) can be substituted by the numerical fluxes at the elemental interfaces denoted by H^{BR1} and H^{aux}, respectively. Here, the central flux or Bassi Rebay (BR1) scheme is employed for the viscous and auxiliary numerical fluxes to calculate the flux at elemental interfaces.

$$w_h \cdot n \equiv H^{aux}\left(w_h^-, w_h^+\right) = \frac{1}{2}\left[w_h^- + w_h^+\right],$$
$$F(\Theta_h)\cdot n \equiv H^{BR1}\left(\Theta_h^-, \Theta_h^+\right) = \frac{1}{2}\left[F\left(\Theta_h^-\right) + F\left(\Theta_h^+\right)\right]. \tag{11.15}$$

Here, the superscripts $(-)$ and $(+)$ are the left and right states of the elemental interface. As a result, the DG weak formulation is obtained as

$$\int_{\Omega_e} \Theta_h b_h \, dV + \int_{\Omega_e} \nabla b_h \cdot w_h \, dV - \int_{\partial\Omega_e} b_h H^{aux} \, d\Gamma = 0,$$
$$\frac{\partial}{\partial t}\int_{\Omega_e} w_h b_h \, dV - \int_{\Omega_e} \nabla b_h \cdot F(\Theta_h) dV + \int_{\partial\Omega_e} b_h H^{BR1} \, d\Gamma = \int_{\Omega_e} b_h S(w_h) \, dV. \tag{11.16}$$

In the expression equation (11.16), the emerging volume and surface integrals are approximated by the Gaussian–Legendre quadrature rule within the elements to ensure the high-order accuracy (Cockburn and Shu, 2001). Furthermore, a nonlinear TVB limiter (Cockburn and Shu, 1998) is utilized to remove the artificial oscillations arising in the numerical solution.

Finally, the DG spatial discretization equation (11.16) can be expressed in semi-discrete ordinary differential equations form as

$$\mathbf{M}\frac{dw_h}{dt} = \mathbf{R}(w_h), \tag{11.17}$$

where $\mathbf{M}$ and $\mathbf{R}(w_h)$ are the orthogonal mass matrix and the residual function, respectively. Here, an explicit form of Strongly Stability Preserving Runge–Kutta method with third-order accuracy (Shu and Osher, 1988) is adopted as the time marching scheme.

$$w_h^{(1)} = w_h^n + \Delta t \mathbf{M}^{-1}\mathbf{R}(w_h),$$
$$w_h^{(2)} = \frac{3}{4}w_h^n + \frac{1}{4}w_h^{(1)} + \frac{1}{4}\Delta t \mathbf{M}^{-1}\mathbf{R}\left(w_h^{(1)}\right), \tag{11.18}$$
$$w_h^{n+1} = \frac{1}{3}w_h^n + \frac{2}{3}w_h^{(2)} + \frac{2}{3}\Delta t \mathbf{M}^{-1}\mathbf{R}\left(w_h^{(2)}\right),$$

where $\mathbf{R}\left(w_h^n\right)$ is the residual approximation at time t_n, and Δt is the suitable time step value.

11.3 Numerical Results

In this section, four test problems of Fisher–KPP equations are examined by simulating numerical results with the proposed mixed modal DG scheme.

PROBLEM 11.1

For testing the accuracy of the present DG scheme, we consider a one-dimensional Fisher's KPP problem (Wazwaz and Gorguis, 2004)

$$\frac{\partial w}{\partial t} = \lambda_1 \frac{\partial^2 w}{\partial x^2} + \mu w(1-w), \tag{11.19}$$

subject to the initial condition and boundary conditions

$$w(x,0) = \frac{1}{\left(1+e^x\right)^2}, \qquad \lim_{x\to-\infty} w(x,t) = 1.0, \quad \lim_{x\to\infty} w(x,t) = 0. \tag{11.20}$$

The analytical solution of the standard form of Fisher–KPP equation is given by

$$w(x,t) = \left[1 + e^{\sqrt{\frac{\mu}{6}}x - \frac{5}{6}\mu t}\right]^{-2}. \tag{11.21}$$

The numerical solution of one-dimensional Fisher–KPP problem equation (11.19) has been obtained for $\lambda_1 = 1.0$ with $N = 200$ elements and $\Delta t = 0.0001$. Figure 11.2 illustrates the numerical and exact solutions at different time instants $t = 0, 0.2, 0.4, 0.6, 0.8$ and 1. The computed results are depicted graphically in such a way that the numerical result can be compared with the analytical solutions. The provided solutions are found in good match with the analytical solutions. In addition, Table 11.1 shows the comparison between computed DG and analytical solutions, and the maximum error (L_∞) at two different time instants $t = 0.5$, and 1.

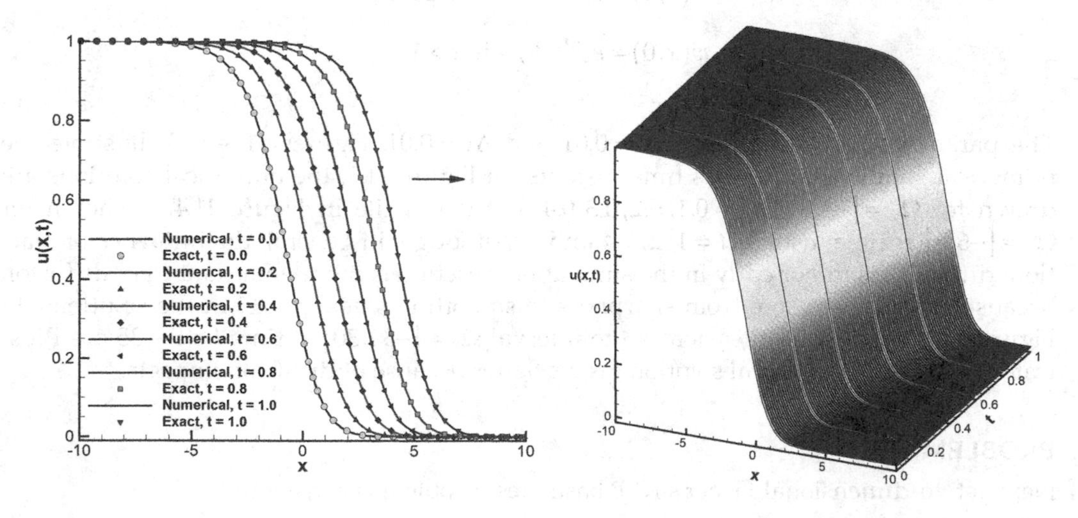

FIGURE 11.2
Profiles and space-time plots for one-dimensional Fisher–KPP Problem 1 at early time instants. Arrow symbol represents the direction of traveling wave solutions.

TABLE 11.1

Comparison of Exact and Present Numerical Solutions of Problem 1 at Two Different Time Level st=0.5 and 1.0.

	t=0.5			t=1.0		
x	Present	Exact	Error	Present	Exact	Error
-10	0.10000E+1	0.99999E+0	0.42469E-4	0.10001E+1	0.99999E+0	0.49076E-4
–8	0.99999E+0	0.99994E+0	0.47085E-4	0.10001E+1	0.99999E+0	0.49529E-4
–6	0.99962E+0	0.99957E+0	0.46834E-4	0.10000E+1	0.99996E+0	0.49533E-4
–4	0.99689E+0	0.99681E+0	0.44718E-4	0.99979E+0	0.99974E+0	0.49400E-4
–2	0.97708E+0	0.97704E+0	0.36362E-4	0.99814E+0	0.99809E+0	0.49064E-4
0	0.85748E+0	0.85744E+0	0.47593E-4	0.98604E+0	0.98599E+0	0.50665E-4
2	0.40232E+0	0.40213E+0	0.18926E-3	0.91172E+0	0.91165E+0	0.70946E-4
4	0.36165E-1	0.36101E-1	0.64248E-4	0.54906E+0	0.54889E+0	0.17092E-3
6	0.94896E-3	0.94672E-3	0.22402E-5	0.77927E-1	0.77833E-1	0.93739E-4
8	0.18347E-4	0.18301E-4	0.46535E-7	0.24806E-2	0.24761E-2	0.45552E-5
10	0.54037E-6	0.27650E-6	0.26387E-7	0.80924E-4	0.40598E-4	0.40325E-4

PROBLEM 11.2

Consider a one-dimensional Fisher's KPP equation (Tang and Weber, 1991)

$$\frac{\partial w}{\partial t} = \lambda_1 \frac{\partial^2 w}{\partial x^2} + \mu w(1-w), \tag{11.22}$$

with the parameters $\lambda_1 = 1,\ \mu = 1$ and the initial conditions are provided by

$$\begin{aligned} w(x,0) &= e^{10(x+1)}, && \text{if } x < -1 \\ w(x,0) &= 1, && \text{if } -1 \le x \le -1 \\ w(x,0) &= e^{-10(x-1)}, && \text{if } x > 1. \end{aligned} \tag{11.23}$$

The parameters are selected as $\Delta t = 0.01$ and $\Delta x = 0.01$. Figures 11.3–11.5 illustrate the numerical solutions at various time instants. In Figure 11.3, the numerical solutions are drawn for $\Omega_x = [-3,3]$ at $t = 0.1, 0.2, 0.3, 0.4$ and 0.5, while in Figure 11.4, is shown for $\Omega_x = [-6,6]$ at time instants $t = 1, 2, 3, 4$ and 5. Although, in general, the influence of reaction–diffusion is minor early in the simulation, reaction is more dominant than diffusion because there is a change from sharpness to smoothness near $x = \pm 1$ in the solutions. In Figure 11.5, solution profiles across the interval $\Omega_x = [-30,30]$ at time $t = 0 - 35$ are illustrated, where the numerical solution grows flatter because of diffusion impacts.

PROBLEM 11.3

Next, a two-dimensional Fisher's KPP based test problem is considered.

$$\frac{\partial w}{\partial t} = \lambda_1 \frac{\partial^2 w}{\partial x^2} + \lambda_2 \frac{\partial^2 w}{\partial y^2} + \mu w(1-w), \tag{11.24}$$

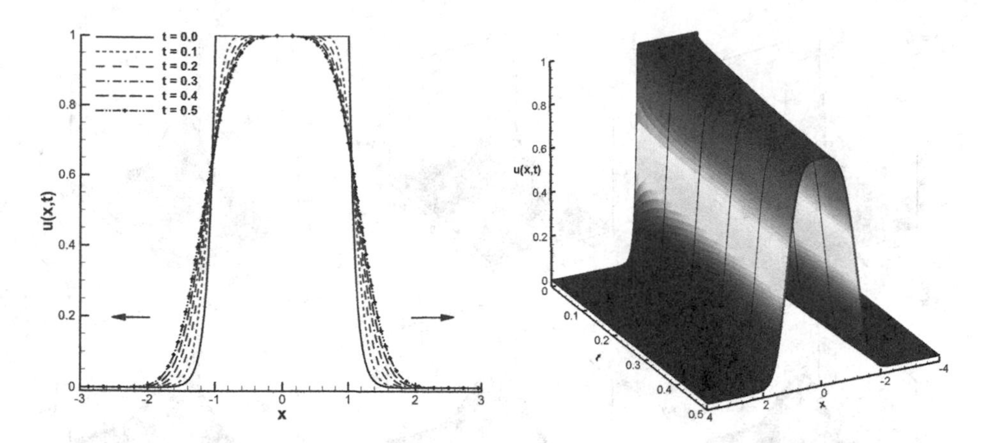

FIGURE 11.3
Profiles and space-time plots for one-dimensional Fisher–KPP Problem 2 at early time instants. Arrow symbol represents the direction of traveling wave solutions.

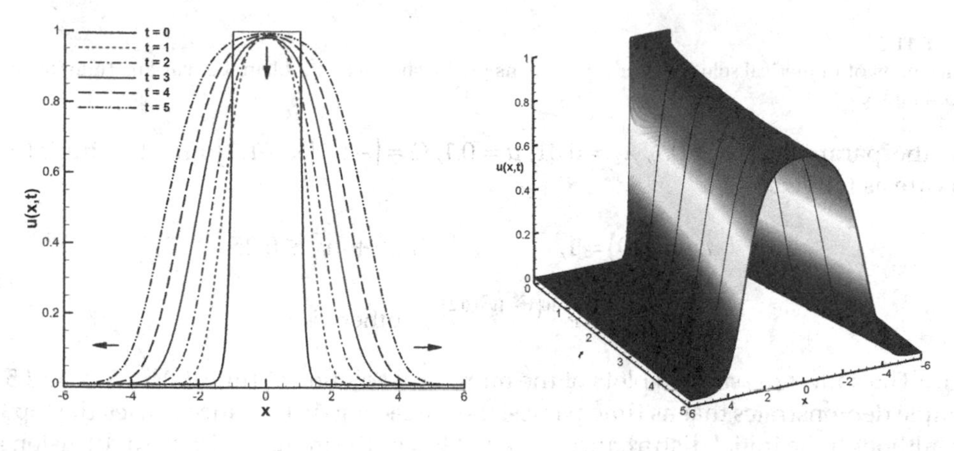

FIGURE 11.4
Profiles and space-time plots for one-dimensional Fisher–KPP Problem 2 at middle time instants. Arrow symbol represents the direction of traveling wave solutions.

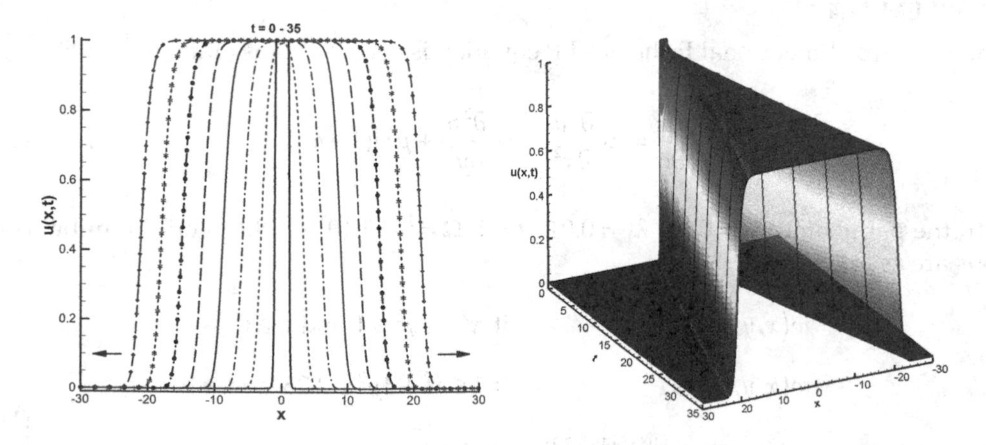

FIGURE 11.5
Profiles and space-time plots for one-dimensional Fisher–KPP Problem 2 at longer time instants. Arrow symbol represents the direction of traveling wave solutions.

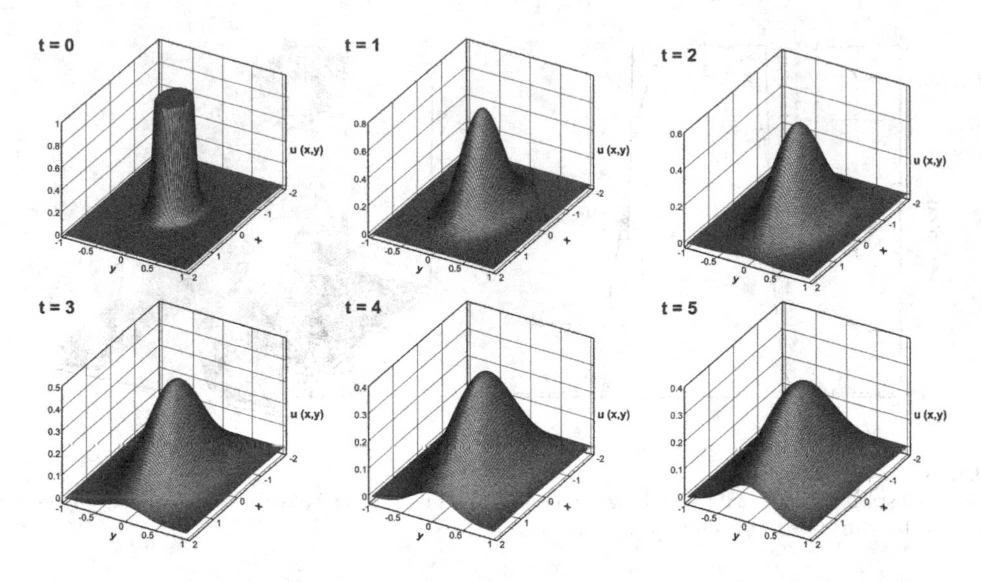

FIGURE 11.6
Surface plots of numerical solutions for two-dimensional Fisher–KPP Problem 3 at various time instants $t=0$, 1, 2, 3, 4 and 5.

with the parameters $\lambda_1 = 0.1$, $\lambda_2 = 0.01$, $\mu = 0.1$, $\Omega = [-2,2]\times[-1,1]$ and the initial conditions are as follows:

$$\begin{aligned} w(x,y,0) &= 1, && \text{if } x^2 + 4y^2 \le 0.25 \\ w(x,y,0) &= e^{-10(x^2+4y^2-0.25)}, && \text{otherwise.} \end{aligned} \tag{11.25}$$

Figure 11.6 shows the surface plots of the numerical solution at time 0, 1, 2, 3, 4, and 5. This example demonstrates that as time passes, the spreading disturbance causes the top to fall first, although the initial disturbance has a flat top in the middle with zero diffusion and a significant diffusion at the edge.

PROBLEM 11.4

Finally, a two-dimensional Fisher's KPP equation is considered as (Tang et al. 1993)

$$\frac{\partial w}{\partial t} = \lambda_1 \frac{\partial^2 w}{\partial x^2} + \lambda_2 \frac{\partial^2 w}{\partial y^2} + \mu w(1-w), \tag{11.26}$$

with the parameters $\lambda_1 = 0.02$, $\lambda_1 = 0.01$, $\mu = 1$, $\Omega = [-10,10]\times[-8,8]$ and the initial conditions are as following:

$$\begin{aligned} w(x,y,0) &= x^2 + 4y^2, && \text{if } x^2 + 4y^2 \le 1 \text{ and } x \ge 0 \\ w(x,y,0) &= 1, && \text{if } 1 \le x^2 + 4y^2 \le 2.25 \text{ and } x \ge 0 \\ w(x,y,0) &= e^{-10(x^2+4y^2-2.25)}, && \text{if } x^2 + 4y^2 \ge 2.25 \text{ and } x \ge 0 \\ w(x,y,0) &= 0, && \text{if } x > 0. \end{aligned} \tag{11.27}$$

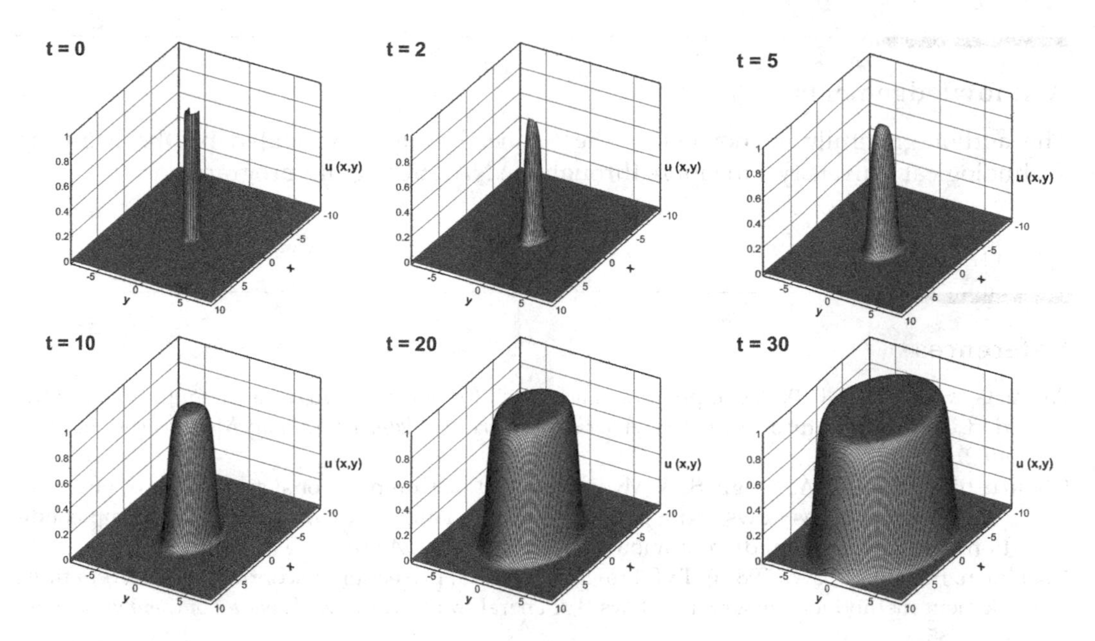

FIGURE 11.7
Surface plots of numerical solutions for two-dimensional Fisher–KPP Problem 4 at various time instants $t=0$, 2, 5, 10, 20 and 30.

The parameters are selected to depict the concave and convex distributions across a vast area. Figure 11.7 illustrates the surface plots of the numerical solutions at several time instants $t = 0, 2, 5, 10,$ 20 and 30. It can be observed that as time passes, the concave part of w and its isopleths will become convex. The two-dimensional quasi-traveling waves are obtained as shown and discussed by Tang and Weber (1991).

11.4 Concluding Remarks

The aim of this chapter is to find approximate solutions to the nonlinear Fisher–KPP reaction-diffusion equation emerging in biology. A mixed modal DG scheme is developed to simulate the two-dimensional Fisher–KPP equation. This numerical approach is based on the concept of addressing an additional auxiliary unknown in the high-order derivative diffusion term. The scaled Legendre polynomials with third-order accuracy are used for spatial discretization, while, a third-order Strongly-Stability-Preserving Runge-Kutta (SSP-RK33) scheme is used for temporal discretization. This numerical approach is widely applicable for several nonlinear reaction–diffusion problems. To verify the accuracy and reliability of the DG scheme, four well-known one-dimensional and two-dimensional numerical problems in literature are solved. The derived numerical solutions and errors show that the results are in good match with the analytical solutions. In the case of two-dimensional work, surface plots are also shown, which is identical to the results reported graphically in the literature. This proposed DG approach demonstrates that it is an efficient scheme to find numerical solutions in a variety of linear and nonlinear physical models. It may thus be extended to tackle higher-dimensional problems efficiently.

Acknowledgements

The author gratefully acknowledges the financial support provided by the Nanyang Technological University, Singapore through the NAP-SUG grant program.

References

Arora, G. & Joshi, V. (2021). A computational approach for one and two dimensional Fisher's equation using quadrature technique. *American Journal of Mathematical and Management Sciences, 40(2),* 145–162.

Chourushi, T., Rahimi, A., Singh, S., & Myong, R.S. (2020). Computational simulation of near continuum external gas flows using the Navier-Stokes-Fourier equations with slip/jump conditions based on a modal discontinuous Galerkin method, *Advances in Aerodynamics, 2(8),* 1–37.

Cockburn, B., & Shu, C.-W. (1989a). TVB Runge-Kutta local projection discontinuous Galerkin finite element method for conservation laws II, General framework. *Mathematics of Computation, 52,* 411–435.

Cockburn, B., & Shu, C.-W. (1989b). TVB Runge-Kutta local projection discontinuous Galerkin finite element method for conservation laws III: one-dimensional systems. *Journal of Computational Physics, 84,* 90–113

Cockburn, B., & Shu, C.-W. (1998). The Runge-Kutta discontinuous Galerkin method for conservation laws V: Multidimensional systems. *Journal of Computational Physics, 141(2),* 199–224.

Cockburn, B., & Shu, C.-W. (2001). Runge–Kutta discontinuous Galerkin methods for convection-dominated problems. *Journal of Scientific Computing, 16(3),* 173–261.

Dehghan, M. & Abbaszadeh, M. (2017). Numerical investigation based on direct meshless local Petrov Galerkin (direct MLPG) method for solving generalized Zakharov system in one and two dimensions and generalized Gross-Pitaevskii equation. *Engineering Computations, 34(4),* 983–996.

Dehghan, M. & Abbaszadeh, M. (2018). Solution of multi-dimensional Klein-Gordon-Zakharov and Schrödinger/Gross-Pitaevskii equations via local radial basis functions-differential quadrature (RBF-DQ) technique on non-rectangular computational domains. *Engineering Analysis with Bundary Elements, 92,* 156–170.

Evans, D.J., & Sahimi, M.S. (1989). The alternating group explicit iterative method to solve parabolic and hyperbolic partial differential equations. *Annual Review of Numerical FluidMechanics and Heat Transfer, 2,* 283–389.

Fisher, R.A. (1937). The wave of advance of advantageous genes. *Annals of Eugenics, 7(4),* 355–369.

Gazdag, J., & Canosa, J. (1974). Numerical solution of Fisher's equation. *Journal of Applied Probability, 11(3),* 445–457.

Hagstrom, T., & Keller, H.B. (1986). The numerical calculation of travelling wave solutions of nonlinear parabolic equations. *SIAM Journal on Scientific and Statistical Computing, 7(3),* 978–988.

José, C. (1969). Diffusion in nonlinear multiplicative media. *Journal of Mathematical Physics, 10(10),* 1862–1868.

Kolmogorov, A.N. (1937). Étude de l'équation de la diffusion avec croissance de la quantité de matière et son application à un problème biologique. *Moskow University Bulletin, International Series, Section A, 1,* 1–25.

Le, N.T.P., Xiao, H., & Myong, R.S. (2014). A triangular discontinuous Galerkin method for non-Newtonian implicit constitutive models of rarefied and microscale gases. *Journal of Computational Physics, 273,* 160–184.

Mittal, R.C. & Kumar, S. (2006). Numerical study of Fisher's equation by wavelet Galerkin method. *International Journal of Computer Mathematics 83(3),* 287–298.

Mittal, R.C., & Arora, G. (2010). Efficient numerical solution of Fisher's equation by using B-spline method. *International Journal of Computer Mathematics, 87(13)*, 3039–3051.

Mo, J. Q., Zhang, W. J., & He, M. (2007). Asymptotic method of travelling wave solutions for aclass of nonlinear reaction diffusion equation. *Acta Mathematica Scientia, 27(4)*, 777–780.

Parekh, N., & Puri, S. (1990). A new numerical scheme for the Fisher's equation. *Journal of Physics A: Mathematical and General, 23*, 1085–1091.

Raj, L.P., Singh, S., Karchani, A., & Myong, R.S. (2017). A super-parallel mixed explicit discontinuous Galerkin method for the second-order Boltzmann-based constitutive models of rarefied and microscale gases. *Computers & Fluids, 157*, 146–163.

Reed, W.H., & Hill, T.R. (1973). Triangular mesh methods for the neutron transport equation. *Los Alamos Scientific Laboratory Report LA-UR, 13*(7), 73–79

Roessler, J. & Hüssner, H. (1997).Numerical solution of the 1+ 2 dimensional Fisher's equation by finite elements and the Galerkin method. *Mathematical and Computer Modelling, 25(3)*, 57–67.

Shu, C.-W., & Osher, S. (1988). Efficient implementation of essentially non-oscillatory shock-capturing schemes. *Journal of Computational Physics, 77*(2), 439–471.

Singh, S. (2018). Development of a 3D discontinuous Galerkin method for the second-order Boltzmann-Curtiss based hydrodynamic models of diatomic and polyatomic gases. (Doctoral dissertation, Gyeongsang National University South Korea. Department of Mechanical and Aerospace Engineering.)

Singh, S. (2020). Role of Atwood number on flow morphology of a planar shock-accelerated square bubble: A numerical study. *Physics of Fluids, 32*, 126112.

Singh, S. (2021a). Numerical investigation of thermal non-equilibrium effects of diatomic and polyatomic gases on the shock-accelerated square light bubble using a mixed-type modal discontinuous Galerkin method. *International Journal of Heat and Mass Transfer, 169*, 121708.

Singh, S. (2021b). Contribution of Mach number on the evolution of Richtmyer-Meshkov instability induced by a shock-accelerated square light bubble. *Physical Review Fluids 6(10)*, 104001.

Singh, S. (2021c). Mixed-type discontinuous Galerkin approach for solving the generalized FitzHugh-Nagumo reaction-diffusion model. *International Journal of Applied and Computational Mathematics, 7*, 207.

Singh, S. (2021d). A mixed-type modal discontinuous Galerkin approach for solving nonlinear reaction diffusion equations. *AIP Conference Proceedings* (accepted).

Singh, S., & Battiato, M. (2020a). Effect of strong electric fields on material responses: The Bloch oscillation resonance in high field conductivities. *Materials, 13*, 1070.

Singh, S., & Battiato, M. (2020b). Strongly out-of-equilibrium simulations for electron Boltzmann transport equation using explicit modal discontinuous Galerkin method. *International Journal of Applied and Computational Mathematics, 6*,133.

Singh, S., & Battiato, M. (2021a). An explicit modal discontinuous Galerkin method for Boltzmann transport equation under electronic nonequilibrium conditions. *Computers & Fluids, 224*, 104972.

Singh, S., & Battiato, M. (2021b). Behavior of a shock-accelerated heavy cylindrical bubble under nonequilibrium conditions of diatomic and polyatomic gases. *Physical Review Fluids, 6*, 044001.

Singh, S., & Myong, R.S. (2017). A computational study of bulk viscosity effects on shock-vortex interaction using discontinuous Galerkin method. *Journal of Computational Fluids Engineering, 22*, 86–95.

Singh, S., Battiato, M. & Myong, R.S. (2021). Impact of bulk viscosity on flow morphology of shock-accelerated cylindrical light bubble in diatomic and polyatomic gases. *Physics of Fluids, 33*, 066103.

Tan, Y., Xu, H., & Liao, S.J. (2007). Explicit series solution of travelling waves with a front of Fisher equation. *Chaos, Solitons and Fractals, 31(2)*, 462–472.

Tang, S., Qin, S. & Weber, R.O. (1993). Numerical studies on 2-dimensional reaction-diffusion equations. *The Journal of the Australian Mathematical Society. Series B. Applied Mathematics, 35(2)*, 223–243.

Tang, S. & Weber, R.O. (1991). Numerical study of Fisher's equation by a Petrov–Galerkin finite element method. *The Journal of the Australian Mathematical Society. Series B. Applied Mathematics, 33(1)*, 27–38.

Wazwaz, A.M. (2004). The tanh method for traveling wave solutions of nonlinear equations. *Applied Mathematics and Computation, 154(3)*, 713–723.

Wazwaz, A.-M. & Gorguis, A. (2004). An analytic study of Fisher's equation by using Adomian decomposition method. *Applied Mathematics and Computation, 154(3)*, 609–620.

Zheng, S. (2004). Nonlinear evolution equations. *Monographs and surveys in pure and applied Mathematics*. CRC Press, Boca Raton.

12

A Numerical Approach on Unsteady Mixed Convection Flow with Temperature-Dependent Variable Prandtl Number and Viscosity

Govindaraj N and Iyyappan G
Hindustan Institute of Technology and Science

A. K. Singh
VIT University

S. Roy
Indian Institute of Technology Madras

P. Shukla
VIT University

CONTENTS

12.1 Introduction 185
12.2 Formation of Governing Equations 186
12.3 Numerical Computation with Quasilinearization Technique 189
12.4 Results and Discussion 190
12.5 Conclusions 194
References 195

12.1 Introduction

The unsteady heat transfer fluid flow is very important for many real-life applications where fluid has direct contact with solid such as plastic sheet extrusion, wire and fiber coating, metal and glass spinning, polymer sheets, the boundary layer along with material handlings conveyers, etc. The boundary layer fluid flow is very thin and sometimes is visible to the naked eye. For example, near the ship, there is a narrow band of water and the relative velocity of the ship is less than that of water. Water is used for working in many industrial and engineering fields and hence the current study is based on water. The study on convection pattern flow over various surfaces was performed with constant physical parameters of fluids. Cortell [1] presented the numerical study of the fluid over moving surface and discussed the direction of wall shear and temperature variations. Fang [2] discussed the unsteady flow over moving plate, and the characteristics of the heat transfer and fluid motion are found. Patil and Roy [3] numerically analyzed fluid flow over moving vertical plates along with parallel to the free stream. They discussed heat absorption or generation within the boundary

DOI: 10.1201/9781003291916-12

layer. Bhattacharyya et al. [4] analyzed the slip impact on the flow over a flat plate. They have found that velocity and temperature overshoot acting on near the plate. The fluid flow over moving a plate is theoretically investigated with viscous dissipation term by Bachok et al. [5]. The results indicated that the solution's nature of the similarity equations depends on the velocity ratio parameter, mixed convection parameter and the Eckert number. Aman and Ishak [6] discussed the mixed convection flow toward a plate. The results indicate that dual and unique solutions are based on opposing and assisting flow.

In general, various studies of boundary layer flow are based on constant Prandtl number and variable viscous fluid. But in practical, liquids say water, are particularly sensitive to temperature changes as for as Prandtl number and viscosity are concerned. It is assumed that the Prandtl and viscosity are varied on inverse of linear function of temperature [7,8]. Pop et al. [9] examined the impact on variable viscosity of the fluid flow over moving flat plate. The nonuniform slot suction can be used for delayed separation, but the opposite effect is observed in the case of nonuniform slot injection. Saikrishnan and Roy [10] presented the forced convection flow impact on Prandtl number and viscosity with temperature. They discussed the effects of skin friction and heat transfer rate on viscous dissipation. Roy et al. [11] studied the laminar flow and heat transfer over a sphere and analyzed the impact of injection and suction. They have found that nonuniform slot suction can be used for delaying the separation but the opposite effect is observed in the case of nonuniform slot injection. Mureithi et al. [12] performed the numerical study of fluid flow on moving flat plates with the streaming flow with temperature-dependent viscosity. Das et al. [13] discussed unsteady fluid flow over a parallel plate with different temperature and mass diffusion in the presence of thermal radiation. Govindaraj et al. [14,15] study the convection fluid flow over a stretching exponential surface with a porous plate. Maitil and Mandal [16] have investigated impacts of thermal radiation and buoyancy force over an infinite vertical flat plate with slip conditions. Talha Anwar et al. [17] discussed magneto hydrodynamics effects on natural convection nanofluid flow over a porous media with time-dependent and heat source/Sink. The effects of thermal radiation heat absorption over an accelerated infinite vertical plate and ramped plate of convective flow of nanofluids with time-dependent are discussed in Refs. [18] and [19].

The objective of the current work is to focus on the unsteady flow over a vertical plate with temperature-dependent and physical parameters of fluids. This investigation of the current problem is considered mainly focused on the time-dependent, various values of Reynolds number, Prandtl number, and viscosity of fluid flow moving on a flat plate. The nonlinear governing equations have been converted into linear equation using the Quasi-linearization technique and further obtained the system of equations based on the finite difference method. The effects of physical parameters on velocity, temperature, heat transfer, and skin friction have been investigated. The unsteadiness in the flow and temperature fields is induced by the free stream and moving plate velocity. The validation of current results has been done by comparing them with previously published results that are shown in Table 12.2.

12.2 Formation of Governing Equations

Consider a vertical plate with an unsteady laminar water-based mixed convection flow with variable viscosity and temperature. In the range of 0°C–45°C, Prandtl number (Pr) and dynamic viscosity μ are considered too different as an inverse linear function of

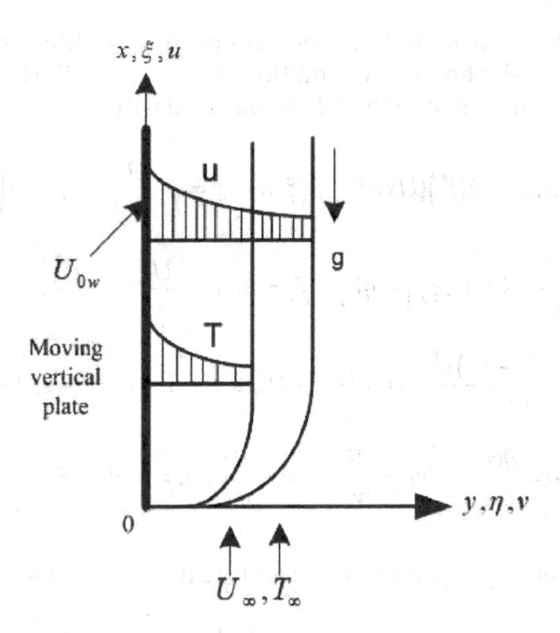

FIGURE 12.1
Physical coordination of model figure.

temperature. At constant strain, the variance of both specific heat (C_p) and density (ρ) is assumed to be less than 1%, so they are treated as constants. Reference [7] provided the values for thermophysical properties of water at various temperatures. Buoyancy force rise to the fluid properties to relate the density changes to temperature, and the thermophysical properties of water with various temperature has been shown in [20]. In an incompressible fluid, the contribution of heat due to the contraction of the fluid is very small and it is ignored. The fluid at the surface is kept at a constant temperature T_w, and the boundary layer's edge is kept at the same temperature T_∞.

The variable Prandtl number and variable viscosity are defined as

$$Pr = \frac{1}{c_1 + c_2 T} \quad \text{and} \quad \mu = \frac{1}{b_1 + b_2 T}$$

where b_1, b_2, c_1 and c_2 are the best approximation of water and the numerical data presented [20] (Figure 12.1).

The governing equations of the current problem are given as

$$\nabla . V = 0 \tag{12.1}$$

$$(\nabla u)_t + u(\nabla u)_i + v(\nabla u)_j = (\nabla U_e)_t + \frac{1}{\rho}(\nabla)_j\left[\mu(\nabla u)_j\right] + g\left[\beta(T_w - T_\infty)\right] \tag{12.2}$$

$$(\nabla T)_t + u(\nabla T)_i + v(\nabla T)_j = \frac{1}{\rho}(\nabla)_j\left[\frac{\mu}{Pr}(\nabla T)_j\right] \tag{12.3}$$

The boundary constraints:

$$\begin{aligned} y = 0: \; & u(t,x,y) = U_w(t) = U_{0w}\phi(t^*), \; T(t,x,y) = T_w, \; v(t,x,y) = 0, \\ y \to \infty: \; & u(t,x,y) \to U_e(t) \to U_{\infty}\phi(t^*), \; T(t,x,\infty) \to T_\infty \end{aligned} \tag{12.4}$$

To determine the velocity field and temperature profiles within the boundary layer, the problem is well posted with boundary condition equations (12.1)–(12.4) grouping into non-dimensional fluid parameters with the following similarity transformation:

$$\xi = \frac{x}{L},\ \psi(x,y) = \phi(t^*)(Uxv)^{1/2} f(\xi,\eta),\ \varepsilon = \frac{U_\infty}{U_\infty + U_{0w}},\ \eta = \left(\frac{U}{xv}\right)^{1/2} y;$$

$$v = -\frac{(Uvx)^{1/2}}{2x}\{f + 2\xi f_\eta - \eta F\},\ f_\eta = F,\ t^* = \frac{U}{x}t;$$

$$Gr_L = \frac{g\beta_T(T_w - T_\infty)L^3}{v^2},\ U = U_{0w} + U_\infty,\ T - T_\infty = (T_w - T_\infty)\theta(t^*,\xi,\eta)$$

$$\lambda = \frac{Gr_L}{Re_L^2},\ u = \frac{\partial\psi}{\partial y},\quad Re_L = \frac{UL}{v},\ v = -\frac{\partial\psi}{\partial x},\ u = U\phi(t^*)F \tag{12.5}$$

The transformed governing equations (12.2) and (12.3) are as follows:

$$(NF_\eta)_\eta + \phi(t^*)\left(\frac{f}{2}\right)F_\eta + Re_L\xi\phi(t^*)^{-1}\frac{d\phi(t^*)}{dt^*}(\epsilon - F) - Re_L\xi\frac{dF}{dt^*} + \lambda\theta\xi\phi(t^*)^{-1}$$

$$= \xi\phi(t^*)(FF_\xi - f_\xi F_\eta) \tag{12.6}$$

$$(NPr^{-1}\theta_\eta)_\eta + \theta_\eta\phi(t^*)\left(\frac{f}{2}\right) - Re_L\xi\frac{d\theta}{dt^*} = \xi\phi(t^*)(F\theta_\xi - \theta_\eta f_\xi) \tag{12.7}$$

The transformed boundary constrains are

$$\begin{aligned} F &= 1 - \varepsilon,\ \theta = 1,\ at\ \eta = 0 \\ F &= \varepsilon,\ \theta = 0,\quad at\ \eta = \eta_\infty \end{aligned} \tag{12.8}$$

where $f = \int_0^\eta F d_\eta + f_w$, $f_w = 0$ (Table 12.1)

TABLE 12.1

The Values of Thermo-Physical Properties at Different Temperature. Values of Water as Given Below [7]

Temperature (T) (C^o)	Density (g/m^3)	Specific Heat ($J.10^7$/kg K)	Thermal Conductivity ($erg.10^5$/cm.s.K)	Viscosity ($g.10^{-2}$-cm.s)	Prandtl Number
0	1.00228	4.2176	0.5610	1.7930	13.48
10	0.99970	4.1921	0.5800	1.3070	9.45
20	0.99821	4.1818	0.5984	1.0060	7.03
30	0.99565	4.1784	0.6154	0.7977	5.12
40	0.99222	4.1785	0.6305	0.6532	4.32
50	0.98803	4.1806	0.6435	0.5470	3.55

i.e., $f_w = 0$ is impermeable plate. Noted that $\phi(t^*) = 1 + \alpha t^*; \alpha > 0$ or $\alpha < 0$, where the time $t^* = 0$ is steady-state flow are obtained from equations (12.6) and (12.7) by substituting $\frac{d\varnothing(t^*)}{dt^*} = \frac{dF}{dt^*} = 0, \varnothing(t^*) = 1$ and if $t^* > 0$ become unsteady flow with free stream velocity $[U_e(t) = U_\infty \phi(t^*)]$ and velocity at wall $[U_w(t) = U_{0w}\phi(t^*)]$.

The skin friction coefficient (C_{fx}) as below:

$$C_{fx} = \frac{2\mu\left(\frac{\partial u}{\partial y}\right)_{y=0}}{\rho U_w^2} = 2\phi(t^*)F_\eta(\xi,0)(Re_L\xi)^{-1/2} \tag{12.9}$$

$$i.e., \backslash dsC_{fx}\left((Re_L\xi)^{1/2}\right) = 2\phi(t^*)F_\eta(\xi,0)$$

The Nusselt number (Nu_x) as below:

$$Nu_x = \frac{-x\left(\frac{\partial T}{\partial y}\right)_{y=0}}{(T_w - T_\infty)} = -\theta_\eta(\xi,0)(Re_L \sim \xi)^{1/2} \tag{12.10}$$

$$i.e., \backslash dsNu_x\left((Re_L\xi)^{-1/2}\right) = -\theta_\eta(\xi,0).$$

12.3 Numerical Computation with Quasilinearization Technique

In the process of numerical computation, the set of nonlinear governing equations are playing a vital role given its efficiency methods and accuracy in yielding the solutions. The nonlinear equations (12.6) and (12.7) with boundary conditions (12.8) have been linearized numerically by using quasi-linearization technique, to obtain the following coupled PDEs.

$$X_1^s F_{\eta\eta}^{s+1} + X_2^s F_\eta^{s+1} + X_3^s F^{s+1} + X_4^s F_\xi^{s+1} + X_5^s F_{t^*}^{s+1} + X_6^s \theta_\eta^{s+1} + X_7^s \theta^{s+1} = X_8^s \tag{12.11}$$

$$Y_1^s \theta_{\eta\eta}^{s+1} + Y_2^s \theta_\eta^{s+1} + Y_3^s \theta^{s+1} + Y_4^s \theta_\eta^{s+1} + Y_5^s \theta_{t^*}^{s+1} + Y_6^s F^{s+1} = Y_7^s \tag{12.12}$$

The coefficient functions are determined by an iterative method, subject to constraints,

$$F^{s+1} = 1 - \varepsilon,\ \theta^{s+1} = 1 \quad at \quad \eta = 0$$

$$F^{s+1} = \epsilon,\ \theta^{s+1} = 0 \quad at \quad \eta = \eta_\infty \tag{12.13}$$

The coefficients in the equations (12.11) and (12.12) are given by

$$X_1^s = N$$

$$X_2^s = -a_1 N^2 \theta_\eta + \phi(t^*)\frac{f}{2} + \phi(t^*)\xi f_\xi$$

$$X_3^s = -\xi Re\phi(t^*)^{-1} \backslash ds\frac{d\phi(t^*)}{dt^*} - \xi\phi(t^*)F_\xi$$

$$X_4^s = -\xi\phi(t^*)F$$

$$X_5^s = -\xi Re$$

$$X_6^s = -a_1 N^2 F_\eta$$

$$X_7^s = -a_1 N^2 F_{\eta\eta} + 2a_1^2 N^3 F_\eta \theta_\eta + \lambda\xi\phi(t^*)^{-1}$$

$$X_8^s = -a_1 N^2 F_\eta \theta_\eta - a_1 N^2 F_{\eta\eta}\theta + 2a_1^2 N^3 F_\eta \theta_\eta \theta - \xi\epsilon Re\phi(t^*)^{-1\backslash} ds\frac{d\phi(t^*)}{dt^*} - \xi\phi(t^*)FF_\xi$$

$$Y_1^s = NPr^{-1}$$

$$Y_2^s = -2a_1 N^2 Pr^{-1}\theta_\eta + 2a_3\theta_\eta N + \phi(t^*)\frac{f}{2} + \phi(t^*)\xi f_\xi$$

$$Y_3^s = a_3 N\theta_{\eta\eta} - a_1 N^2 Pr^{-1}\theta_{\eta\eta} - 2a_1 a_3 N^2 \theta_\eta^2 + 2a_1^2 N^3 Pr^{-1}\theta_\eta^2$$

$$Y_4^s = -\xi\phi(t^*)F$$

$$Y_5^s = -\xi Re$$

$$Y_6^s = -\xi\phi(t^*)\theta_\xi$$

$$Y_7^s = a_3\theta_\eta^2 N - a_1\theta_\eta^2 Pr^{-1}N^2 + a_3\theta_{\eta\eta}\theta N - a_1\theta_{\eta\eta}\theta Pr^{-1}N^2$$
$$- 2a_1 a_3\theta_\eta^2\theta N^2 + 2Pr^{-1}a_1^2\theta_\eta^2\theta N^3 - \xi\phi(t^*)F\theta_\xi$$

12.4 Results and Discussion

To obtain the nonsimilar solutions, the numerical procedure in the above section is implemented. Unsteady mixed convection flow with variable viscous fluid and variable

TABLE 12.2

Comparison of the Steady State Results

Pr	2	5	7	10	100
Ali [21]	–	–	–	1.6713	–
Chen [22]	0.68324	–	1.38619	1.68008	5.54450
A.K. Singh [23]	0.6830	1.151	1.38600	1.6801	5.5450
Present work	0.6832	1.149	1.385	1.6800	5.5449

Prandtl number is considered. Numerical computation process has been carried out for different parameters values of $\alpha(0 \leq \alpha \leq 2.5)$, $\epsilon(0.1 \leq \epsilon \leq 0.9)$, $\xi(0 \leq \xi \leq 1)$, $\lambda(0 \leq \lambda \leq 4)$, and $t^*(0 \leq t^* \leq 2)$.

Comparison of the steady state results $-\theta_\eta(\xi,0)$ for $\lambda = 0, \xi = 0, \varepsilon = 0$ with those of Ali and Yousef [21], Chen [22] and A.K. Singh et al. [23].

Figure 12.2 displays the influence of accelerating fluid flow $[\phi(t^*)]$, time parameter (t^*), and buoyancy force (λ) on the velocity profile. Noted that, near the wall high overshoot is presents in the velocity profile as increase λ with decrease t^*. Also, the velocity profile enhanced near the wall due to the increase of buoyancy force (λ). In case of time parameter increase, the steepness of velocity profile is almost absent, which means that the intensity of the fluid flow goes down. For example, when λ increases from 2 to 4 the fluid accelerated near the wall and increased the velocity profile for $\epsilon = 0.4, \alpha = 0.5, \xi = 0.5, t^* = 1$ and Re=1.0.

Figure 12.3 shows the impact of t^*, ε and accelerating flow $[\varnothing(t^*)]$ on θ (temperature profile). It is noted that intensity of temperature profile θ increases with increase of ε and its followed decreasing trends with the increase of time t^*. However, due increase of ε the fluid gets accelerated and the magnitude of temperature is increased. As $[\varnothing(t^*)]$ increase the thickness of the thermal boundary layer is close to the lower segment of the plate. The temperature gradient increased when t^* decreases from 1 to 0 and hence thermal thickness are much smaller for $t = 1$ as compared to $t = 0$. For example, the intensity of θ increases by 36% as ε varies from 0.1 to 0.7 at $t^* = 0, \xi = 0, \lambda = 2, \alpha = 0.5$ and Re=1.

Figure 12.4 depicts the impact of the accelerating flow α and t^* on the velocity profile. The impacts of velocity profile indicate that it is strongly dependent on α and t^*. In F, the decreasing trend is observed for all values of α. Physically, furthermore accelerating parameter is directly related to α and more significant effects are observed in velocity profiles. However, velocity profiles are enhanced for low values of $\alpha(\alpha = 0.5)$ and time $t^* = 1$. In particular, for $\varepsilon = 0.2, \eta = 1, \xi = 0, \lambda = 2.0$ and Re = 100, the magnitude of $F(\eta, \xi)$ increases by 20% with decreases of α from 2.5 to 0.5 at $t^* = 1.0$.

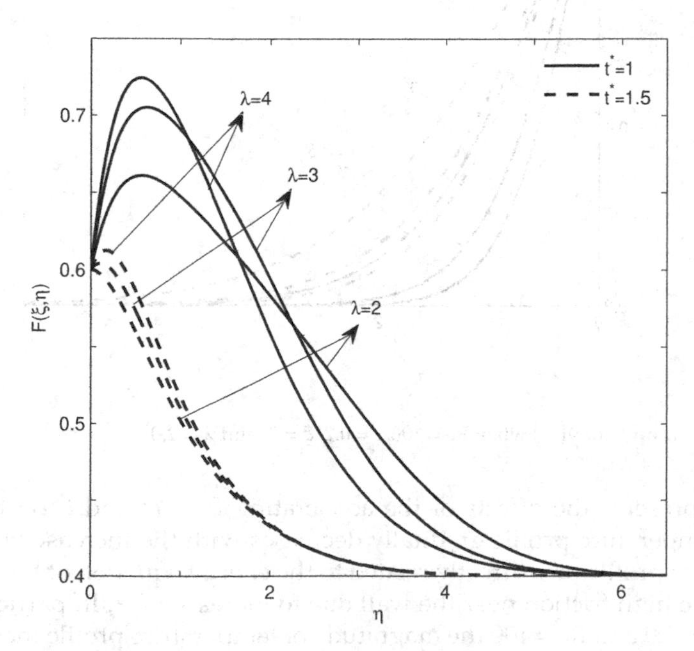

FIGURE 12.2
Variations of t^* and α on F for $\phi(t^*)$ when Re=100, $\epsilon = 0.2, \xi = 0$, and $\lambda = 2.0$.

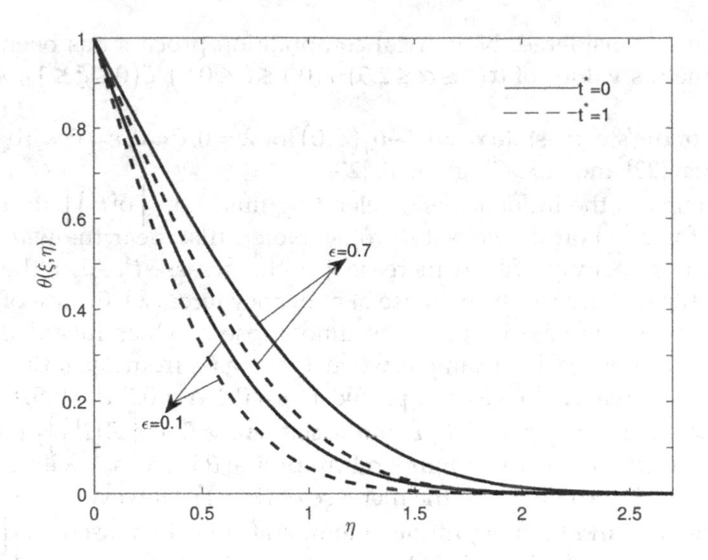

FIGURE 12.3
Variations of t^* and α on F for $\phi(t^*)$ when Re = 100, $\epsilon = 0.2$, $\xi = 0$, and $\lambda = 2.0$.

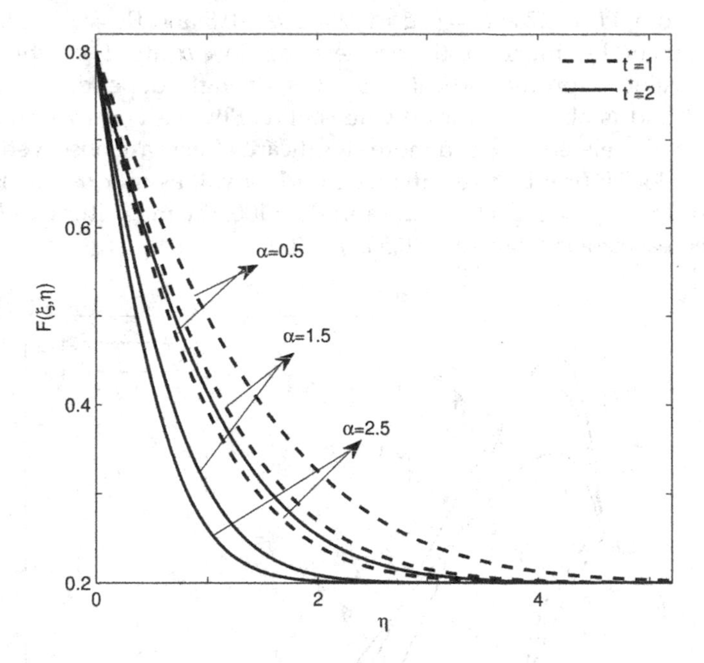

FIGURE 12.4
Variations of t^* and α on F for $\phi(t^*)$ when Re = 100, $\epsilon = 0.2$, $\xi = 0$, and $\lambda = 2.0$.

Figure 12.5 presents the effects of the accelerating flow α and t^* on the temperature profile. The temperature profile gradually decreases with the increase of α and time (t^*). The temperature profiles are directly related to the energy equation of the right-hand side, which gives the high friction near the wall due to increase of α. In particular, for $\varepsilon = 0.2$, $\varepsilon = 0.2, \xi = 0, \lambda = 2.0$ and Re = 100 the magnitude of temperature profile increases by 26% as α decreases 2.3 to 0.4 at $t^* = 1.5$.

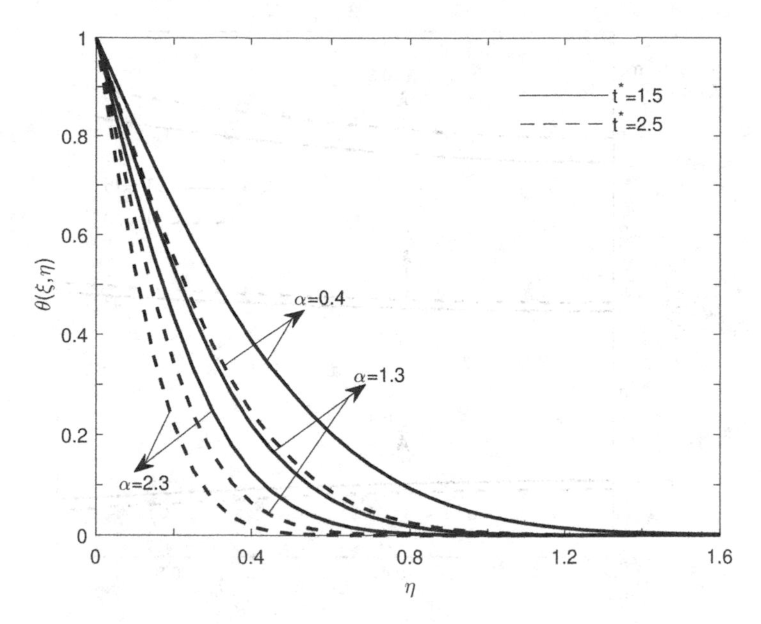

FIGURE 12.5

Variations of α and t^* on θ for $\phi(t^*)$ when $\epsilon = 0.2$, Re = 100, $\xi = 0$ and $\lambda = 2.0$.

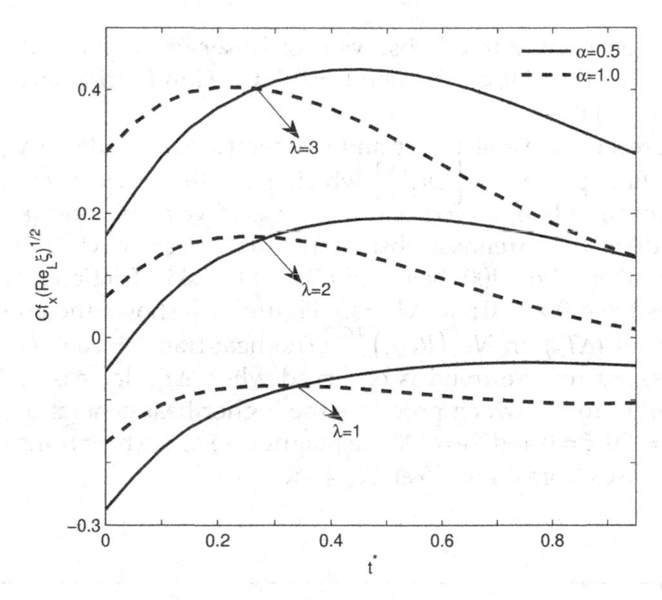

FIGURE 12.6

Variations of α and λ on $\left(C_{fx}\left(Re_L\xi\right)^{1/2}\right)$ for $\phi(t^*)$ when $\epsilon = 0.2$, $\xi = 0.5$ and Re = 1.0.

Figure 12.6 depicts the impacts of λ and accelerating flow α on the skin-friction. For the higher values of (λ >0), the fluid moves faster near the wall and increase the thermal and acceleration of the fluid. Due to the fluid acceleration, the fluid gets hot and moves towards the momentum boundary layers. In fact, the shear stress increased near the wall and the magnitude of the $C_{fx}\left(Re_L\xi\right)^{-1/2}$ is increased. Initially ($t^* \leq 0.4$), the magnitude of $C_{fx}\left(Re_L\xi\right)^{-1/2}$ is very high for higher values of α = 1.0, whereas, the magnitude

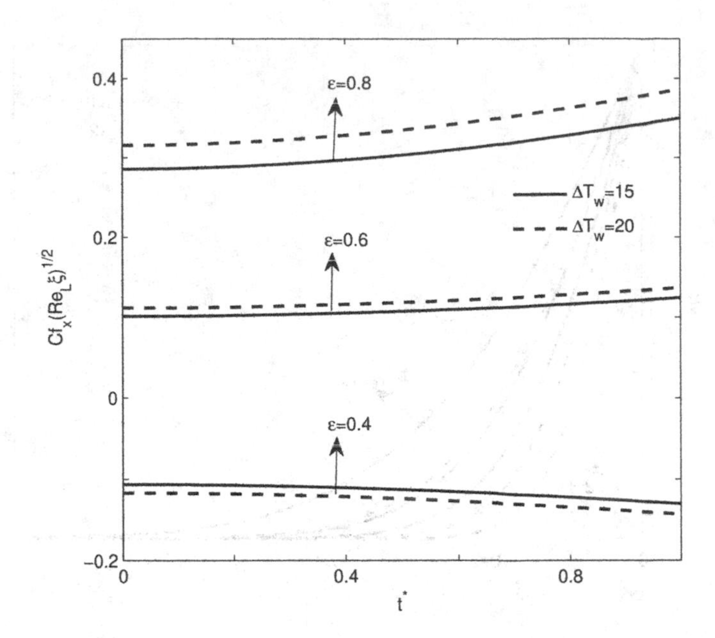

FIGURE 12.7
Variations of ΔT and ϵ on $\left(C_{fx}\left(Re_L\xi\right)^{1/2}\right)$ for $\phi\left(t^*\right)$, $\alpha = 0.5$ when $\xi = 0$, $Re = 100$, and $\lambda = 1.0$.

of $C_{fx}\left(Re_L\xi\right)^{-1/2}$ is decreasing trend observed for lesser value of $\alpha = 0.5$ when $t^* \geq 0.5$. In particular, for $\varepsilon = 0.2$, Re = 1.0, $\xi = 0.5$, and $t^* = 0.2$, the skin-friction increases by 179% as λ increase from 2 to 3 at $\alpha = 0.5$.

Figure 12.7 represents the effects of ε and temperature dependent (ΔT_w) on $C_{fx}\left(Re_L\xi\right)^{1/2}$. The accelerating flow parameter $\left[\varnothing\left(t^*\right)\right]$, which gives the acceleration to the fluid due to this the increasing trends observed with increase of velocity ratio (ε). In case of lesser values of (ε) the decreasing trends is observed with increase of ΔT_w. In particular, for $\xi = 0$, $\alpha = 0.5$, $\xi = 0$, $\lambda = 1.0$ and Re = 100 the magnitude of the skin-friction increases about 195% when ε increases from 0.4 to 0.6 at $\Delta T_w = 15$. Figure 12.8 shows the effects of ε and temperature dependent (ΔT_w) on $Nu_x\left(Re_L\xi\right)^{-1/2}$. The heat-transfer coefficient increases with the increase of ε, but reverse trend is observed when ΔT_w decreases. The velocity ratio (ε) increase from 0.1 to 0.5, which produces the higher heat generation. In particular, for $\alpha = 0.5$, $t^* = 0.3$, $\lambda = 1.0$, $\xi = 0$ and Re = 100 the magnitude of the heat transfer increases about 93%, when ε increases from 0.4 to 0.6 at $\Delta T_w = 15$.

12.5 Conclusions

This work investigated numerically an unsteady laminar fluid flow of stretching sheets the following results are summarized.

- The velocity profile overshoots with an increase of buoyancy parameters (λ) and undershoot velocity profile with an increase of t^*. In particular, for $\varepsilon = 0.4$, $\alpha = 0.5$, Re = 1.0, $t^* = 1$ and $\xi = 0.5$, the magnitude of velocity profile maximum overshoot by 21% between $\lambda = 2$ and $\lambda = 4$.

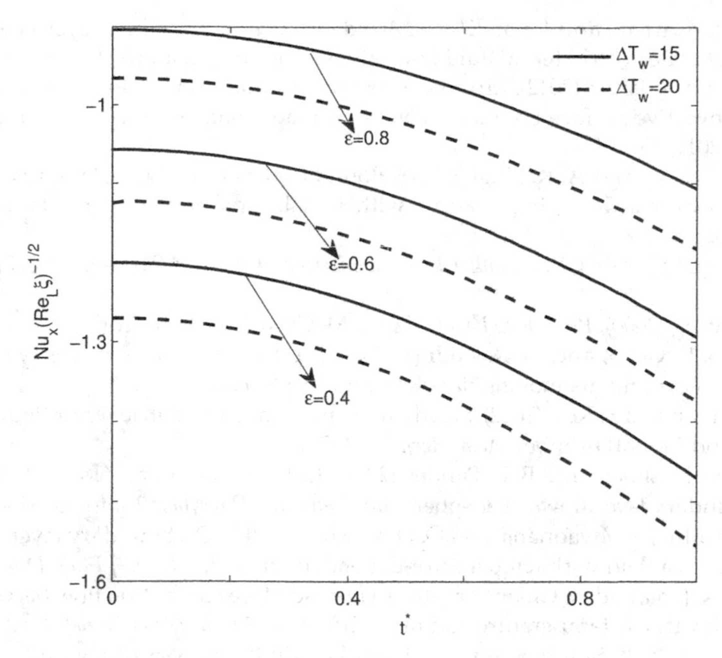

FIGURE 12.8
Variations ΔT and ϵ on $\left(Nu_x\left(Re_L\xi\right)^{-1/2}\right)$ for $\phi\left(t^*\right)$, $\alpha = 0.5$, when Re = 100, $\xi = 0$ and $\lambda = 1.0$.

- The temperature distribution increases with an increase of composite velocity (ε) and decreases with an increase of t^*. For example, the temperature (θ) increases by 10% as t^* decreases from 1 to 0 for $\varepsilon = 0.2, \xi = 0$ and Re=1.0.
- The α acts as suction/injection factor on the velocity and temperature profiles. In particular, for $\varepsilon = 0.2$, Re = 100, $t^* = 1$ and $\xi = 0$, the velocity (F) and temperature (θ) decreased by 23% and 90%, respectively as α increases from $\alpha = 0.5$ to $\alpha = 2.5$ at $\eta = 1$.
- The magnitude of skin friction ($C_{fx}\left(Re_L\xi\right)^{-1/2}$) and heat transfer. ($Nu_x\left(Re_L\xi\right)^{-1/2}$) are increased with an increase of ε. In particular $t^* = 1$ the skin friction coefficient by 129% and heat transfer rate is increased by 117%.

References

1. R. Cortell (2007), Flow and heat transfer in moving fluid over a moving flat surface. *Theor Comput Fluid Dyn* 21, 435–446.
2. T. Fang (2008), A note on the unsteady boundary layer over a flat plate, I. *J. Non-linear Mech* 43, 1007–1011.
3. P. M. Patil and S. Roy (2010), Unsteady mixed convection flow from a vertical plate in parallel free stream: influence of heat generation or absorption, *Int J Heat Mass Transf* 53(21–22), 4749–4756.
4. K. Bhattacharyya, S. Mukhopadhyay, and G. C. Layek (2013), Similarity solution of mixed convective boundary layer slip flow over a vertical plate, *Ain Shams Eng J* 4, 299–305.

5. N. Bachok, A. Ishak, and I. Pop (2013), Mixed convection boundary layer flow over a moving vertical flat plate in an external fluid flow with viscous dissipation effect, *PLoS One* 8(4), e60766.
6. F. Aman and A. Ishak (2012), Mixed convection boundary layer flow towards a vertical plate with a convective surface boundary condition mathematical problems in engineering, *Math Probl Eng* 2012, 453457.
7. I. Govindhasamy and A. K. Singh (2020), Boundary layer flow and stability analysis of forced convection over a diverging channel with variable properties of fluids, *Heat Transf* 49(8), 5050–5065.
8. D. R. Lide (Ed.) (1990), *CRC Handbook of Chemistry and Physics*, 71st ed., CRC Press, BocaRaton, FL.
9. H. Schlichting (2000), *Boundary Layer Theory*, McGraw Hill, New York.
10. I. Pop, R. S. R. Gorla, and M. Rashidi (1992), The effect of variable viscosity on flow and heat transfer to a continuous moving flat plate, *Int J Eng Science* 30, 1–6.
11. P. Saikrishnan and S. Roy (2002), Steady nonsimilar axisymmetric water boundary layer with variable and Prandtl number, *Acta Mech* 157 187–199.
12. S. Roy, P. Saikrishnan, and B. D. Pandey (2009), Influence of double slot suction (injection) into water boundary layer flows over sphere, *Int Commun Heat Mass Transf* 36, 646–650.
13. E. M. Mureithi, J. J. Mwaonanji, and O. D. Makinde (2013), On boundary layer flow over a moving surface in a fluid with temperature-dependent viscosity, *Open J Fluid Dyn* 3, 135–140.
14. S. Das, R. N. Jana, and A. Chamkha (2015), Unsteady free convection flow between two vertical plates with variable temperature and mass diffusion, *J Heat Mass Transf Res* 2 49–58.
15. N. Govindaraj, A. K. Singh, S. Roy, and P. Shukla (2019), Analysis of a boundary layer flow over moving an exponentially stretching surface with variable viscosity and Prandtl number, *Heat Transf* 48(7), 2736–2751.
16. N. Govindaraj, A. K. Singh, and S. Roy (2018), Water boundary layer flow over an exponentially permeable stretching sheet with variable viscosity and Prandtl number, In *International Conference on Mathematical Modelling and Scientific Computation* (pp. 207–217). Springer, Singapore.
17. D. K. Maiti and H. Mandal (2019), Unsteady slip flow past an infinite vertical plate with ramped plate temperature and concentration in the presence of thermal radiation and boyancy, *J Eng Thermophys* 28(3), 431–452.
18. T. Anwar, P. Kumam, Z. Shah, W. Watthayu, and P. Thounthong (2020), Unsteady radiative natural convective MHD nanofluid flow past a porous moving vertical plate with heat source/sink, *Molecules* 25, 854.
19. M. V. Krishna, N. A. Ahamad, and A. J. Chamkha (2021), Radiation absorption on MHD convective flow of nanofluids through vertically travelling absorbent plate, *Ain Shams Eng J* 12, 3043–3056.
20. Y. D. Reddy, B. S. Goud, and M. A. Kumar (2021), Radiation and heat absorption effects on an unsteady MHD boundary layer flow along an accelerated infinite vertical plate with ramped plate temperature in the existence of slip condition, *Partial Diff Equ Appl Math* 4, 100166.
21. M. Ali and F. Al-Yousef (1998), Laminar mixed convection from a moving vertical surface with suction or injection, *Heat Mass Transf* 33(4), 301–306.
22. C. H. Chen (1998), Laminar mixed convection adjacent to vertical, continuously stretching sheets, *Heat Mass Transf* 33, 471–476.
23. A. K. Singh, A. K. Singh, and S. Roy (2019), Analysis of mixed convection in water boundary layer flows over a moving vertical plate with variable viscosity and Prandtl number, *Int. J. Numer. Methods Heat Fluid Flow* 29(2), 602–616.

13

Study of Incompressible Viscous Flow Due to a Stretchable Rotating Disk Through Finite Element Procedure

Anupam Bhandari
University of Petroleum & Energy Studies (UPES)

CONTENTS

13.1 Introduction 197
13.2 Mathematical Formulation 198
13.3 Finite Element Procedure for Numerical Solution 200
13.4 Results and Discussion 203
13.5 Conclusion 211
References 212

13.1 Introduction

The study of three-dimensional incompressible flow in a rotating system has been researched owing to its significant role in rotating equipment. Such types of interesting problems have been solved using the similarity transformation. It reduces the three-dimensional Navier–Stokes equations into nondimensional nonlinear ordinary differential equations. Then the converted nondimensional equations represent the nondimensional velocity distribution for the nondimensional similarity variable of the flow.

Cochran implemented the power series approximations method for the solution of nondimensional differential equations (Cochran, 1934). In 1966, Benton tried to obtain a more accurate solution than Cochran (Benton, 1966). In magnetohydrodynamic flow, Hassan and Attia used a finite difference scheme for solving problems of rotational flow (Aboul-Hassan & Attia, 1997). Turkyilmazoglu used the fourth-order Runge–Kutta scheme for solving the nonlinear coupled differential equations for magnetohydrodynamic stagnation revolving flow with the help of MATHEMATICA Software (Turkyilmazoglu, 2012). Similarity solutions have been obtained through the homotopy analysis method for different types of steady flow in the rotating system considering the influence of variable thickness and slip effects (Rashidi et al., 2011, 2012; Sravanthi, 2019).

Due to the importance of the finite element method over other numerical methods, this method is being commercialized (Reddy & Gartling, 2010). Gupta et al. developed the finite element procedure to investigate the micropolar flow over a shrinking sheet (Gupta et al., 2018). Anjos et al. implemented the finite element method procedure for the flow over a rotating plate (Anjos et al., 2014). Takhar et al. presented the finite element solution for micropolar fluid flow from an enclosed rotating plate (Takhar et al., 2001). Yoseph

DOI: 10.1201/9781003291916-13

and Olek described the solution of compressible flow over a rotating porous disk using asymptotic and finite element approximation (Bar-Yoseph & Olek, 1984). Sheikholeslami and Ganji used the control volume finite element technique to investigate ferrofluid flow and heat transfer in a semiannulus enclosure (Sheikholeslami & Ganji, 2014). Vasu et al. demonstrated the magnetohydrodynamic flow through a stenosed coronary artery using finite element analysis (Vasu et al., 2020). Feistauer et al. demonstrated the finite element solution for nonviscous subsonic irrotational flow (Feistauer et al., 1992).

This work investigates the steady, incompressible, and axisymmetric flow of viscous fluid due to rotating and stretching disks. The influence of rotation and stretching of the disk on the velocity profiles is graphically presented. The finite element procedure is explained for solving nonlinear coupled differential equations of the flow due to a rotating and stretching disk. The problem formulation is presented over a typical element of the solution domain. In the present analysis, the solution of nonlinear differential equations is demonstrated through finite element procedure in COMSOL Multiphysics Software and is compared with the previously reported solutions. The present solution finite element solution is corrected up to ten decimal places.

13.2 Mathematical Formulation

Figure 13.1 demonstrates the flow configuration of viscous over a rotating and stretching plate. Let (u,v,w) be the velocity components in the cylindrical coordinates (r,φ,z) direction, respectively. It is assumed that the disk uniformly rotates with the angular velocity ω about z-axis. Along with the rotation, the disk is stretching at a uniform rate s in the direction of. The governing equations for the steady and axisymmetric flow due to rotating disk are as follows (Benton, 1966; Cochran, 1934; Turkyilmazoglu, 2014):

$$\frac{\partial u}{\partial r}+\frac{u}{r}+\frac{\partial w}{\partial z}=0 \tag{13.1}$$

$$\rho\left[u\frac{\partial u}{\partial r}+w\frac{\partial u}{\partial z}-\frac{v^2}{r}\right]=-\frac{\partial p}{\partial r}+\mu\left[\frac{\partial^2 u}{\partial z^2}+\frac{1}{r}\frac{\partial u}{\partial r}-\frac{u}{r^2}+\frac{\partial^2 u}{\partial r^2}\right] \tag{13.2}$$

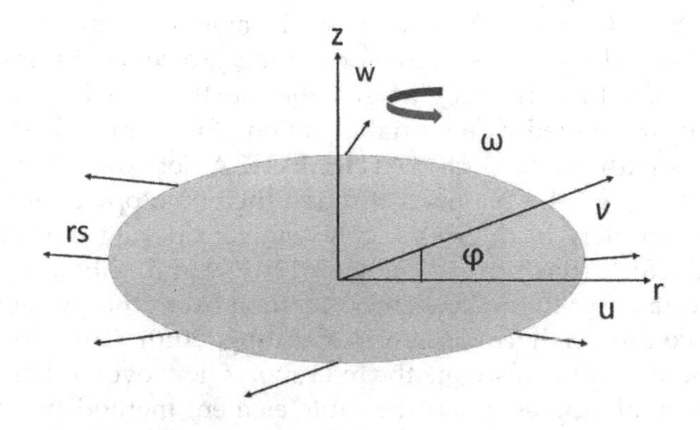

FIGURE 13.1
Flow configuration of viscous fluid due to rotating and stretching disk.

$$\rho\left[u\frac{\partial v}{\partial r}+w\frac{\partial v}{\partial z}+\frac{uv}{r}\right]=\mu\left[\frac{\partial^2 v}{\partial z^2}+\frac{1}{r}\frac{\partial v}{\partial r}+\frac{\partial^2 v}{\partial r^2}-\frac{v}{r^2}\right] \tag{13.3}$$

$$\rho\left[u\frac{\partial w}{\partial r}+w\frac{\partial w}{\partial z}\right]=-\frac{\partial p}{\partial z}+\mu\left[\frac{\partial^2 w}{\partial z^2}+\frac{1}{r}\frac{\partial w}{\partial r}+\frac{\partial^2 w}{\partial r^2}\right] \tag{13.4}$$

With boundary conditions:

$$z=0;\ u=rs,\ v=r\omega,\ w=0;\quad z\to\infty;\ u\to 0,\ v\to 0 \tag{13.5}$$

The following similarity transformation reduces equations (13.1)–(13.5) into nondimensional nonlinear differential equations (Bhandari, 2020; Cochran, 1934; Ram & Bhandari, 2013; Benton, 1966):

$$u=r\omega F(\eta),\ v=r\omega G(\eta),\ w=\sqrt{v\omega}\,H(\eta),\ p-p_\infty=-\rho v\omega P(\eta),\ \eta=\sqrt{\frac{\omega}{v}}\,z \tag{13.6}$$

The reduced nondimensional equations are as follows:

$$\frac{dH}{d\eta}+2F=0 \tag{13.7}$$

$$\frac{d^2F}{d\eta^2}-H\frac{dF}{d\eta}-F^2+G^2=0 \tag{13.8}$$

$$\frac{d^2G}{d\eta^2}-H\frac{dG}{d\eta}-2FG=0 \tag{13.9}$$

With boundary conditions:

$$F(0)=C,\ G(0)=1,\ H(0)=0,\ F(\infty)=0,\ G(\infty)=0 \tag{13.10}$$

where $C=\frac{s}{\omega}$ denotes the stretching strength parameter.

A case $C=0$ demonstrates that there is no stretching in the disk and the problem reduces into the flow due to rotating disk only. However, the present problem can also be investigated for different speeds of rotation of the disk. In this case, the following similarity transformation can be used:

$$u=rsF(\eta),\ v=rsG(\eta),\ w=\sqrt{v\omega}\,H(\eta),\ p-p_\infty=-\rho v\omega P(\eta),\ \eta=\sqrt{\frac{s}{v}}\,z \tag{13.11}$$

In this case, the boundary conditions can be written as

$$F(0)=1,\ G(0)=\Omega,\ H(0)=0,\ F(\infty)=0,\ G(\infty)=0 \tag{13.12}$$

where $\Omega = \dfrac{\omega}{s}$ represents the rotation strength parameter.

A case $\Omega = 0$ represents there is no rotation of the disk and the flow of viscous fluid is appeared due to only stretching. If only stretching of the plate is present, then the three-dimensional flow reduces into the two-dimensional flow.

13.3 Finite Element Procedure for Numerical Solution

Figure 13.2(a) demonstrates the entire domain of the solution and, Figure 13.2(b) represents the element-wise discretization of the domain. The weighted integral form of the above differential equation over a typical element $\Omega^e = \left(\eta_a^e, \eta_b^e\right)$ is (Reddy et al., 2020; Reddy & Gartling, 2010):

$$\int_{\eta_a^e}^{\eta_b^e} W\left\{\frac{dH}{d\eta} + 2F\right\} d\eta = 0 \tag{13.13}$$

$$\int_{\eta_a^e}^{\eta_b^e} W\left\{\frac{d^2F}{d\eta^2} - H\frac{dF}{d\eta} - F^2 + G^2\right\} d\eta = 0 \tag{13.14}$$

$$\int_{\eta_a^e}^{\eta_b^e} W\left\{\frac{d^2G}{d\eta^2} - H\frac{dG}{d\eta} - 2FG\right\} d\eta = 0 \tag{13.15}$$

(a)

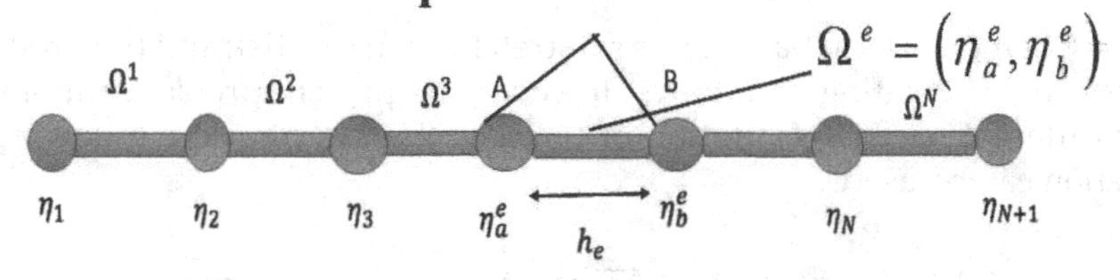

(b)

FIGURE 13.2
(a) Domain of the solution and (b) finite element mesh of the domain

Equations (13.14) and (13.15) can be written as

$$\int_{\eta_a^e}^{\eta_b^e}\left(\frac{dW}{d\eta}\frac{dF}{d\eta}+W\overline{H}\frac{dF}{d\eta}+W\overline{F}F-W\overline{G}G\right)d\eta=\left(W\frac{dF}{d\eta}\right)_{\eta_a^e}^{\eta_b^e} \tag{13.16}$$

$$\int_{\eta_a^e}^{\eta_b^e}\left(\frac{dW}{d\eta}\frac{dG}{d\eta}+W\overline{H}\frac{dG}{d\eta}-2W\overline{F}G\right)d\eta=\left(W\frac{dG}{d\eta}\right)_{\eta_a^e}^{\eta_b^e} \tag{13.17}$$

The above equations, $\overline{F},\overline{G}$ and $\overline{H}$ are nonlinear and considered the known quantity using Lagrange's interpolation formula. Therefore, the system of equations becomes linear to obtain the unknown dependent variables.

For a typical element $\Omega^e=\left(\eta_a^e,\eta_b^e\right)$, we consider the quadratic Lagrange's interpolation functions (J. N. Reddy, 2019; Reddy & Gartling, 2010)

$$F(\eta)\approx F_h^e(\eta)=\sum_{j=1}^{3}F_j^e\psi_j^e(\eta) \tag{13.18}$$

$$G(\eta)\approx G_h^e(\eta)=\sum_{j=1}^{3}G_j^e\psi_j^e(\eta) \tag{13.19}$$

$$H(\eta)\approx H_h^e(\eta)=\sum_{j=1}^{3}H_j^e\psi_j^e(\eta) \tag{13.20}$$

where $\psi_j^e(\eta)$ are the Lagrange's interpolation functions. In the present analysis, quadratic interpolation functions are used, which are defined as follows (J. N. Reddy, 2019):

$$\psi_1^e(\overline{\eta})=\left(1-\frac{\overline{\eta}}{h_e}\right)\left(1-\frac{2\overline{\eta}}{h_e}\right),\ \psi_2^e(\overline{\eta})=\frac{4\overline{\eta}}{h_e}\left(1-\frac{\overline{\eta}}{h_e}\right),\ \psi_3^e(\overline{\eta})=\frac{\overline{\eta}}{h_e}\left(1-\frac{2\overline{\eta}}{h_e}\right) \tag{13.21}$$

where $\overline{\eta}$ denotes the local coordinates. Therefore, the local coordinate $\overline{\eta}$ and global coordinate η are related to each other with the linear relation $\overline{\eta}=\eta-\eta_a^e$.

Consider the weight function $W=\psi_i^e(\eta)$, equations (13.13), (13.16), and (13.17) can be written as

$$\int_{\eta_a^e}^{\eta_b^e}\left\{\psi_1^e\frac{d}{d\eta}\left(H_1^e\psi_1^e+H_2^e\psi_2^e+H_3^e\psi_3^e\right)+2\psi_1^e\left(F_1^e\psi_1^e+F_2^e\psi_2^e+F_3^e\psi_3^e\right)\right\}d\eta=0 \tag{13.22}$$

$$\int_{\eta_a^e}^{\eta_b^e}\left\{\psi_2^e\frac{d}{d\eta}\left(H_1^e\psi_1^e+H_2^e\psi_2^e+H_3^e\psi_3^e\right)+2\psi_2^e\left(F_1^e\psi_1^e+F_2^e\psi_2^e+F_3^e\psi_3^e\right)\right\}d\eta=0 \tag{13.23}$$

$$\int_{\eta_a^e}^{\eta_b^e}\left\{\psi_3^e\frac{d}{d\eta}\left(H_1^e\psi_1^e+H_2^e\psi_2^e+H_3^e\psi_3^e\right)+2\psi_3^e\left(F_1^e\psi_1^e+F_2^e\psi_2^e+F_3^e\psi_3^e\right)\right\}d\eta=0 \tag{13.24}$$

$$\int_{\eta_a^e}^{\eta_b^e}\left\{\frac{d\psi_1^e}{d\eta}\frac{d}{d\eta}\left(\sum_{i=1}^{3}F_i^e\psi_i^e\right)+\psi_1^e\overline{H}\frac{d}{d\eta}\left(\sum_{i=1}^{3}F_i^e\psi_i^e\right)+\psi_1^e\overline{F}\left(\sum_{i=1}^{3}F_i^e\psi_i^e\right)-\psi_1^e\overline{G}\left(\sum_{i=1}^{3}G_i^e\psi_i^e\right)\right\}d\eta$$

$$=\psi_1^e\left(\eta_a^e\right)Q_1^e+\psi_1^e\left(\eta_b^e\right)Q_2^e \tag{13.25}$$

$$\int_{\eta_a^e}^{\eta_b^e}\left\{\frac{d\psi_2^e}{d\eta}\frac{d}{d\eta}\left(\sum_{i=1}^{3}F_i^e\psi_i^e\right)+\psi_2^e\overline{H}\frac{d}{d\eta}\left(\sum_{i=1}^{3}F_i^e\psi_i^e\right)+\psi_2^e\overline{F}\left(\sum_{i=1}^{3}F_i^e\psi_i^e\right)-\psi_2^e\overline{G}\left(\sum_{i=1}^{3}G_i^e\psi_i^e\right)\right\}d\eta$$

$$=\psi_2^e\left(\eta_a^e\right)Q_1^e+\psi_2^e\left(\eta_b^e\right)Q_2^e \tag{13.26}$$

$$\int_{\eta_a^e}^{\eta_b^e}\left\{\frac{d\psi_3^e}{d\eta}\frac{d}{d\eta}\left(\sum_{i=1}^{3}F_i^e\psi_i^e\right)+\psi_3^e\overline{H}\frac{d}{d\eta}\left(\sum_{i=1}^{3}F_i^e\psi_i^e\right)+\psi_3^e\overline{F}\left(\sum_{i=1}^{3}F_i^e\psi_i^e\right)-\psi_3^e\overline{G}\left(\sum_{i=1}^{3}G_i^e\psi_i^e\right)\right\}d\eta$$

$$=\psi_3^e\left(\eta_a^e\right)Q_1^e+\psi_3^e\left(\eta_b^e\right)Q_2^e \tag{13.27}$$

$$\int_{\eta_a^e}^{\eta_b^e}\left\{\frac{d\psi_1^e}{d\eta}\frac{d}{d\eta}\left(\sum_{i=1}^{3}G_i^e\psi_i^e\right)+\psi_1^e\overline{H}\frac{d}{d\eta}\left(\sum_{i=1}^{3}G_i^e\psi_i^e\right)-2\psi_1^e\overline{F}\left(\sum_{i=1}^{3}G_i^e\psi_i^e\right)\right\}d\eta$$

$$=\psi_1^e\left(\eta_a^e\right)Q_3^e+\psi_1^e\left(\eta_b^e\right)Q_4^e \tag{13.28}$$

$$\int_{\eta_a^e}^{\eta_b^e}\left\{\frac{d\psi_2^e}{d\eta}\frac{d}{d\eta}\left(\sum_{i=1}^{3}G_i^e\psi_i^e\right)+\psi_2^e\overline{H}\frac{d}{d\eta}\left(\sum_{i=1}^{3}G_i^e\psi_i^e\right)-2\psi_2^e\overline{F}\left(\sum_{i=1}^{3}G_i^e\psi_i^e\right)\right\}d\eta$$

$$=\psi_2^e\left(\eta_a^e\right)Q_3^e+\psi_2^e\left(\eta_b^e\right)Q_4^e \tag{13.29}$$

$$\int_{\eta_a^e}^{\eta_b^e}\left\{\frac{d\psi_3^e}{d\eta}\frac{d}{d\eta}\left(\sum_{i=1}^{3}G_i^e\psi_i^e\right)+\psi_3^e\overline{H}\frac{d}{d\eta}\left(\sum_{i=1}^{3}G_i^e\psi_i^e\right)-2\psi_3^e\overline{F}\left(\sum_{i=1}^{3}G_i^e\psi_i^e\right)\right\}d\eta$$

$$=\psi_3^e\left(\eta_a^e\right)Q_3^e+\psi_3^e\left(\eta_b^e\right)Q_4^e \tag{13.30}$$

In the above equations Q_1^e, Q_2^e, Q_3^e and Q_4^e represents the secondary variables. These are defined as follows:

$$Q_1^e=-\left(\frac{dF}{d\eta}\right)_{\eta_a^e},\ Q_2^e=-\left(\frac{dF}{d\eta}\right)_{\eta_b^e},\ Q_3^e=-\left(\frac{dG}{d\eta}\right)_{\eta_a^e},\ Q_4^e=-\left(\frac{dG}{d\eta}\right)_{\eta_a^e} \tag{13.31}$$

The matrix form of the algebraic equations can be written as:

$$\begin{bmatrix} [A_{ij}^e] & [B_{ij}^e] & [C_{ij}^e] \\ [D_{ij}^e] & [E_{ij}^e] & [F_{ij}^e] \\ [M_{ij}^e] & [N_{ij}^e] & [O_{ij}^e] \end{bmatrix} \begin{bmatrix} F \\ G \\ H \end{bmatrix} = \begin{bmatrix} 0 \\ Q_1^e + Q_2^e \\ Q_3^e + Q_4^e \end{bmatrix} \quad (13.32)$$

$$A_{ij}^e = 2\int_{\eta_a^e}^{\eta_b^e} \psi_i^e \psi_j^e d\eta,\ C_{ij}^e = \int_{\eta_a^e}^{\eta_b^e} \psi_i^e \frac{d\psi_j^e}{d\eta} d\eta,\ D_{ij}^e = \int_{\eta_a^e}^{\eta_b^e} \left(\frac{d\psi_i^e}{d\eta}\frac{d\psi_j^e}{d\eta} + \bar{H}\psi_i^e \frac{d\psi_j^e}{d\eta} + \bar{F}\psi_i\psi_j \right) d\eta,$$

$$E_{ij}^e = -\int_{\eta_a^e}^{\eta_b^e} \bar{G}\psi_i\psi_j d\eta,\ N_{ij}^e = \int_{\eta_a^e}^{\eta_b^e} \left(\frac{d\psi_i^e}{d\eta}\frac{d\psi_j^e}{d\eta} + \bar{H}\psi_i^e \frac{d\psi_j^e}{d\eta} - 2\bar{F}\psi_i\psi_j \right) d\eta \quad (13.33)$$

$$B_{ij}^e = 0,\ F_{ij}^e = 0,\ M_{ij}^e = 0,\ O_{ij}^e = 0,$$

After the assembly of element equations, a set of nonlinear equations can be obtained. We assume the functions for $\bar{F}$ and $\bar{H}$ at a lower iteration level to obtain a system of linear equations. The present domain of the solution is considered (0,10) to obtain the asymptotic solution. If we consider the element length 0.001, then 10,000 element equations need to assemble through the finite element method. This attempt is used in COMSOL Multiphysics Software and then automatic Newton is used to obtain the numerical results.

13.4 Results and Discussion

Current results describe finite element solution of similarity equations for flow over a rotating and stretching plate. In the previous section, the mathematical procedure of the finite element procedure is presented. The finite element equations are obtained here for a typical element Ω^e. The size and geometry of each element same, therefore equations of each element can be assembled in a matrix form using quadratic Lagrange's interpolation. The graphical representation of the present model is obtained using COMSOL Multiphysics Software and is compared with the previously published solutions.

Figures 13.3–13.5 represent the radial, tangential, and axial velocity profiles with the variation of the parameter C. Cochran (1934) solved the similarity equations for $C = 0$ and $\Omega = 1$ using power series approximation and obtained the solution up to three decimal places. In the literate (Benton, 1966; Cochran, 1934), the radial, tangential, and axial velocities are demonstrated in the domain $(0, 4.4)$. However, in the present solution, the domain of the similarity solution is $(0, 10)$ for the variation of stretching and rotation parameters. The present solution represents that the velocities reach asymptotically zero and this trend of the graph was absent in the previous study due to its small domain. Escalating the strength of the stretching parameter enhances the radial and axial velocities. Stretching of the disk favors the two-dimensional flow in the radial and axial directions. Therefore, the radially stretch of the disk diminishes the azimuthal velocity profile. Figures 13.6 and 13.7 represents the radial and axial velocity profiles for different values of stretching strength parameter. In this case, the rotation of the disk is considered zero. The stretching of the disk without rotation produces only a two-dimensional flow. The comparative study of rotation and stretching indicates that the stretching parameter influences the radial and axial velocities more than the rotation parameter. Stretching of the disk increases the axial velocity more than the rotation of the disk as shown in Figures 13.5, 13.7, and 13.10. It represents the dominance of stretching effects on the velocity.

Figures 13.8–13.10 represent the radial, tangential, and axial velocity profiles, respectively for variation of the strength of rotation parameter along with constant radial stretching. Increasing the rotation parameter enlarges the momentum boundary layer. At $\Omega = 0$, the tangential velocity becomes zero which reduces the three-dimensional boundary layer flow into the flow in the radial and axial directions only.

Table 13.1 represents the comparison of the finite element method with previous results (Bachok et al., 2011; Benton, 1966; Cochran, 1934; Kelson & Desseaux, 2000; Turkyilmazoglu, 2014). In references (Benton, 1966; Cochran, 1934), the similarity equations were solved using power series approximation and recurrence relation method and the solution was not achieved to a higher degree of accuracy due to the unavailability of a high computational computer. However, using computational software, the similarity equations were numerically solved using the finite difference scheme, shooting method, and Runge–Kutta fourth-order method (Bachok et al., 2011; Kelson & Desseaux, 2000; Turkyilmazoglu, 2014). Using the finite element method, we obtained the solution with better accuracy in comparison to the previous results. The derivative of the radial and tangential velocity profiles are shown in Figures 13.11 and 13.12.

Tables 13.2 and 13.3 show the values of velocities and derivatives of radial and tangential velocities. The values of Tables 13.2 and 13.3 can be compared with the finite element

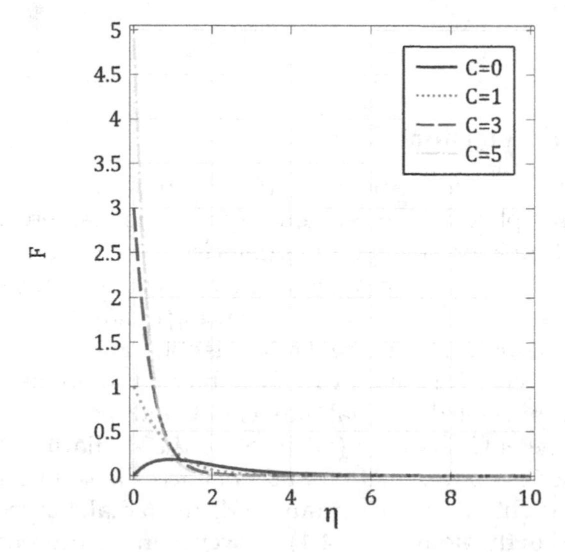

FIGURE 13.3
Influence of stretching on the radial velocity profile.

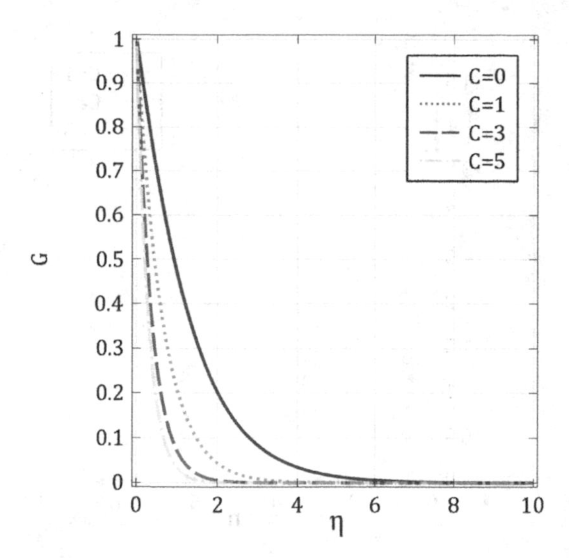

FIGURE 13.4
Influence of stretching on the tangential velocity profile.

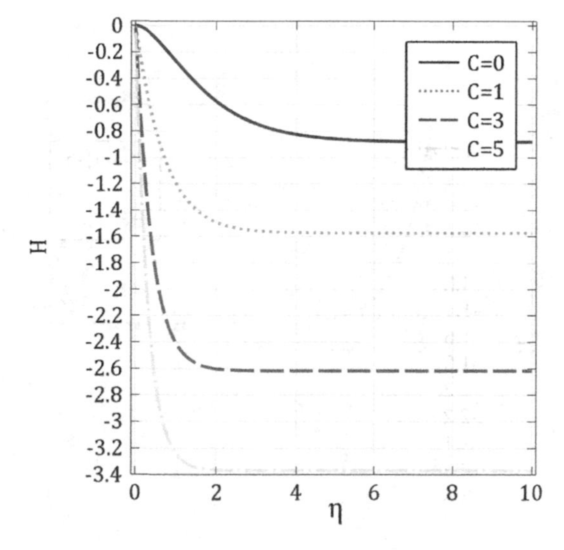

FIGURE 13.5
Influence of stretching on the axial velocity profile.

solution presented in Tables 13.4 and 13.5. The finite element method improves the previous solution of incompressible viscous flow due to the rotating disk. The degree of accuracy is higher as compared to previous methods. The present solution represents better asymptotic behaviors since the domain of the solution is extended.

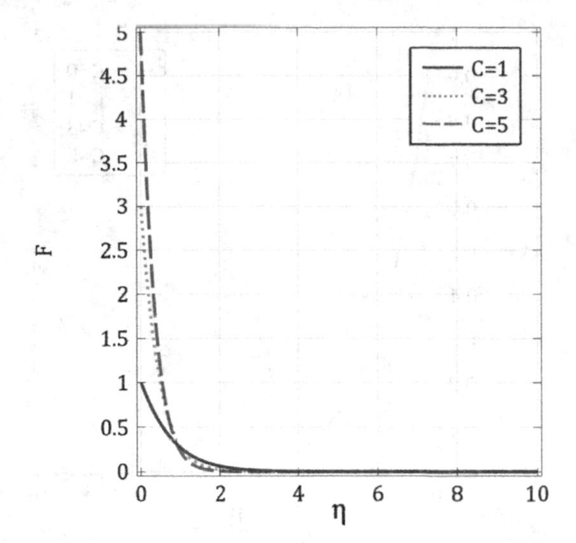

FIGURE 13.6
Influence of stretching in the absence of the rotation of the disk on the radial velocity profile.

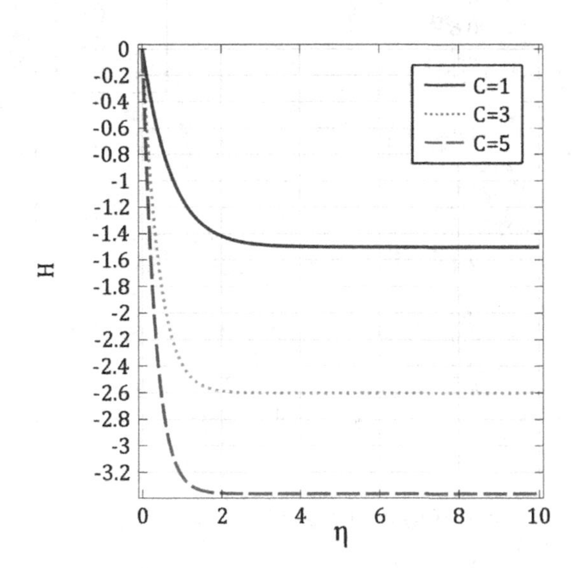

FIGURE 13.7
Influence of stretching in the absence of the rotation of the disk on the axial velocity profile.

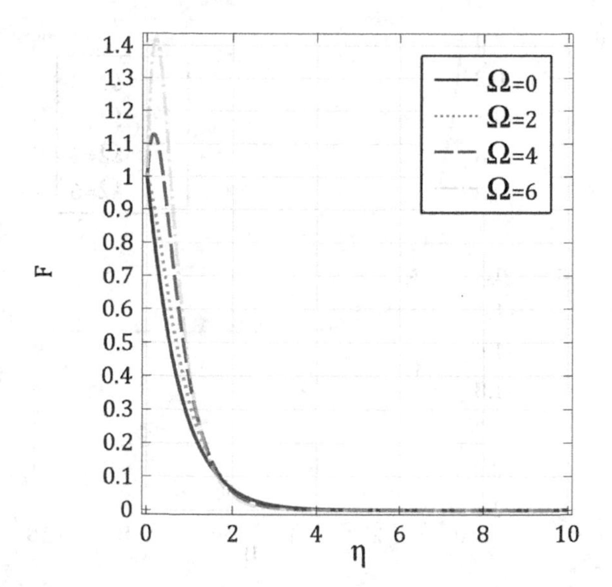

FIGURE 13.8
Influence of rotation of the disk in the presence of constant stretching on the radial velocity profile.

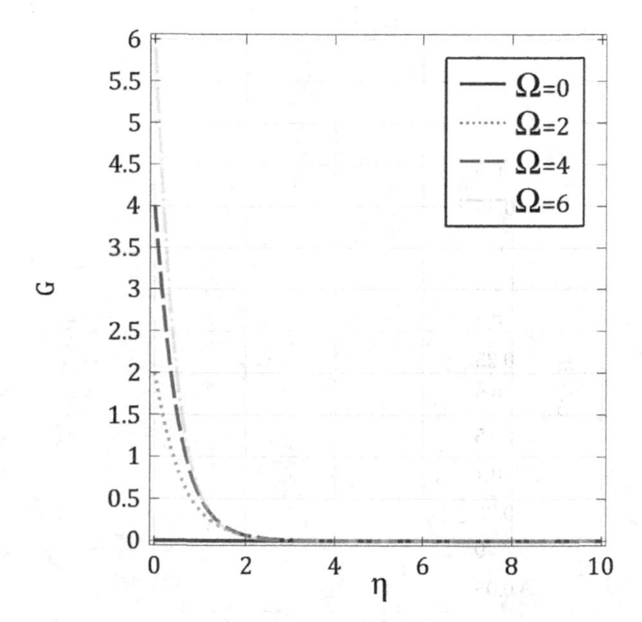

FIGURE 13.9
Influence of rotation of the disk in the presence of constant stretching on the tangential velocity profile.

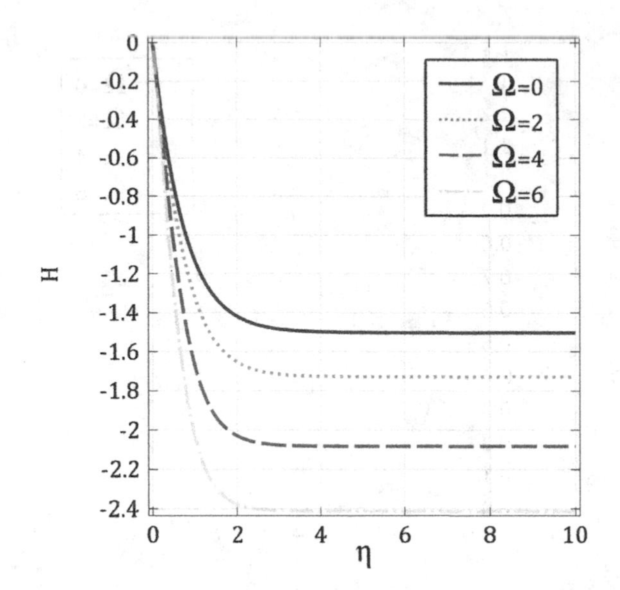

FIGURE 13.10
Influence of rotation of the disk in the presence of constant stretching on the axial velocity profile.

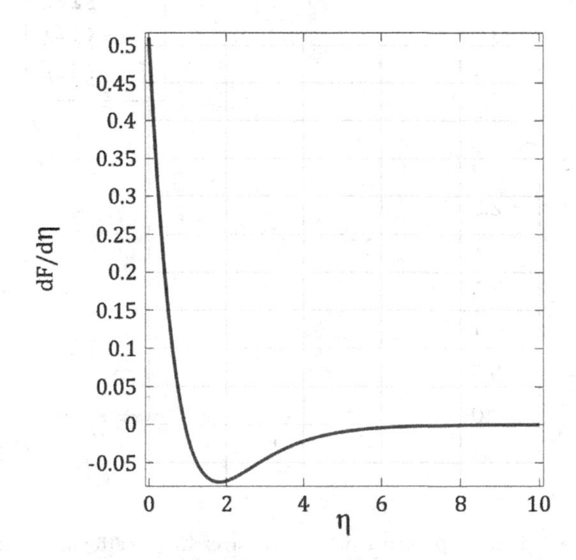

FIGURE 13.11
Derivative of the radial velocity profile at $\Omega = 1$ and $C = 0$.

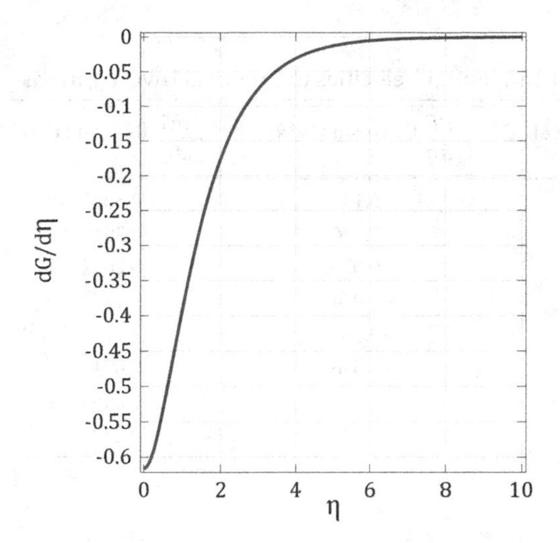

FIGURE 13.12
Derivative of the tangential velocity profile at $\Omega = 1$ and $C = 0$.

TABLE 13.1
Comparison of the Finite Element Solution with Previous Results

	$\left[\frac{dF}{d\eta}\right]_{\eta=0}$	$\left[\frac{dG}{d\eta}\right]_{\eta=0}$	H_∞
Cochran (1934)	0.510	–0.616	–0.844
Benton (1966)	0.5102	–0.6159	–0.8845
Kelson & Desseaux(2000)	0.510233	–0.65922	–0.884474
Bachok et al. (2011)	0.5102	–0.6159	–
Turkyilmazoglu (2014)	0.51023262	–0.61592201	–0.88447411
Present Result	0.5102134911	–0.6159097466	–0.8822962810

TABLE 13.2
Steady-State Radial, Tangential and Axial Velocities in Previous Studies

η	Benton (1966)	Cochran (1934)	G Benton (1966)	G Cochran (1934)	Benton (1966)	$-H$ Cochran (1934)
0.50	0.1536	0.154	0.7076	0.708	0.0919	0.092
1.0	0.1802	0–180	0.4766	0.468	0.2655	0.266
1.5	0.1559	0–156	0.3132	0.313	0.4357	0.435
2	0.1189	0–118	0.2033	0.203	0.5732	0.572
2.5	0.0848	0–084	01313	0.131	0.6745	0.674
3	0.0581	0–058	0.0845	0.083	0.7452	0.746
3.5	0.0389	–	0.0544	–	0.7851	–
4	0.0257	0–026	0.0349	0.035	0.8251	0.826

TABLE 13.3

Derivative of Radial and Tangential Velocities in Previous Investigations

η	$\frac{dF}{d\eta}$ Benton (1966)	$\frac{dF}{d\eta}$ Cochran (1934)	$-\frac{dG}{d\eta}$ Benton (1966)	$-\frac{dG}{d\eta}$ Cochran (1934)
0.50	0.1467	0.147	0.5321	0.532
1.0	–0.0157	–0.016	0.3911	0.391
1.5	–0.0693	–0.070	0.2677	0.268
2	–0.0739	–0.074	0.1771	0.177
2.5	–0.0643	–0.061	0.1153	0.116
3	–0.0455	–0.046	0.0745	0.075
3.5	–0.0319	–	0.0480	–
4	–0.0216	–0.022	0.0309	0.031

TABLE 13.4

Steady-State Radial, Tangential, and Axial Velocities Obtained Through Finite Element Procedure

η	F	G	H
0.50	0.1536131403	0.7075852568	–0.0918753802
1.0	0.1801341021	0.4766341550	–0.2654525478
1.5	0.1558260932	0.3129478876	–0.3189416011
2	0.1187269872	0.2031634678	–0.5733460188
2.5	0.0846300712	0.1311306716	–0.6744852792
3	0.0579977303	0.0844028956	–0.7451407153
3.5	0.0387932911	0.0542459886	–0.7929703003
4	0.0255481732	0.0348269501	–0.8247142674
4.5	0.0166514271	0.0223339711	–0.8455125851
5	0.0107715388	0.0142996297	–0.8590187475
5.5	0.0069307523	0.00913324178	–0.8677299880
6	0.0044236713	0.00581107831	–0.8733141462
6.5	0.0028043468	0.0036747368	–0.8768694802
7	0.0017585143	0.0023008742	–0.8791128449
7.5	0.0010841529	0.0014173033	–0.8805095641
8	6.497563E-4	8.4901977E-4	–0.8813606353
8.5	3.701088E-4	4.8349759E-4	–0.8818603319
9	1.901497E-4	2.483802E-4	–0.8821338701
9.5	7.436672E-5	9.71374E-5	–0.8822618862
10	0	0	–0.8822962810

TABLE 13.5

Derivative of Radial and Tangential Velocities Through Finite Element Procedure

η	$\frac{dF}{d\eta}$	$-\frac{dG}{d\eta}$
0.50	0.1467227786	0.5321338435
1.0	−0.0157293502	0.3911387287
1.5	−0.0693942529	0.2675140128
2	−0.0739028582	0.1770084427
2.5	−0.0611822145	0.1152706230
3	−0.0454564158	0.0745204320
3.5	−0.0318893113	0.0480213081
4	−0.0216399180	0.0309029877
4.5	−0.0143921656	0.0198760889
5	−0.0094532799	0.0127814181
5.5	−0.0061608755	0.0082188276
6	−0.0039953788	0.0052850668
6.5	−0.0025829536	0.0033986920
7	−0.0016665482	0.0021857371
7.5	−0.001073940	0.0014057558
8	−6.915250E-4	9.041639E-4
8.5	−4.450734E-4	5.815783E-4
9	−2.863757E-4	3.741021E-4
9.5	−1.842369E-4	2.406515E-4
10	−1.1862427E-4	1.549458E-4

13.5 Conclusion

The present investigation describes the flow of viscous fluid over a rotating and stretching plate. Without rotation of the plate, the three-dimensional flow reduces into a two-dimensional flow case. However, in the absence of the stretching, the flow remains three-dimensional. The present results show that the influence of stretching on the radial and axial velocity profiles is more dominant than the rotation of the disk. The results in the present solution are improved to a higher degree of accuracy as compared to the previous studies.

References

Aboul-Hassan, A. L., & Attia, H. A. (1997). Flow due to a rotating disk with Hall effect. *Physics Letters, Section A: General, Atomic and Solid State Physics, 228*(4–5), 286–290. https://doi.org/10.1016/S0375–9601(97)00086–8

Anjos, G. R., Mangiavacchi, N., & Pontes, J. (2014). Three-dimensional finite element method for rotating disk flows. *Journal of the Brazilian Society of Mechanical Sciences and Engineering, 36*(4), 709–724. https://doi.org/10.1007/s40430-013-0120–0

Bachok, N., Ishak, A., & Pop, I. (2011). Flow and heat transfer over a rotating porous disk in a nanofluid. *Physica B: Physics of Condensed Matter, 406*(9), 1767–1772. https://doi.org/10.1016/j.physb.2011.02.024

Bar-Yoseph, P., & Olek, S. (1984). Asymptotic and finite element approximations for heat transfer in rotating compressible flow over an infinite porous disk. *Computers and Fluids, 12*(3), 177–197. https://doi.org/10.1016/0045–7930(84)90003–3

Benton, E. R. (1966). On the flow due to a rotating disk. *Journal of Fluid Mechanics, 24*(4), 781–800. https://doi.org/10.1017/S0022112066001009

Bhandari, A. (2020). Effect of magnetic field dependent viscosity on the unsteady ferrofluid flow due to a rotating disk. *International Journal of Applied Mechanics and Engineering, 25*(2), 22–39. https://doi.org/10.2478/ijame–2020–0018

Cochran, W. G. (1934). The flow due to a rotating disc. *Mathematical Proceedings of the Cambridge Philosophical Society, 30*(3), 365–375. https://doi.org/10.1017/S0305004100012561

Feistauer, M., Felcman, J., Rokyta, M., & Vlášek, Z. (1992). Finite-element solution of flow problems with trailing conditions. *Journal of Computational and Applied Mathematics, 44*(2), 131–166. https://doi.org/10.1016/0377–0427(92)90008–L

Gupta, D., Kumar, L., Anwar Bég, O., & Singh, B. (2018). Finite element analysis of MHD flow of micropolar fluid over a shrinking sheet with a convective surface boundary condition. *Journal of Engineering Thermophysics, 27*(2), 202–220. https://doi.org/10.1134/S1810232818020078

Kelson, N., & Desseaux, A. (2000). Note on porous rotating disk flow. *ANZIAM Journal, 42*, 837. https://doi.org/10.21914/anziamj.v42i0.624

Ram, P., & Bhandari, A. (2013). Negative viscosity effects on ferrofluid flow due to a rotating disk. *International Journal of Applied Electromagnetics and Mechanics, 41*(4), 467–478. https://doi.org/10.3233/JAE–121637

Rashidi, M. M., Hayat, T., Erfani, E., Mohimanian Pour, S. A., & Hendi, A. A. (2011). Simultaneous effects of partial slip and thermal-diffusion and diffusion-thermo on steady MHD convective flow due to a rotating disk. *Communications in Nonlinear Science and Numerical Simulation, 16*(11), 4303–4317. https://doi.org/10.1016/j.cnsns.2011.03.015

Rashidi, M. M., Mohimanian Pour, S. A., Hayat, T., & Obaidat, S. (2012). Analytic approximate solutions for steady flow over a rotating disk in porous medium with heat transfer by homotopy analysis method. *Computers and Fluids, 54*(1), 1–9. https://doi.org/10.1016/j.compfluid.2011.08.001

Reddy, J. N. (2019). *Introduction to the Finite Element Method, Fourth Edition.* https://www.accessengineeringlibrary.com/content/book/9781259861901

Reddy, J. N., & Gartling, D. K. (2010). The finite element method in heat transfer and fluid dynamics. In *The Finite Element Method in Heat Transfer and Fluid Dynamics, Third Edition.* https://books.google.co.in/books?hl=en&lr=&id=sv0VKLL5lWUC&oi=fnd&pg=PP1&dq=Finite+element+method+in+Fluid+Mechanics+JN+Reddy&ots=r-MY7AjsSs&sig=OrMQdhVV45vXpSoYkzaOa_z-qIU&redir_esc=y#v=onepage&q=Finite element method in Fluid Mechanics JN Reddy&f=false

Reddy, J. N., Nampally, P., & Srinivasa, A. R. (2020). Nonlinear analysis of functionally graded beams using the dual mesh finite domain method and the finite element method. *International Journal of Non-Linear Mechanics, 127.* https://doi.org/10.1016/j.ijnonlinmec.2020.103575

Sheikholeslami, M., & Ganji, D. D. (2014). Ferrohydrodynamic and magnetohydrodynamic effects on ferrofluid flow and convective heat transfer. *Energy, 75*, 400–410. https://doi.org/10.1016/j.energy.2014.07.089

Sravanthi, C. S. (2019). Effect of nonlinear thermal radiation on silver and copper water nanofluid flow due to a rotating disk with variable thickness in the presence of nonuniform heat source/sink using the homotopy analysis method. *Heat Transfer - Asian Research, 48*(8), 4033–4048. https://doi.org/10.1002/htj.21581

Takhar, H. S., Bhargava, R., & Agarwal, R. S. (2001). Finite element solution of micropolar fluid flow from an enclosed rotating disc with suction and injection. *International Journal of Engineering Science, 39*(8), 913–927. https://doi.org/10.1016/S0020-7225(00)00075-6

Turkyilmazoglu, M. (2012). Three dimensional MHD stagnation flow due to a stretchable rotating disk. *International Journal of Heat and Mass Transfer, 55*(23–24), 6959–6965. https://doi.org/10.1016/j.ijheatmasstransfer.2012.05.089

Turkyilmazoglu, M. (2014). Nanofluid flow and heat transfer due to a rotating disk. *Computers and Fluids, 94*, 139–146. https://doi.org/10.1016/j.compfluid.2014.02.009

Vasu, B., Dubey, A., & Bég, O. A. (2020). Finite element analysis of non-Newtonian magnetohemodynamic flow conveying nanoparticles through a stenosed coronary artery. *Heat Transfer - Asian Research, 49*(1), 33–66. https://doi.org/10.1002/htj.21598

14

Instability of a Viscoelastic Cylindrical Jet: The VCVPF Theory

Mukesh Kumar Awasthi
Babasaheb Bhimrao Ambedkar University

Sudhir Kumar Pundir
S. D. (P.G.) College

Manu Devi
Motherhood University

Dhananjay Yadav
University of Nizwa

Vivek Kumar
Shri Guru Ram Rai (PG) College

A. K. Singh
VIT University

CONTENTS

14.1 Introduction 215
14.2 Formulation of the Problem 217
14.3 Viscoelastic Potential Flow (VPF) Analysis 218
14.4 Viscoelastic Contribution for the Viscoelastic Potential Flow (VCVPF) Analysis ... 219
14.5 Dissipation Calculation for the Cylindrical Jet of Viscoelastic Fluid 222
14.6 Dimensionless Form of the Dispersion Relation 224
14.7 Results and Discussions 227
14.8 Conclusions 229
References 229

14.1 Introduction

It is necessary to know the breakup behavior of liquid jet of non-Newtonian fluids because this phenomenon has many engineering applications, including fertilizers, inkjets, paint leveling, and roll coating, etc. The cylindrical free surface containing jets of viscoelastic fluids can be applied in many engineering processes such as oil-drilling fiber spinning, bottle-filling, etc.

DOI: 10.1201/9781003291916-14

Viscoelastic fluids are non-Newtonian fluids that can display a response that resembles that of a viscous liquid under some circumstances or the response of an elastic solid under other circumstances. The most common type of fluids with visco-elasticity is polymeric liquids: melts and solutions of polymers. Middleman [1] considered the stability of a jet in the cylindrical flow of three-constant linearized Oldroyd fluid for axisymmetric disturbances in an inviscid medium. Goldin et al. [2] extended the study of Middleman [1] for the generalized viscoelastic material. Bousfield et al. [3] considered the combined effect of nonlinearity and surface tension on the breakup of viscoelastic filament. Later, Chang et al. [4] extended the outcomes of Bousfield et al. [3] by performing an extended wave investigation of the stretching dynamics of bead string filament for FENE as well as Oldroyd-B fluids. Brenn et al. [5] analyzed the stability of non-Newtonian liquid linearly taking the effects of aerodynamic forces as well as surface tension. El-Dib and Moatimid [6] investigated the instability of a cylindrical interface of viscoelastic fluids which supports free-surface currents. Moatimid [7] also considered the nonlinear instability of two dielectric viscoelastic fluids. The linear stability investigation of a viscoelastic liquid jet has been studied by Liu and Liu [8,9] taking both symmetrical as well as unsymmetrical disturbances.

In 1994, Joseph and Liao [10] have given the idea of viscoelastic flow theory. In the viscoelastic flow concept, tangential elements of stresses which are irrotational assumed to be zero and normal stresses balance has to include viscosity. This concept does not require a no-slip condition at rigid boundaries. Several vorticity and theorems of circulation of inviscid potential flow are also valid in viscous potential flow theory. At some finite value of Reynolds number, viscous potential flow theory produces excellent physical results in many cases. A theory, in which both pressure and motion are irrotational and normal stresses balance equation was used to include viscosity at the interface, is termed as viscoelastic potential flow (VPF). The stability of viscous liquid jets into incompressible and viscous gases and liquids using viscous flow theory has been measured by Funada et al. [11]. They showed that instability may be motivated by Kelvin–Helmholtz instability (KHI) because of the difference in the velocity at the interface and a neck-down because of capillary instability.

At the high Weber number, the Rayleigh–Taylor stability of viscoelastic fluids was examined by Joseph et al. [12]. They found that the majority of unstable waves, are the functions of the retardation time and it fits into experimental data excellently if the ratio of the relaxation time to that of retardation time is of order 10^3. Funada and Joseph [13] study capillary instability of viscoelastic fluid of Maxwell-type and observe that the growth is bigger for the viscoelastic fluid as compared to the similar Newtonian fluids. The stability of a cylindrical jet of viscoelastic fluid has been considered by Awasthi et al. [14] using potential theory and observed that the elastic property enhances the instability.

If the analysis is based on VPF theory; normal stresses balance has to include viscosity and there is no contribution of the tangential element of viscous stresses. An additional way to analyze the stability of viscous fluids by using potential flow is the dissipation approximation. The dissipation approximation is based on an assessment of the equations that govern the progress of energy in fields. The coefficient of the dissipation integral includes viscosity and estimates, certain viscous stress at the boundaries. This technique has been used by Lamb [15] and by Funada et al. [16] to study the stability of cylindrical incompressible jet into incompressible fluids, and also by Wang et al. [17,18] to analyze the capillary instability of two liquids.

These two theories, VPF and DM, are not the same and, therefore, it is important to study the difference between the VPF and dissipation approximation. It was established by Joseph et al. [19] that the above two theories could provide identical observations if extra viscous stress in terms of a viscoelastic modification of the extra irrotational pressure, which was computed through a technique based on the concept that this extra contribution is developed in a very thin boundary layer. The theory based on added extra pressure, which gives almost identical results as given by dissipation approximation, is termed VCVPF (viscoelastic contribution of VPF). Therefore, VCVPF is the VPF adding extra pressure. In this paper, we use VCVPF to study the stability of a cylindrical jet of viscoelastic fluid moving in a viscous medium.

14.2 Formulation of the Problem

A liquid cylindrical jet of viscoelastic fluid of density ρ_1, viscosity μ_1, and of signifying radius R moves with an unvarying axial velocity U in an infinite viscous fluid of density ρ_2, viscosity μ_2 with a polar cylindrical reference frame (r,θ,z). The viscoelastic fluid cylinder is lying in the region $0 \le r < R+\eta$ and $-\infty < z < \infty$ where $\eta = \eta(\theta,z,t)$ represents the interfacial displacement. It is assumed that σ denotes surface tension at the interface of two fluids.

If we impose small asymmetric perturbations to the balanced state, the equation of the interface can be denoted as

$$f(r,\theta,z,t) = r - R - \eta(\theta,z,t) = 0. \tag{14.1}$$

the outward unit normal at the interface is given by

$$\boldsymbol{n} = \left(\boldsymbol{e}_r - \frac{1}{r}\frac{\partial\eta}{\partial\theta}\boldsymbol{e}_\theta - \frac{\partial\eta}{\partial z}\boldsymbol{e}_z\right)\left\{1+\left(\frac{1}{r}\frac{\partial\eta}{\partial\theta}\right)^2+\left(\frac{\partial\eta}{\partial z}\right)^2\right\}^{-1/2} \tag{14.2}$$

where $\boldsymbol{e}_z, \boldsymbol{e}_r$ and $\boldsymbol{e}_\theta$ are unit vectors along the z, r, and θ directions, respectively.

In the present analysis, the velocity can be denoted as the gradient of some potential function and the potential functions always satisfy the Laplace equation because the fluids are incompressible, i.e.,

$$\nabla^2\phi_j = 0, \qquad (j = 1,2) \tag{14.3}$$

where $\nabla^2 = \dfrac{\partial^2}{\partial r^2} + \dfrac{1}{r}\dfrac{\partial}{\partial r} + \dfrac{1}{r^2}\dfrac{\partial^2}{\partial\theta^2} + \dfrac{\partial^2}{\partial z^2}$.

We assume that every particle, which is on the interface currently, will remain on the interface throughout the analysis, we derive the following kinematic conditions:

$$\frac{\partial\eta}{\partial t} + U\frac{\partial\eta}{\partial z} = \frac{\partial\phi_1}{\partial r} \quad \text{at } r = R \tag{14.4}$$

$$\frac{\partial\eta}{\partial t} = \frac{\partial\phi_2}{\partial r} \quad \text{at } r = R \tag{14.5}$$

Using normal mode technique and taking the help of kinematics boundary conditions equations (14.4) and (14.5), the solution of equation (14.3) can be expressed as

$$\eta = Ce^{(ikz+in\theta-i\omega t)} + c.c. \tag{14.6}$$

$$\phi^{(1)} = (ikU - i\omega)CE_1(kr)e^{(ikz+in\theta-i\omega t)} + c.c. \tag{14.7}$$

$$\phi^{(2)} = -i\omega CE_2(kr)e^{(ikz+in\theta-i\omega t)} + c.c. \tag{14.8}$$

where C denotes complex constant, m is the positive integer, k represents real wave number. ω represents growth rate and $E_1(kr) = \frac{I_n(kr)}{I'_n(kR)}$, $E_2(kr) = \frac{K_n(kr)}{K'_n(kR)}$. $I_n(kr)$ and $K_n(kr)$ are the modified Bessel's function of first and second kind of order n, respectively, and *c.c.* denotes the complex conjugate of the previous term.

Here, we consider the fluid is of the Oldroyd-B type [13] because it has a feature that combines the effects of relaxation and nonlinearity, with a relative case of execution, better than any other viscoelastic fluid. The governing equation of the linear viscoelastic fluid of Oldroyd-B type is written as

$$\left[1 + \lambda_1 \frac{\partial}{\partial t}\right]\tau = 2\mu_1\left[1 + \lambda_2 \frac{\partial}{\partial t}\right]\gamma, \tag{14.9}$$

where τ denotes viscous stress tensor, μ_1 represents viscosity, γ denotes strain tensor, and λ_2 and λ_1 are the retardation and relaxation time, respectively.

14.3 Viscoelastic Potential Flow (VPF) Analysis

As per dynamical condition, we know that the normal components of stresses must be uninterrupted at the interface, i.e.,

$$p_1 - p_2 - 2\mu_1 \frac{\partial^2 \phi_1}{\partial r^2} + 2\mu_2 \frac{\partial^2 \phi_2}{\partial r^2} = -\sigma\left(\frac{\partial^2 \eta}{\partial z^2} + \frac{1}{r^2}\frac{\partial^2 \eta}{\partial \theta^2} + \frac{\eta}{R^2}\right) \tag{14.10}$$

where $p_j(j = 1,2)$ represent the pressure for the inner and outer fluid, respectively. If we use the equation given by Bernoulli for irrotational flow and linearize it, we have

$$-\rho_1\left(\frac{\partial \phi_1}{\partial t} + U\frac{\partial \phi_1}{\partial z}\right) + \rho_2\left(\frac{\partial \phi_2}{\partial t}\right) - 2\mu_1 \frac{\partial^2 \phi_1}{\partial r^2} + 2\mu_2 \frac{\partial^2 \phi_2}{\partial r^2}$$
$$= -\sigma\left(\frac{\partial^2 \eta}{\partial z^2} + \frac{1}{r^2}\frac{\partial^2 \eta}{\partial \theta^2} + \frac{\eta}{R^2}\right) \tag{14.11}$$

Putting the values of η, ϕ_1 and ϕ_2 in equation (14.11), we get the expression

$$\rho_1(\omega - kU)^2 E_1(kR) - \omega^2 \rho_2 E_2(kR) - 2i\mu_1 k^2 (kU - \omega)F_1(kR) - 2i\mu_2 k^2 \omega F_2(kR) = \sigma\left(k^2 + \frac{n^2 - 1}{R^2}\right) \tag{14.12}$$

where $F_1(kR) = \frac{I_n''(kR)}{I_n'(kR)} = \left(1 + \frac{n^2}{k^2R^2}\right)E_1(kR) - \frac{1}{kR}$

$$F_2(kR) = \frac{K_n''(kR)}{K_n'(kR)} = \left(1 + \frac{n^2}{k^2R^2}\right)E_2(kR) - \frac{1}{kR}$$

Since the inside fluid is viscoelastic fluid of Oldroyd-B type, viscosity μ_1 is modified as $\frac{1 - i\lambda_2\omega}{1 - i\lambda_1\omega}\mu_1$.

Using the expression for μ_1 in equation (14.12), the dispersion relation for the VPF analysis can be written as

$$a_3\omega^3 + (a_2 + ib_2)\omega^2 + (a_1 + ib_1)\omega + (a_0 + ib_0) = 0 \tag{14.13}$$

where

$$a_3 = \lambda_1\left(\rho_1 E_1(kR) - \rho_2 E_2(kR)\right)$$

$$a_2 = -2k\lambda_1 U\rho_1 E_1(kR)$$

$$b_2 = \left(\rho_1 E_1(kR) - \rho_2 E_2(kR)\right) + 2k^2\left(\lambda_2\mu_1 F_1(kR) - \lambda_1\mu_2 F_2(kR)\right)$$

$$a_1 = \lambda_1\left(k^2U^2\rho_1 E_1(kR) - \sigma k(k^2 + \frac{n^2 - 1}{R^2})\right) - 2k^2\left(\mu_1 F_1(kR) - \mu_2 F_2(kR)\right)$$

$$b_1 = -2kU\rho_1 E_1(kR) - 2k^3U\lambda_2\mu_1 F_1(kR)$$

$$a_0 = 2k^3U\mu_1 F_1(kR)$$

$$b_0 = k^2U^2\rho_1 E_1(kR) - \sigma k(k^2 + \frac{n^2 - 1}{R^2})$$

14.4 Viscoelastic Contribution for the Viscoelastic Potential Flow (VCVPF) Analysis

Taking viscoelastic pressure as an extra pressure into the normal stress balance equation, we can fluidly take account of the effect of shearing components of stresses in the VPF analysis of cylindrical jet.

In this study, we are not considering the small deformation η found in the linear analysis. We assume that $\boldsymbol{n}_1 = \boldsymbol{e}_r$ denotes the outward unit normal for inner fluid at the interface and, therefore, $\boldsymbol{n}_2 = -\boldsymbol{n}_1$ represents the outward unit normal for outer fluid in the jet, and also $\boldsymbol{t} = \boldsymbol{e}_z$ is the unit tangent vector at the interface. Here, the superscripts "i" shows irrotational and "v" represents viscous. The tangential and normal components of the viscous stresses are denoted by τ^s and τ^n, respectively.

The mechanical energy balance equations for outer and inner fluids are as follows;

$$\begin{aligned}\frac{d}{dt}\int_{V_2}\frac{\rho_2}{2}|\boldsymbol{u}_2|^2 dV &= \int_A(\boldsymbol{u}_2\cdot\boldsymbol{T}\cdot\boldsymbol{n}_2)dA - \int_{V_2}2\mu_2\boldsymbol{D}_2 : \boldsymbol{D}_2\, dV \\ &= -\int_A\left(\boldsymbol{u}_2\cdot\boldsymbol{n}_1(-p_2^i+\tau_2^n)+\boldsymbol{u}_2\cdot\boldsymbol{t}\,\tau_2^s\right)dV - \int_{V_2}2\mu_2\boldsymbol{D}_2 : \boldsymbol{D}_2\, dV\end{aligned} \tag{14.14}$$

$$\begin{aligned}\frac{d}{dt}\int_{V_1}\frac{\rho_1}{2}|\boldsymbol{u}_1|^2 dV &= \int_A(\boldsymbol{u}_1\cdot\boldsymbol{T}\cdot\boldsymbol{n}_1)dA - \int_{V_1}2\mu_1\boldsymbol{D}_1 : \boldsymbol{D}_1\, dV \\ &= \int_A\left(\boldsymbol{u}_1\cdot\boldsymbol{n}_1(-p_1^i+\tau_1^n)+\boldsymbol{u}_1\cdot\boldsymbol{t}\,\tau_1^s\right)dV - \int_{V_1}2\mu_1\boldsymbol{D}_1 : \boldsymbol{D}_1\, dV\end{aligned} \tag{14.15}$$

where $\boldsymbol{D}_j (j = 1,2)$ represents the symmetric component of the rate of strain tensor for inner and outer fluids, respectively.

As the normal components of velocities are continuous at the interface, we have

$$\boldsymbol{u}_2\cdot\boldsymbol{n}_1 = \boldsymbol{u}_1\cdot\boldsymbol{n}_1 = u_n,$$

Adding equations (14.14) and (14.15), we have

$$\begin{aligned}\frac{d}{dt}\int_{V_2}\frac{\rho_2}{2}|\boldsymbol{u}_2|^2 dV + \frac{d}{dt}\int_{V_1}\frac{\rho_1}{2}|\boldsymbol{u}_1|^2 dV &= \int_A\left(u_n(-p_1^i+\tau_1^n+p_2^i-\tau_2^n)+\boldsymbol{u}_2\cdot\boldsymbol{t}\,\tau_2^s-\boldsymbol{u}_1\cdot\boldsymbol{t}\,\tau_1^s\right)dA \\ &\quad - \int_{V_2}2\mu_2\boldsymbol{D}_2 : \boldsymbol{D}_2\, dV - \int_{V_1}2\mu_1\boldsymbol{D}_1 : \boldsymbol{D}_1\, dV\end{aligned} \tag{14.16}$$

If we introduce two viscoelastic pressure terms, namely, p_1^v and p_2^v for the inner and outer flow, the horizontal velocity and shear stress will be continuous at the interface, and, therefore,

$$\boldsymbol{u}_2\cdot\boldsymbol{t} = \boldsymbol{u}_1\cdot\boldsymbol{t} = u_s \quad \text{and} \quad \tau_1^s = \tau_2^s = \tau^s$$

Here, we have taken an assumption that the boundary layer estimate has an insignificant outcome on the flow in the bulk liquid, but it disturbs the continuity and pressure conditions at the interface. Therefore, the equation (14.16) can be written as

$$\begin{aligned}\frac{d}{dt}\int_{V_2}\frac{\rho_2}{2}|\boldsymbol{u}_2|^2 dV + \frac{d}{dt}\int_{V_1}\frac{\rho_1}{2}|\boldsymbol{u}_1|^2 dV &= \int_A\left(u_n(-p_1^i-p_1^v+\tau_1^n+p_2^i+p_2^v-\tau_2^n)\right)dA \\ &\quad - \int_{V_2}2\mu_2\boldsymbol{D}_2 : \boldsymbol{D}_2\, dV - \int_{V_1}2\mu_1\boldsymbol{D}_1 : \boldsymbol{D}_1\, dV\end{aligned} \tag{14.17}$$

Now we can obtain a relation between irrotational shear stresses and viscoelastic pressure from equations (14.16) and (14.17)

$$\int_A \left(u_n(-p_1^v + p_2^v)\right)dA = \int_A \left(\boldsymbol{u}_1 \cdot \boldsymbol{t}\tau_1^s - \boldsymbol{u}_2 \cdot \boldsymbol{t}\tau_2^s\right)dA \tag{14.18}$$

The equation which governs viscoelastic pressure is written as

$$\nabla^2 p_i^v = 0 \qquad \text{for } (i = 1,2) \tag{14.19}$$

Solving the above equation, the expressions for viscoelastic pressure can be written as

$$p_1^v = -C_k I_n(kr)e^{(ikz+in\theta-i\omega t)} \tag{14.20}$$

$$p_2^v = -D_k K_n(kr)e^{(ikz+in\theta-i\omega t)} \tag{14.21}$$

Here C_k, D_k are the constants. At the undisturbed interface $r = R$ the variation in the viscoelastic pressure is written as

$$-p_1^v + p_2^v = \left[C_k I_n(kR) + D_k K_n(kR)\right]e^{(ikz+in\theta-i\omega t)} \tag{14.22}$$

We consider here that an extra pressure termed as viscoelastic pressure will be present in the normal stress balance equation with irrotational pressures and, therefore, the normal stress balance equation can be written as

$$p_1^i + p_1^v - p_2^i - p_2^v - 2\mu_1 \frac{\partial^2 \phi_1}{\partial r^2} + 2\mu_2 \frac{\partial^2 \phi_2}{\partial r^2} = -\sigma\left(\frac{\partial^2 \eta}{\partial z^2} + \frac{1}{r^2}\frac{\partial^2 \eta}{\partial \theta^2} + \frac{\eta}{R^2}\right) \tag{14.23}$$

where p_j^i denote irrotational pressure, p_j^v representing viscous pressure for inner and outer fluid $(j = 1,2)$, respectively.

To find irrotational pressure, we use Bernoulli's equation and then linearizing it, we have

$$-\rho_1\left(\frac{\partial \phi_1}{\partial t} + U\frac{\partial \phi_1}{\partial z}\right) + p_1^v + \rho_2\left(\frac{\partial \phi_2}{\partial t}\right) - p_2^v - 2\mu_1 \frac{\partial^2 \phi_1}{\partial r^2} + 2\mu_2 \frac{\partial^2 \phi_2}{\partial r^2} = -\sigma\left(\frac{\partial^2 \eta}{\partial z^2} + \frac{1}{r^2}\frac{\partial^2 \eta}{\partial \theta^2} + \frac{\eta}{R^2}\right) \tag{14.24}$$

On solving equation (14.22) along with equation (14.18) we get

$$\left[C_k I_n(kR) - D_k K_n(kR)\right] = 2kC\left[\mu_1(ikU - i\omega)E_1(kR) - \mu_2 i\omega E_2(kR)\right] \tag{14.25}$$

Putting the values of η, ϕ_1, ϕ_2 and contribution of viscous pressure in equation (14.24) we get the expression

$$\begin{aligned} &\rho_1(\omega - kU)^2 E_1(kR) - \omega^2 \rho_2 E_2(kR) - 2i\mu_1 k^2(kU - \omega) \times \\ &\left(F_1(kR) + E_1(kR)\right) - 2i\mu_2 k^2 \omega\left(F_2(kR) + E_2(kR)\right) = \sigma\left(k^2 + \frac{n^2 - 1}{R^2}\right) \end{aligned} \tag{14.26}$$

Since the inside fluid is viscoelastic fluid of Oldroyd-B type so viscosity μ_1 is modified as $\frac{1-i\lambda_2\omega}{1-i\lambda_1\omega}\mu_1$.

Using the expression for μ_1 in equation (14.26), the dispersion relation for VCVPF solution is expressed as:

$$a_3\omega^3 + (a_2 + ib_2)\omega^2 + (a_1 + ib_1)\omega + (a_0 + ib_0) = 0 \tag{14.27}$$

where

$$a_3 = \lambda_1(\rho_1 E_1(kR) - \rho_2 E_2(kR))$$

$$a_2 = -2k\lambda_1 U\rho_1 E_1(kR)$$

$$b_2 = (\rho_1 E_1(kR) - \rho_2 E_2(kR))$$
$$+2k^2(\lambda_2\mu_1(F_1(kR) + E_1(kR)) - \lambda_1\mu_2(F_2(kR) + E_2(kR)))$$

$$a_1 = \lambda_1\left(k^2U^2\rho_1 E_1(kR) - \sigma k(k^2 + \frac{n^2-1}{R^2})\right)$$
$$-2k^2(\mu_1(F_1(kR) + E_1(kR)) - \mu_2(F_2(kR) + E_2(kR)))$$

$$b_1 = -2kU\rho_1 E_1(kR) - 2k^3U\lambda_2\mu_1(F_1(kR) + E_1(kR))$$

$$a_0 = 2k^3U\mu_1(F_1(kR) + E_1(kR))$$

$$b_0 = k^2U^2\rho_1 E_1(kR) - \sigma k(k^2 + \frac{n^2-1}{R^2})$$

14.5 Dissipation Calculation for the Cylindrical Jet of Viscoelastic fluid

Here, we are going to derive dispersion relation for liquid cylindrical jet using dissipation approximations. As we know that dissipation approximation is a technique to take account of viscous effects into the stability of fluid layer taking flow as irrotational.

The addition of the mechanical energy balance equations for inner and outer fluids can be expressed as

$$\frac{d}{dt}\int_{V_2}\frac{\rho_2}{2}|\boldsymbol{u}_2|^2 dV + \frac{d}{dt}\int_{V_1}\frac{\rho_1}{2}|\boldsymbol{u}_1|^2 dV = \int_A\left[u_n(-p_1^i + \tau_1^n + p_2^i - \tau_2^n) + \boldsymbol{u}_2\cdot\boldsymbol{t}\ \tau_2^s - \boldsymbol{u}_1\cdot\boldsymbol{t}\ \tau_1^s\right]dA$$
$$-\int_{V_2}2\mu_2\boldsymbol{D}_2 : \boldsymbol{D}_2\, dV - \int_{V_1}2\mu_1\boldsymbol{D}_1 : \boldsymbol{D}_1\, dV \tag{14.28}$$

Here, we consider the annular cylinder together with the surface $r = R$; the length of the cylinder is taken as λ. Therefore, the volume of the inner cylinder is written as $V_1 = \int_0^{2\pi}\int_0^{R}\int_z^{z+\lambda} r\, d\theta\, dr\, dz.$

Similarly, the volume of the outer cylinder is given by $V_2 = \int_0^{2\pi}\int_R^{\infty}\int_z^{z+\lambda} r\, d\theta\, dr\, dz$

The conservation equation of momentum in linear form is expressed as

$$p_2 - \tau_2^n - p_1 + \tau_2^n = \sigma\left(\frac{\partial^2 \eta}{\partial z^2} + \frac{1}{R^2}\frac{\partial^2 \eta}{\partial \theta^2} + \frac{\eta}{R^2}\right) \tag{14.29}$$

and also we know that horizontal velocity and stresses are continuous at the interface, i.e.,

$$\boldsymbol{u}_2 \cdot \boldsymbol{t} = \boldsymbol{u}_1 \cdot \boldsymbol{t} = u_s, \qquad \tau_2^s = \tau_1^s = \tau^s \tag{14.30}$$

Here, the potential flow theory approximates the flows of the fluids, and therefore, we can use Gauss's divergence theorem, i.e.,

$$\int_V 2\mu \boldsymbol{D} : \boldsymbol{D}\, dV = \int_A \boldsymbol{n} \cdot 2\mu \boldsymbol{D} \cdot \boldsymbol{u}\, dA \tag{14.31}$$

where V is the volume bounded by the surface A and $\boldsymbol{n}$ represents outward unit normal vector. Putting equations (14.29)–(14.31) into equation (14.28), we obtain

$$\begin{aligned}\frac{d}{dt}\int_{V_2} \frac{\rho_2}{2}|\boldsymbol{u}_2|^2 dV + \frac{d}{dt}\int_{V_1} \frac{\rho_1}{2}|\boldsymbol{u}_1|^2 dV &= \int_A \left[u_n \sigma\left(\frac{\partial^2 \eta}{\partial z^2} + \frac{1}{R^2}\frac{\partial^2 \eta}{\partial \theta^2} + \frac{\eta}{R^2}\right)\right] dA \\ &+ \int_A \boldsymbol{n}_1 \cdot 2\mu_2 \boldsymbol{D}_2 \cdot \boldsymbol{u}_2\, dA - \int_A \boldsymbol{n}_1 \cdot 2\mu_1 \boldsymbol{D}_I \cdot \boldsymbol{u}_1\, dA\end{aligned} \tag{14.32}$$

The integrals in equation (14.32) are solved as

$$\begin{aligned}\frac{d}{dt}\int_{V_2} \frac{\rho_2}{2}|\boldsymbol{u}_2|^2 dV &= \int_0^{2\pi}\int_z^{z+\lambda}\int_R^{\infty} \frac{\rho_2}{2}\frac{\partial}{\partial t}|\boldsymbol{u}_2|^2\, r\, d\theta\, dz\, dr \\ &= 2i\pi\omega\rho_2\lambda\frac{R}{k}\left(\frac{\alpha}{\rho_2} - i\omega\right)\left(\frac{\alpha}{\rho_2} + i\bar{\omega}\right)|C|^2 E_2(kR)\exp\left[i(\bar{\omega} - \omega)t\right]\end{aligned} \tag{14.33}$$

$$\begin{aligned}\frac{d}{dt}\int_{V_1} \frac{\rho_1}{2}|\boldsymbol{u}_1|^2 dV &= \int_0^{2\pi}\int_z^{z+\lambda}\int_0^{R} \frac{\rho_1}{2}\left(\frac{\partial}{\partial t} + U\frac{\partial}{\partial z}\right)|\boldsymbol{u}_1|^2\, r\, d\theta\, dz\, dr \\ &= \pi\rho_1\lambda\frac{R}{k}(2ikU - 2i\omega)\left(\frac{\alpha}{\rho_1} + ikU - i\omega\right)\left(\frac{\alpha}{\rho_1} - ikU + i\bar{\omega}\right)|C|^2 E_1(kR)\exp\left[i(\bar{\omega} - \omega)t\right]\end{aligned} \tag{14.34}$$

$$\int_A \left[u_n \sigma \left(\frac{\partial^2 \eta}{\partial z^2} + \frac{1}{R^2}\frac{\partial^2 \eta}{\partial \theta^2} + \frac{\eta}{R^2} \right) \right] dA$$

$$= \int_0^{2\pi} d\theta \int_z^{z+\lambda} \sigma \left(-k^2 - \frac{n^2}{R^2} + \frac{1}{R^2} \right) \left(\frac{\alpha}{\rho_1} - ikU + i\bar{\omega} \right) |C|^2 E_1'(kR) \exp\left[i(\bar{\omega} - \omega)t \right] r dz \quad (14.35)$$

$$= 2\pi\lambda R \sigma \left(-k^2 - \frac{n^2}{R^2} + \frac{1}{R^2} \right) \left(\frac{\alpha}{\rho_1} - ikU + i\bar{\omega} \right) |C|^2 \exp\left[i(\bar{\omega} - \omega)t \right]$$

$$\int_A \boldsymbol{n}_1 \cdot 2\mu_2 \boldsymbol{D}_2 \cdot \boldsymbol{u}_2 \, dA$$

$$= 4\pi\lambda\mu_2 kR \left(\frac{\alpha}{\rho_2} - i\omega \right) \left(\frac{\alpha}{\rho_2} + i\bar{\omega} \right) |C|^2 \left(F_2(kR) + E_2(kR) \right) \exp\left[i(\bar{\omega} - \omega)t \right] \quad (14.36)$$

$$\int_A \boldsymbol{n}_1 \cdot 2\mu_1 \boldsymbol{D}_1 \cdot \boldsymbol{u}_1 \, dA$$

$$= 4\pi\lambda\mu_1 kR \left(\frac{\alpha}{\rho_1} + ikU - i\omega \right) \left(\frac{\alpha}{\rho_1} - ikU + i\bar{\omega} \right) |C|^2 \left(F_1(kR) + E_1(kR) \right) \exp\left[i(\bar{\omega} - \omega)t \right] \quad (14.37)$$

Putting the values from equations (14.33)–(14.37) into equation (14.32) and using the concept that normal velocity is continuous at the interface, we have

$$\rho_1 \left(\frac{\alpha}{\rho_1} + ikU - i\omega \right) (ikU - i\omega) E_1(kR) - \rho_2 \left(\frac{\alpha}{\rho_2} - i\omega \right) \times (-i\omega) E_2(kR) = k\sigma \left(-k^2 - \frac{n^2}{R^2} + \frac{1}{R^2} \right)$$

$$+ 2\mu_2 k^2 \left(\frac{\alpha}{\rho_2} - i\omega \right) \times \left(F_2(kR) + E_2(kR) \right) - 2\mu_1 k^2 \left(\frac{\alpha}{\rho_1} + ikU - i\omega \right) \left(F_1(kR) + E_1(kR) \right) \quad (14.38)$$

By putting $\frac{1 - i\lambda_2\omega}{1 - i\lambda_1\omega}\mu_1$ at the place of μ_1, we are getting the same dispersion relation as we obtain in VCVPF analysis.

14.6 Dimensionless Form of the Dispersion Relation

If we consider the non-dimensional variables;

$$D = 2R, \quad \Lambda_1 = \frac{\lambda_1 V}{D}, \quad \Lambda_2 = \frac{\lambda_2 V}{D}, \quad \hat{\omega} = \frac{\omega D}{V}, \quad \hat{\rho} = \frac{\rho^{(2)}}{\rho^{(1)}}, \quad \hat{\mu} = \frac{\mu^{(2)}}{\mu^{(1)}}$$

$$\hat{R} = \frac{R}{D} = \frac{1}{2}, \quad \hat{k} = kD, \quad \mathrm{Re} = \frac{\rho^{(1)} V D}{\mu^{(1)}}, \quad W = \frac{\rho^{(1)} D V^2}{\sigma}, \quad \hat{U} = \frac{U}{V}$$

where Re denotes the Reynolds number that is defined as the ratio of inertia forces to viscous forces. The ratio of the inertia force to the surface tension force is known as Weber number and it is denoted by W. Here, Λ_1 denotes The Deborah number and is defined as the ratio of relaxation time to the characteristic time scale. $\hat{\rho}$ and $\hat{\mu}$ represents the density ratio and viscosity ratio, respectively.

The dimensionless form of the dispersion relation for the VPF analysis is written as

$$\hat{a}_3\hat{\omega}^3 + \left(\hat{a}_2 + i\hat{b}_2\right)\hat{\omega}^2 + \left(\hat{a}_1 + i\hat{b}_1\right)\hat{\omega} + \left(\hat{a}_0 + i\hat{b}_0\right) = 0 \tag{14.39}$$

where

$$\hat{a}_3 = \Lambda_1\left(E_1(\hat{k}/2) - \hat{\rho}E_2(\hat{k}/2)\right)$$

$$\hat{a}_2 = -2\hat{k}\hat{U}\Lambda_1 E_1(\hat{k}/2)$$

$$\hat{b}_2 = \left(E_1(\hat{k}/2) - \hat{\rho}E_2(\hat{k}/2)\right) + \frac{2\hat{k}^2}{\text{Re}}\left(\Lambda_2 F_1(\hat{k}/2) - \Lambda_1\hat{\mu}F_2(\hat{k}/2)\right)$$

$$\hat{a}_3 = \Lambda_1\left(\hat{k}^2\hat{U}^2 E_1(\hat{k}/2) - W^{-1}\hat{k}(\hat{k}^2 + 4(n^2 - 1))\right) - \frac{2\hat{k}^2}{\text{Re}}\left(F_1(\hat{k}/2) - \hat{\mu}F_2(\hat{k}/2)\right)$$

$$\hat{b}_1 = -2\hat{k}\hat{U}E_1(\hat{k}/2) - \Lambda_2\frac{2\hat{k}^3}{\text{Re}}\hat{U}F_1(\hat{k}/2)$$

$$\hat{a}_0 = \frac{2k^3}{\text{Re}}\hat{U}F_1(\hat{k}/2)$$

$$\hat{b}_0 = \hat{k}^2\hat{U}^2 E_1(\hat{k}/2) - W^{-1}\hat{k}(\hat{k}^2 + 4(n^2 - 1))$$

The dimensionless dispersion relation for VCVPF analysis can be written as

$$\hat{a}_3\hat{\omega}^3 + \left(\hat{a}_2 + i\hat{b}_2\right)\hat{\omega}^2 + \left(\hat{a}_1 + i\hat{b}_1\right)\hat{\omega} + \left(\hat{a}_0 + i\hat{b}_0\right) = 0 \tag{14.40}$$

where $\hat{a}_3 = \Lambda_1\left(E_1(\hat{k}/2) - \hat{\rho}E_2(\hat{k}/2)\right)$

$$\hat{a}_2 = -2\hat{k}\hat{U}\Lambda_1 E_1(\hat{k}/2)$$

$$\hat{b}_2 = \left(E_1(\hat{k}/2) - \hat{\rho}E_2(\hat{k}/2)\right)$$
$$+ \frac{2\hat{k}^2}{\text{Re}}\left(\Lambda_2\left(F_1(\hat{k}/2) + E_1(\hat{k}/2)\right) - \Lambda_1\hat{\mu}\left(F_2(\hat{k}/2) + E_2(\hat{k}/2)\right)\right)$$

$$\hat{a}_1 = \Lambda_1\left(\hat{k}^2\hat{U}^2E_1(\hat{k}/2) - W^{-1}\hat{k}(\hat{k}^2 + 4(n^2 - 1))\right)$$

$$-\frac{2\hat{k}^2}{\text{Re}}\left(\left(F_1(\hat{k}/2) + E_1(\hat{k}/2)\right) - \hat{\mu}\left(F_2(\hat{k}/2) + E_2(\hat{k}/2)\right)\right)$$

$$\hat{b}_1 = -2\hat{k}\hat{U}E_1(\hat{k}/2) - \Lambda_2\frac{2\hat{k}^3}{\text{Re}}\hat{U}\left(F_1(\hat{k}/2) + E_1(\hat{k}/2)\right)$$

$$\hat{a}_0 = \frac{2k^3}{\text{Re}}\hat{U}\left(F_1(\hat{k}/2) + E_1(\hat{k}/2)\right)$$

$$\hat{b}_0 = \hat{k}^2\hat{U}^2E_1(\hat{k}/2) - W^{-1}\hat{k}(\hat{k}^2 + 4(n^2 - 1))$$

In the case of inviscid viscoelastic fluid ($\text{Re} \rightarrow \infty$), the VPF and VCVPF solution gives the same dispersion relation. From the above two expressions, it can also be observed that the VCVPF and VPF give the same solution at the higher values of W^{-1}, while the difference between the two theories increases at the smaller values of W^{-1}.

For axisymmetric disturbances, i.e., for $n = 0$, the dimensionless dispersion relation becomes

$$\begin{aligned}
&\left[\Lambda_1(\alpha_1 + \hat{\rho}\alpha_2)\right]\hat{\omega}^3 + \hat{\omega}^2\left[-2\hat{k}\hat{U}\Lambda_1\alpha_1 + i\left((\alpha_1 + \hat{\rho}\alpha_2) + \frac{2\hat{k}^2}{\text{Re}}(\Lambda_1\hat{\mu}(\beta_2 + \alpha_2) + \Lambda_2(\beta_1 + \alpha_1))\right)\right] \\
&+\left[\Lambda_1\left(\hat{k}^2\hat{U}^2\alpha_1 - W^{-1}\hat{k}(\hat{k}^2 - 4)\right) - \frac{2k^2}{\text{Re}}((\beta_1 + \alpha_1) + \hat{\mu}(\beta_2 + \alpha_2)) - 2i\hat{k}\hat{U}\alpha_1 - i\frac{2\hat{k}^3}{\text{Re}}\Lambda_2\hat{U}(\beta_1 + \alpha_1)\right]\hat{\omega} \\
&+\left[\frac{2k^3}{\text{Re}}\hat{U}(\beta_1 + \alpha_1) + i\left(\hat{k}^2\hat{U}^2\alpha_l - W^{-1}\hat{k}(\hat{k}^2 - 4)\right)\right] = 0
\end{aligned} \tag{14.41}$$

If we put Deborah number $\Lambda_1 = \Lambda_2$ and $\hat{U} = 1$ in the equation (14.41), the dispersion relation becomes

$$\begin{aligned}
&(\alpha_l + \hat{\rho}\alpha_a)\hat{\omega}^2 + \left[-2\hat{k}\alpha_l + i\frac{2\hat{k}^2}{\text{Re}}((\beta_1 + \alpha_1) + \hat{\mu}(\beta_2 + \alpha_2))\right]\hat{\omega} \\
&+\left[\hat{k}^2\alpha_l - i\frac{2k^3}{\text{Re}}(\beta_1 + \alpha_1) - W^{-1}\hat{k}(\hat{k}^2 - 4)\right] = 0
\end{aligned} \tag{14.42}$$

It is the same expression as obtained by Funada et al. [16].

14.7 Results and Discussions

The previous study of the stability of a viscoelastic cylindrical jet has been made by Awasthi et al. [14]. They used VPF theory without taking an additional viscoelastic pressure (VCVPF). In the literature, we have seen that the VCVPF solution is more stable than the VPF solution, and therefore, we recommend VCVPF theory to study the stability problem studied here.

As we have shown that the discrepancy between VCVPF and VPF vanishes in the inviscid limiting case. In the case of inviscid viscoelastic fluid ($\mathrm{Re} \to \infty$), The VPF and VCVPF solution gives the same dispersion relation. The instability is the liquid jet is driven by capillary instability because of surface tension and KHI because there is a velocity difference at the interface. For Weber number, $W \to \infty$ the effect of surface tension vanishes, so the instability in the liquid jet becomes pure KHI. If Weber number $W \to 0$, the instability in the liquid jet is driven by pure capillary instability. It has also been shown that the discrepancy between the above two theories vanishes when capillary instability dominates (small value of W) and is reasonable when KHI dominates (large value of W).

The dispersion relation for viscoelastic cylindrical jets for both VCVPF and VPF solution is cubic in terms of growth rate parameter and instability arises due to a positive imaginary part of the growth rate parameter (i.e., $\omega_I > 0$). The viscoelastic fluid parameters and their properties used here are the same as taken by Awasthi et al. [14] and given in Table 14.1.

First, we compare the growth rate curves for axisymmetric disturbance. In Figure 14.1, the highest growth rate curves versus Weber number obtained for the VPF solution and VCVPF solution for the cylindrical jet of viscoelastic fluid have been compared.

TABLE 14.1
Viscoelastic Fluid Parameters and Their Properties

	2% PAA	2% PO
$\rho_1\,(\mathrm{g/cm^3})$	0.99	0.99
$\mu_1\,(\mathrm{P})$	96.0	350.0
$\rho_2\,(\mathrm{g/cm^3})$	1.947×10^{-3}	1.776×10^{-3}
$\mu_2\,(\mathrm{P})$	1.8×10^{-4}	1.8×10^{-4}
$\sigma(\mathrm{dyn\,cm^{-1}})$	45.0	63.0
$\lambda_1\,(\mathrm{s})$	0.039	0.21
$\lambda_2\,(\mathrm{s})$	0.0	0.0
$V(\mathrm{cm/s})$	0.4688	0.18
Re	206.8466	1964.10
W	206.8246	1964.10
$\hat{\rho}$	1.967×10^{-3}	1.794×10^{-3}
$\hat{\mu}$	1.875×10^{-6}	5.143×10^{-7}
Λ_1	0.01828	0.0378
Λ_2	0.0	0.0

In Figure 14.1(a), the viscoelastic fluid (2% PAA) is taken while (2% PO) is considered in Figure 14.1(b). It has been found that the highest growth rates for the VCVPF solution are almost identical to the VPF solution for lower values of Weber number. At the larger values of Weber number, the maximum growth rate obtained in the VCVPF solution is lower as compared to the maximum growth rate obtained in the VPF solution. This concludes that at low Weber number, the VPF and VCVPF solutions are the same but as Weber number increases, the VCVPF solution most stable waves. The VPF solution only contains the effect of normal stresses, while in the VCVPF solution, the effect of both normal and tangential stresses is included. Therefore, it is concluded that the effect of horizontal irrotational stresses resists the breakup of the viscoelastic liquid jet.

Figure 14.2 compares the highest growth rate curves for the VCVPF and VPF analysis for the asymmetric disturbances. Figure 14.2(a) is drawn for (2% PAA) viscoelastic fluid and (2% PO) is considered in Figure 14.2(b). Here, we also obtain that the highest growth rate obtained in the VCVPF solution is nearly the same as obtained for the VPF solution; however, at larger values of Weber number, the VPF solution is less stable.

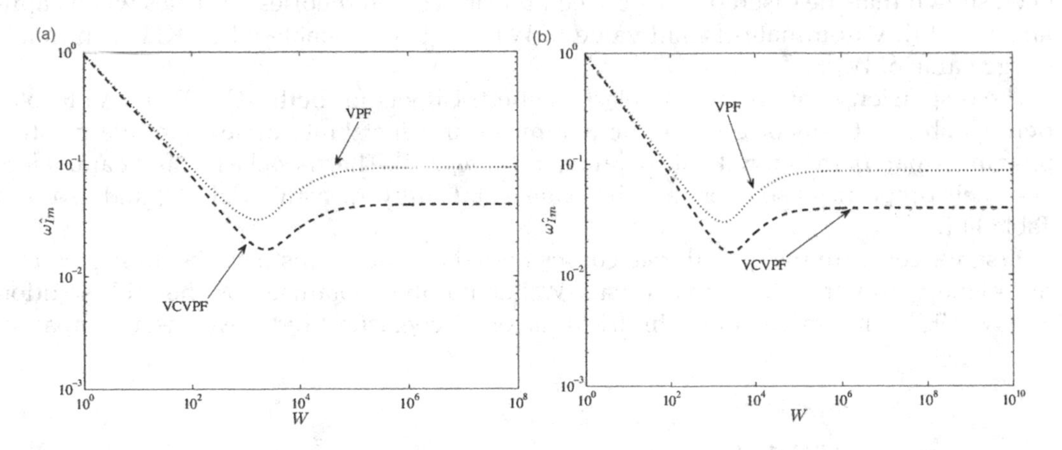

FIGURE 14.1
Comparison of highest growth rate curves versus Weber number for VCVPF solution and VPF solution for axisymmetric disturbances (a) viscoelastic jet (2% PAA) and (b) viscoelastic jet (2% PO).

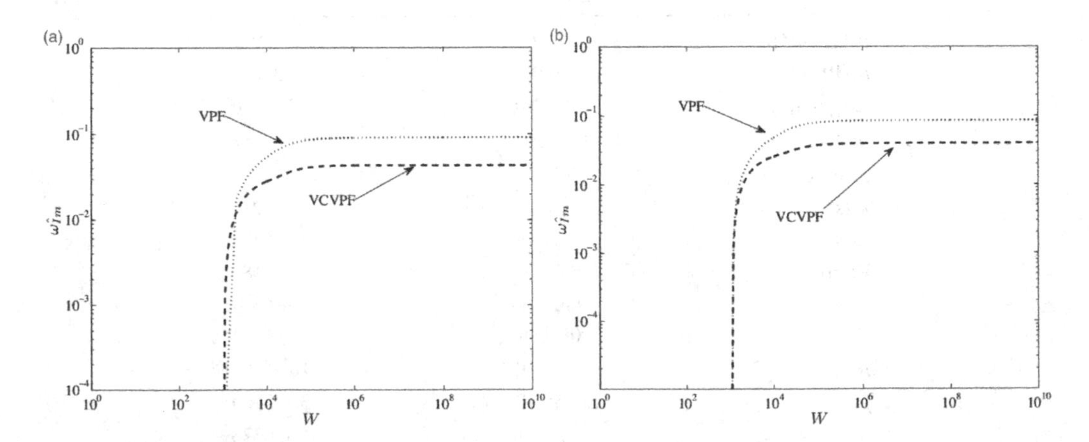

FIGURE 14.2
Comparison of highest growth rate curves versus Weber number for VCVPF solution and VPF solution for asymmetric disturbances (a) viscoelastic jet (2% PAA) and (b) viscoelastic jet (2% PO).

14.8 Conclusions

The temporal stability behavior of viscoelastic cylindrical jet with both asymmetric and axisymmetric perturbations in an infinite viscous fluid has been studied by taking the assumption that the motion in the fluids is irrotational. In the earlier theory (VPF), the stability of the viscoelastic cylindrical jet was investigated taking the assumption that both pressure and motion are irrotational and normal stresses balance equation was used to include viscosity at the interface (Awasthi et al. [14]). Here, a different irrotational theory identified as the VCVPF which incorporates the differences in the shearing stresses and horizontal velocities at the plane interface is used to study the instability of viscoelastic cylindrical jet. The key results are given as follows:

1. The dispersion relation found in the dissipation approximation is the one that comes in the VCVPF solution.
2. In both the theories, the breakup is dominated by capillary instability if Weber's number is low and by KHI if the Weber number is very large.
3. The discrepancy between VCVPF and VPF is significant when the Weber number is large and the Reynolds number is small.
4. The difference between the two theories is irrelevant if capillary instability is dominant and is reasonable when KHI dominates.

References

1. Middleman, S., Stability of a viscoelastic jet, *Chem. Eng. Sci.*, 20 (1965) 1037–1040.
2. Goldin, M., Yerushalmi J., Pfeffer, R. and Shinaar, R., Breakup of laminar capillary jet of viscoelastic fluid, *J. Fluid Mech.* 38 (1969) 689–711.
3. Bousfield, D.W., Keunings, R., Marruci, G. and Denn, M. M., Nonlinear analysis of surface driven breakup of viscoelastic filaments, *J. Non-Newtonian Fluid Mech.* 21(1986) 79–97.
4. Chang, H. C., Demekhin, E. A. and Kalaidin, E., Interated stretching of viscoelastic jets, *Phy. Fluids.* 11(7) (1986) 1717–1737.
5. Brenn, G., Liu, Z. and Durst, F., Linear analysis of temporal instability of axisymmetrical non-Newtonian liquid jets, *Int. J. Multiph. Flow* 26 (2000) 1621–1644.
6. El-Dib, Y. O. and Moatimid, G.M., The instability of a viscoelastic conducting cylindrical interfaces supporting free-surface currents, *Z. Naturforsch A* 57 (2002) 159–176.
7. Moatimid, G. M., Non-linear instability of two dielectric viscoelastic fluids, *Can. J. Phys.* 82 (2004) 1109–1133.
8. Liu, Z. and Liu, Z., Linear analysis of three-dimensional instability of non-Newtonian liquid jets, *J. Fluid Mech.* 559 (2006) 451–459.
9. Liu, Z. and Liu, Z., Instability of viscoelastic liquid jet with axisymmetric and asymmetric disturbances, *Int. J. Multiph. Flow* 34 (2008) 42–60.
10. Joseph, D.D. and Liao, T., Potential flows of viscous and viscoelastic fluids, *J. Fluid Mech.* 256 (1994) 1–23.
11. Funada, T., Joseph, D. D. and Yamashita, S., Stability of liquid jet into incompressible gases and liquids, *Int. J. Multiph. Flow* 30 (2004) 1177–1208.
12. Joseph, D. D., Beavers, G. S. and Funada, T., Rayleigh-Taylor instability of viscoelastic drops at high weber numbers, *J. Fluid Mech.* 453(2002) 109–132.

13. Funada, T. and Joseph, D. D., Viscoelastic potential flow analysis of capillary instability, *J. Non-Newtonian Fluid Mech.* 111(2003) 87–105.
14. Awasthi, M. K., Asthana, R. and Agrawal, G. S., Viscoelastic potential flow analysis of the stability of a cylindrical jet, *Sci. Iran.* 21 (2014) 578–586.
15. Lamb, S. H., *Hydrodynamics*, NewYork: Cambridge University Press (1994).
16. Funada, T., Saitoh, M., Wang, J. and Joseph, D. D., Stability of a liquid jet into incompressible gases and liquids: Part 2. Effects of the irrotational viscous pressure, *Int. J. Multiiph. Flow,* 31 (2005) 1134–1154.
17. Wang, J., Joseph D. D. and Funada, T., Viscous contributions to the pressure for potential flow analysis of capillary instability of two viscous fluids, *Phys. Fluids,* 17 (2005) 052105.
18. Wang, J., Joseph, D. D. and Funada, T., Pressure corrections for potential flow analysis of capillary instability of viscous fluids, *J. Fluid Mech.* 522 (2005) 383–394.
19. Joseph D. D. and Wang, J., The dissipation approximation and viscous potential flow, *J. Fluid. Mech.* 505 (2004) 365–377.

15

Evaporative Kelvin–Helmholtz Instability of a Porous Swirling Annular Layer

Shivam Agarwal and Mukesh Kumar Awasthi
Babasaheb Bhimrao Ambedkar University

CONTENTS

15.1 Introduction 231
15.2 Problem Statement 232
15.3 Boundary and Interfacial Conditions 233
15.4 Dispersion Relationship 235
15.5 Nondimensional Form 237
15.6 Results and Numerical Discussions 239
15.7 Conclusions 243
Acknowledgment 243
References 243

15.1 Introduction

The K-H instability appears when there is a velocity shear at the interface of two fluids. The K-H instability is not only restricted to a water surface but is evident through other natural phenomena such as the ocean, Saturn's band, and Jupiter red spot.

Hsieh [1] first studied the K-H instability between two fluids including heat transport at the interface and found that when the liquid is colder than the vapor, heat transport has a destabilizing nature. Nayak and Chakraborty [2] extended the analysis of the K-H instability in the cylindrical geometry between the vapor–liquid phases admitting heat transport at the interface. They compared results for the plane and cylindrical geometries, which revealed that plane configuration is more stable. Lee [3] analyzed the nonlinear stability of the streaming liquid-vapor interface in the presence of heat transport.

The viscous potential flow (VPF) theory is a theory which deals the fluid motion as irrotational but viscosity is nonzero. Funada and Joseph [4] examined the instability of viscous fluids in a horizontal rectangular channel and they considered the normal irrotational viscous stresses into the model. It was observed that the VPF analysis agrees with the experimental data. Awasthi and Agrawal [5] studied K-H instability in cylindrical geometry. This analysis was also based on the VPF theory. Funada et al. [6] applied the stability of a viscous liquid jet surrounded by a viscous fluid is investigated using the VPF theory. The effect of heat transport in the VPF analysis of K-H instability at the cylindrical interface was included by Asthana et al. [7]. They showed that transport of heat destabilizes the interface but the presence of viscosity delays the instability. Awasthi et al. [8] analyzed

DOI: 10.1201/9781003291916-15

the nonlinear theory of the same problem discussed by Asthana et al. [9] and found that nonlinear effects reduce the range of stability. The effect of the porous medium on the K-H instability was examined by Awasthi et al. [10,11]. It was established that perturbation travels faster in a viscous medium than in a porous medium.

The swirling of cylinders also affects the instability criterion in the cylindrical configuration. Fu et al. [12, 13] studied the swirling effects on interfacial instability admitting heat transport. They considered nonviscous fluids for their study and found that swirling has stabilizing nature. Awasthi et al. [14–16] considered the swirling effect on the VPF analysis of stability problems including heat transport.

The K-H instability in an annular porous zone bounded by two nonflexible cylinders is investigated in this work. The medium porosity and permeability are assumed to be constants. The inner cylinder is stationary, whereas the outer cylinder swirls at a constant angular velocity. The interface is responsible for the transmission of heat along with mass from one phase to the other. The fluids are viscous, and their dynamic viscosities vary. The mathematical equations are solved using the VPF theory, and the stability is tested using the normal mode approach. The marginal stability curves are drawn as well as the stability/instability criteria are explained.

15.2 Problem Statement

The porous annular region is considered with constant porosity ε_1 and permeability k_i. This region is enclosed by two cylinders of radii a_i and b_o $(a_i < b_o)$. The inside cylindrical boundary is immovable while the outside cylinder is swirling with velocity $r\Omega$, where Ω is angular velocity. The fluid with density $\rho^{(i)}$, viscosity $\mu^{(i)}$, and its vapor with density $\rho^{(o)}$, viscosity $\mu^{(o)}$ lie in this porous annular region and is separated by an interface $r = R$. The surface tension and temperature at the interface are denoted by σ and $T^{(o)}$, respectively. The fluid in the region $a_i < r < R$ is flowing with velocity V_i and the temperature $r = a_i$ is represented by $T^{(a)}$. The velocity of the fluid in the region $R < r < b_o$ is V_o and the temperature at $r = b_o$ is represented by $T^{(b)}$.

We use the Darcy–Brinkman model to incorporate the porous medium effects into the analysis. The continuity equation and the Navier–Stokes equation for the considered problem can be expressed as

$$\nabla \cdot \vec{q} = 0 \tag{15.1}$$

$$\frac{\rho}{\varepsilon_1}\left[\frac{\partial \vec{q}}{\partial t} + \frac{1}{\varepsilon_1}(\vec{q} \cdot \nabla)\vec{q}\right] = -\nabla p - \frac{\mu}{k_i}\vec{q} + \frac{\mu}{\varepsilon_1}\nabla^2 \vec{q} \tag{15.2}$$

where μ represents the fluid viscosity, p denotes pressure, and $\vec{q}$ represents the fluid velocity.

We are interested in studying the temporal stability of the considered flow and therefore, a small perturbation is applied to the basic state of flow and interface takes form as $r - R = \xi(\theta, z, t)$, where $\xi(\theta, z, t)$ represents the free-surface deflection.

$$E(r, \theta, z, t) = \xi(\theta, z, t) + R - r = 0 \tag{15.3}$$

The normal unit vector at the interface can be obtained as

$$\hat{n} = \frac{\nabla E}{|\nabla E|} = \frac{\left(\hat{e}_r - \frac{1}{r}\frac{\partial \xi}{\partial \theta}\hat{e}_\theta - \frac{\partial \xi}{\partial z}\hat{e}_z\right)}{\left[\sqrt{1 + \frac{1}{r^2}\left(\frac{\partial \xi}{\partial \theta}\right)^2 + \left(\frac{\partial \xi}{\partial z}\right)^2}\right]} \tag{15.4}$$

In purpose of answering the mathematical problems, the potential flow theory is applied,

$$\vec{q}_j = \nabla \phi^{(j)} \tag{15.5}$$

$$\nabla^2 \phi^{(j)} = 0, \quad \text{where } (j = i, o) \tag{15.6}$$

$$\nabla^2 = \frac{\partial^2}{\partial r^2} + \frac{1}{r^2}\frac{\partial^2}{\partial \theta^2} + \frac{1}{r}\frac{\partial}{\partial r} + \frac{\partial^2}{\partial z^2}$$

The potential functions can also be represented as in their most basic state.

$$\phi_o^{(j)} = V_j z \qquad (j = i, o) \tag{15.7}$$

15.3 Boundary and Interfacial Conditions

Fixed cylindrical surface $r = a_i$ and $r = b_o$, normal velocities vanish and, therefore,

$$\begin{cases} \dfrac{\partial \phi^{(i)}}{\partial r} = 0 & \text{at} \quad r = a_i \\ \dfrac{\partial \phi^{(o)}}{\partial r} = 0 & \text{at} \quad r = b_o \end{cases} \tag{15.8}$$

The interfacial circumstances will be defined at $r = R$. The mass transport equation in mathematical form as

$$\rho^{(i)}\left[\frac{\partial E}{\partial t} + \nabla \phi^{(i)} \cdot \nabla E\right] = \rho^{(o)}\left[\frac{\partial E}{\partial t} + \nabla \phi^{(o)} \cdot \nabla E\right] \tag{15.9}$$

$$\left[\frac{\partial E}{\partial t} + \nabla \phi^{(i)} \cdot \nabla E\right] = \frac{1}{L\rho^{(i)}} S(\xi) \tag{15.10}$$

Here $S(\xi)$ represents the heat flux and L stands for the released latent heat.

The heat flux in the interior phase is achieved as

$$-K_i (T^{(a)} - T^{(i)}) / R(\log(a_i) - \log(R)) \tag{15.11}$$

The heat flux in the outer phase expressed as

$$-K_o(T^{(o)}-T^{(b)})/R(\log(R)-\log(b_o)) \tag{15.12}$$

Hence, the total heat flux is

$$S(\xi)=\frac{1}{(R+\xi)}\left[\frac{K_o(T^{(o)}-T^{(b)})}{\log(b_o)-\log(R+\xi)}-\frac{K_i(T^{(a)}-T^{(i)})}{\log(R+\xi)-\log(a_i)}\right] \tag{15.13}$$

Expanding the total heat flux about $\xi=0$,

$$S(\xi)=S(0)+\xi S'(0)+\frac{\xi^2}{2}S''(0)+\frac{\xi^3}{2}S'''(0)+\ldots\ldots \tag{15.14}$$

Considering the only linear terms, we have

$$S(\xi)=S(0)+\xi S'(0) \tag{15.15}$$

From the symmetric condition $S(0)=0$ and, therefore,

$$\frac{K_o(T^{(o)}-T^{(b)})}{R(\log(b_o)-\log(R))}=\frac{K_i(T^{(a)}-T^{(i)})}{R(\log(R)-\log(a_i))}=P \tag{15.16}$$

equation (15.10) can be expressed in its linear form as

$$\rho^{(i)}\left[\frac{\partial\phi^{(i)}}{\partial r}-\frac{\partial\xi}{\partial t}-\frac{\partial\xi}{\partial z}\cdot\frac{\partial\phi_o^{(i)}}{\partial z}\right]=\beta\xi \tag{15.17}$$

here $\beta=\frac{P}{LR}\left(\frac{\log(b_o/a_i)}{\log(R/a_i)\log(b_o/R)}\right)$

The interfacial normal stress is balanced by the interfacial tension. This equation can be expressed as

$$(p^{(o)}-p^{(i)})-\frac{1}{k_i}\left(\nabla\phi^{(o)}\mu^{(o)}-\nabla\phi^{(i)}\mu^{(i)}\right)-2\left(\mu^{(o)}\frac{\partial^2\phi^{(o)}}{\partial r^2}-\mu^{(i)}\frac{\partial^2\phi^{(i)}}{\partial r^2}\right)-\rho^{(o)}\Omega_o^2R\xi$$

$$=-\sigma\left(\frac{\partial^2\xi^2}{\partial z^2}+\frac{1}{R^2}\frac{\partial^2\xi^2}{\partial\theta^2}+\frac{\xi}{R^2}\right) \tag{15.18}$$

As the flow is irrotational, the pressure can be computed by Bernaoulli's equation and, therefore, above equation takes form as

$$\left[\rho^{(o)}\left(\frac{1}{\varepsilon_1}\frac{\partial\phi^{(o)}}{\partial z}\frac{\partial\phi_o^{(o)}}{\partial z}+\frac{1}{\varepsilon_1}\frac{\partial\phi^{(o)}}{\partial t}\right)+\frac{1}{k_i}\left(\mu^{(o)}\phi^{(o)}-\mu^{(i)}\phi^{(i)}\right)-\rho^{(i)}\left(\frac{1}{\varepsilon_1}\frac{\partial\phi^{(i)}}{\partial z}\frac{\partial\phi_o^{(i)}}{\partial z}+\frac{\partial\phi^{(i)}}{\partial t}\frac{1}{\varepsilon_1}\right)\right.$$

$$\left.+\frac{2}{\varepsilon_1}\left(\mu^{(o)}\frac{\partial^2\phi^{(o)}}{\partial r^2}-\mu^{(i)}\frac{\partial^2\phi^{(i)}}{\partial r^2}\right)+\sigma\left(\frac{\partial^2\xi}{\partial z^2}+\frac{1}{R^2}\frac{\partial^2\xi}{\partial\theta^2}+\frac{\xi}{R^2}\right)=\rho^{(o)}\Omega_o^2R\xi\right] \tag{15.19}$$

15.4 Dispersion Relationship

We are seeking the solution of the above equations in the context of the normal mode technique. Using the normal modes technique, the surface elevation $\xi(\theta,z,t)$ may be written as

$$\xi = B\exp(im\theta - i\omega t)\exp(ikz) + c.c. \tag{15.20}$$

here B represent perturbation's amplitude, k is the wavenumber, m is a positive number, ω is the growth rate, and *c.c.* represents a complex conjugate.

$$\phi^{(i)} = \frac{1}{k}\left(\frac{\beta}{\rho^{(i)}} + ikV_i - i\omega\right)BH^{(i)}(kr)\exp(im\theta - i\omega t)\exp(ikz) \tag{15.21}$$

$$\phi^{(o)} = \frac{1}{k}\left(\frac{\beta}{\rho^{(o)}} + ikV_o - i\omega\right)BH^{(o)}(kr)\exp(im\theta - i\omega t)\exp(ikz) \tag{15.22}$$

here

$$H^{(i)}(kr) = \frac{I_m(kr)K'_m(ka_i) - I'_m(ka_i)K_m(kr)}{I'_m(kR)K'_m(ka_i) - I'_m(ka_i)K'_m(kR)},$$

$$H^{(o)}(kr) = \frac{I_m(kr)K'_m(kb_o) - I'_m(kb_o)K_m(kr)}{I'_m(kR)K'_m(kb_o) - I'_m(kb_o)K'_m(kR)}$$

$$\beta = \frac{P}{LR}\left(\frac{\log(b_o/a_i)}{\log(R/a_i)\log(b_o/R)}\right)$$

Using the expression of equations (15.20)–(15.22) in equation (15.19) we have

$$D(\omega,k) = C_0\omega^2 + (C_1 + iD_1)\omega^2 + (C_2 + iD_2) = 0 \tag{15.23}$$

where $C_0 = \frac{1}{\varepsilon_1}\left[\rho^{(i)}H^{(i)}(kR) - \rho^{(o)}H^{(o)}(kR)\right]$

$$C_1 = \frac{2k}{\varepsilon_1}\left[\rho^{(o)}V_oH^{(o)}(kR) - \rho^{(i)}V_iH^{(i)}(kR)\right]$$

$$D_1 = \frac{\beta}{\varepsilon_1}\left[H^{(i)}(kR) - H^{(o)}(kR)\right] + \frac{2}{\varepsilon_1}\left[\mu^{(i)}k^2G^{(i)}(kR) - \mu^{(o)}k^2G^{(o)}(kR)\right]$$

$$+\frac{1}{k_i}\left[\mu^{(i)}H^{(i)}(kR) - \mu^{(o)}H^{(o)}(kR)\right]$$

$$C_2 = \frac{k^2}{\varepsilon_1}\left[\rho^{(i)}V_i^2H^{(i)}(kR) - \rho^{(o)}V_o^2H^{(o)}(kR)\right] - \frac{\beta}{k_i}\left[\frac{\mu^{(i)}}{\rho^{(i)}}H^{(i)}(kR) - \frac{\mu^{(o)}}{\rho^{(o)}}H^{(o)}(kR)\right]$$
$$+\frac{\sigma k\left(1-m^2-k^2R^2\right)}{R^2} - \frac{2k^2\beta}{\varepsilon_1}\left[\frac{\mu^{(i)}}{\rho^{(i)}}G^{(i)}(kR) - \frac{\mu^{(o)}}{\rho^{(o)}}G^{(o)}(kR)\right] - \rho^{(o)}\Omega_o^2Rk$$

$$D_2 = -2k^3\left[\frac{\mu^{(i)}}{\varepsilon_1}V_iG^{(i)}(kR) - \frac{\mu^{(o)}}{\varepsilon_1}V_oG^{(o)}(kR)\right] - \frac{k}{k_i}\left[\mu^{(i)}V_iH^{(i)}(kR) - \mu^{(o)}V_oH^{(o)}(kR)\right]$$
$$-\beta k\left[\frac{V_i}{\varepsilon_1}H^{(i)}(kR) - \frac{V_o}{\varepsilon_1}H^{(o)}(kR)\right]$$

Here $G^{(i)}(kR) = \left(1+\frac{m^2}{k^2R^2}\right)H^{(i)}(kR) - \frac{1}{kR}$

$$G^{(o)}(kR) = \left(1+\frac{m^2}{k^2R^2}\right)H^{(o)}(kR) - \frac{1}{kR}$$

So, from equation (23)

$$D(\omega,k) = \omega^2 + (P_1 + iQ_1)\omega + (P_2 + iQ_2) = 0 \tag{15.24}$$

where $P_1 = \frac{C_1}{C_o}$, $Q_1 = \frac{D_1}{C_o}$, $P_2 = \frac{C_2}{C_o}$, $Q_2 = \frac{D_2}{C_o}$

Put $\omega = \omega_r + i\omega_i$ in equation (24)

$$(\omega_r^2 - \omega_i^2) + \left(p_1\omega_r - q_1\omega_i\right) + p_2 = 0 \tag{15.25}$$

$$\omega_r\left(2\omega_i + q_1\right) + \left(p_1\omega_i + q_2\right) = 0 \tag{15.26}$$

Eliminating ω_r between equations (15.25) and (15.26), we get

$$B_0\omega_i^4 + B_1\omega_i^3 + B_2\omega_i^2 + B_3\omega_i^1 + B_4\omega_i^0 = 0 \tag{15.27}$$

where $B_0 = 1,\ B_1 = 2q_1,\ B_2 = -p_2 + \frac{5}{4}q_1^2 + \frac{1}{4}p_1^2$

$$B_3 = -p_2q_1 + \frac{1}{4}q_1^3 + \frac{1}{4}p_1^2q_1,\ B_4 = -\frac{1}{4}q_2^2 + \frac{1}{4}p_1q_1q_2 - \frac{1}{4}p_2q_1^2$$

For marginal stability curve $\omega_i = 0$ and, therefore,

$$q_2^2 - p_1q_1q_2 + p_2q_1^2 = 0 \tag{15.28}$$

Then, we get

$$
\begin{aligned}
&\Bigg[4\frac{k^6}{\varepsilon_1^3}\left(\rho^{(i)}\mu^{(o)2}H^{(i)}(kR)\left(G^{(o)}(kR)\right)^2-\rho^{(o)}\mu^{(i)2}H^{(o)}(kR)\left(G^{(i)}(kR)\right)^2\right)+4\frac{\beta k^4}{\varepsilon_1^3}\Big\{H^{(i)}(kR)H^{(o)}(kR)\\
&\times\left(\rho^{(i)}\mu^{(o)}G^{(o)}(kR)-\rho^{(o)}\mu^{(i)}G^{(i)}(kR)\right)\Big\}+4\frac{k^4}{\varepsilon_1^2k_i}\Big\{\rho^{(i)}\mu^{(o)2}H^{(i)}(kR)H^{(o)}(kR)G^{(o)}(kR)-\rho^{(o)}\mu^{(i)2}\\
&\times H^{(1)}(kR)H^{(o)}(kR)G^{(i)}(kR)\Big\}+\frac{\beta^2k^2}{\varepsilon_1^3}\left(\rho^{(i)}H^{(i)}(kR)\left(H^{(o)}(kR)\right)^2-\rho^{(o)}\left(H^{(i)}(kR)\right)^2H^{(o)}(kR)\right)\\
&+\frac{k^2}{\varepsilon_1k_i^2}\left(\rho^{(i)}\mu^{(o)2}H^{(i)}(kR)\left(H^{(o)}(kR)\right)^2-\mu^{(i)2}\rho^{(o)}H^{(o)}(kR)\left(H^{(i)}(kR)\right)^2\right)+2\frac{\beta k^2}{\varepsilon_1^2k_i}\Big\{\rho^{(i)}\mu^{(o)}H^{(i)}(kR)\\
&\times\left(H^{(o)}(kR)\right)^2-\rho^{(o)}\mu^{(i)}H^{(o)}(kR)\left(H^{(i)}(kR)\right)^2\Big\}\Bigg]U_{rel}^2=\Bigg[\frac{\beta}{k_i}\left(\frac{\mu^{(i)}}{\rho^{(i)}}H^{(i)}(kR)-\frac{\mu^{(o)}}{\rho^{(o)}}H^{(o)}(kR)\right)\\
&+2\frac{k^2\beta}{\varepsilon_1}\left(\frac{\mu^{(i)}}{\rho^{(i)}}G^{(i)}(kR)-\frac{\mu^{(o)}}{\rho^{(o)}}G^{(o)}(kR)\right)-\frac{\sigma k\left(1-m^2-k^2R^2\right)}{R^2}+\rho^{(o)}\Omega_o^2Rk\Bigg]\\
&\times\Bigg[2\frac{k^2}{\varepsilon_1}\left(\mu^{(i)}G^{(i)}(kR)-\mu^{(o)}G^{(o)}(kR)\right)+\frac{\beta}{\varepsilon_1}\left(H^{(i)}(kR)-H^{(o)}(kR)\right)\\
&+\left(\frac{\mu^{(i)}}{k_i}H^{(i)}(kR)-\frac{\mu^{(o)}}{k_i}H^{(o)}(kR)\right)\Bigg]^2 \qquad (15.29)
\end{aligned}
$$

Here $U_{rel}=V_2-V_1$.

15.5 Nondimensional Form

If we assume characteristic length $H=b_o-a_i$ and characteristic velocity V, other parameters in the dimensionless form are

$$
\varsigma=kH,\ \hat{\beta}=\frac{\beta H}{V\rho^{(o)}},\ \hat{V}_i=\frac{V_i}{V},\ \hat{V}_o=\frac{V_o}{V},\ \hat{\omega}=\frac{\omega H}{V},\ \rho=\frac{\rho^{(i)}}{\rho^{(o)}},\ \mu=\frac{\mu^{(i)}}{\mu^{(o)}},
$$

$$
\nu=\frac{\mu}{\rho},\ \hat{\sigma}=\frac{\sigma}{\rho^{(o)}HV^2},\ \hat{\Re}_o=\frac{R}{H},\ \hat{P}_i=\frac{H^2}{k_i},\ \mathrm{Re}=\frac{\rho^{(o)}HV}{\mu^{(o)}},\ Ro=\frac{VH}{\Omega_o},
$$

Here Re denotes the Reynolds number, Ro denotes Rossby number, and ν represents the kinematic viscosity.

Hence, the nondimensional form of equation (15.23) is

$$
D(\hat{\omega},\hat{k})=\hat{C}_0\hat{\omega}^2+(\hat{C}_1+i\hat{D}_1)\hat{\omega}^2+(\hat{C}_2+i\hat{D}_2)=0 \qquad (15.30)
$$

where

$$\hat{C}_0 = \frac{\rho}{\varepsilon_1}\left[H^{(i)}(\hat{\Re}_o\varsigma) - \frac{1}{\rho}H^{(i)}(\hat{\Re}_o\varsigma)\right]$$

$$\hat{C}_1 = \frac{2\varsigma}{\varepsilon_1}\left[\hat{V}_o H^{(o)}(\hat{\Re}_o\varsigma) - \rho\hat{V}_i H^{(i)}(\hat{\Re}_o\varsigma)\right]$$

$$\hat{D}_1 = \frac{1}{\varepsilon_1}\left[\hat{\beta}H^{(i)}(\hat{\Re}_o\varsigma) - \hat{\beta}H^{(o)}(\hat{\Re}_o\varsigma)\right] + \frac{2\varsigma^2\hat{\mu}}{\varepsilon_1 \operatorname{Re}}\left[G^{(i)}(\hat{\Re}_o\varsigma) - \frac{1}{\hat{\mu}}G^{(o)}(\hat{\Re}_o\varsigma)\right] + \frac{\hat{P}_i\mu}{\operatorname{Re}}\left[H^{(i)}(\hat{\Re}_o\varsigma) - \frac{1}{\mu}H^{(o)}(\hat{\Re}_o\varsigma)\right]$$

$$\hat{C}_2 = \frac{\varsigma^2}{\varepsilon_1}\left[\rho\hat{V}_i^2 H^{(i)}(\hat{\Re}_o\varsigma) - \hat{V}_o^2 H^{(o)}(\hat{\Re}_o\varsigma)\right] - \frac{\hat{\beta}\hat{P}_i}{\operatorname{Re}}\left[\nu H^{(i)}(\hat{\Re}_o\varsigma) - H^{(o)}(\hat{\Re}_o\varsigma)\right] + \frac{\hat{\sigma}\varsigma\left(1 - m^2 - \hat{\Re}_o^2\varsigma^2\right)}{\hat{\Re}_o^2}$$
$$- \frac{2\varsigma^2\hat{\beta}}{\operatorname{Re}\varepsilon_1}\left[\nu G^{(i)}(\hat{\Re}_o\varsigma) - G^{(o)}(\hat{\Re}_o\varsigma)\right] - \left(\frac{1}{Ro}\right)^2\hat{\Re}_o\varsigma$$

$$\hat{D}_2 = -\frac{2\varsigma^3}{\operatorname{Re}}\left[\frac{\hat{\mu}}{\varepsilon_1}\hat{V}_i G^{(i)}(\hat{\Re}_o\varsigma) - \frac{1}{\varepsilon_1}\hat{V}_o G^{(o)}(\hat{\Re}_o\varsigma)\right] - \frac{\varsigma\hat{P}_i}{\operatorname{Re}}\left[\mu\hat{V}_i H^{(i)}(\hat{\Re}_o\varsigma) - \hat{V}_o H^{(o)}(\hat{\Re}_o\varsigma)\right]$$
$$- \hat{\beta}\varsigma\left[\frac{\hat{V}_i}{\varepsilon_1}H^{(i)}(\hat{\Re}_o\varsigma) - \frac{\hat{V}_o}{\varepsilon_1}H^{(o)}\left(\hat{\Re}_o\varsigma\right)\right]$$

Nondimensional form of equation (15.29) is

$$\left[4\frac{\varsigma^6\rho}{\varepsilon_1^3 \operatorname{Re}^2}\left(\left(G^{(o)}\left(\hat{\Re}_o\varsigma\right)\right)^2 H^{(i)}\left(\hat{\Re}_o\varsigma\right) - \frac{\mu^2}{\rho}\left(G^{(i)}\left(\hat{\Re}_o\varsigma\right)\right)^2 H^{(o)}\left(\hat{\Re}_o\varsigma\right)\right) + 4\frac{\hat{\beta}\varsigma^4}{\varepsilon_1^3 \operatorname{Re}}H^{(i)}\left(\hat{\Re}_o\varsigma\right)H^{(o)}\left(\hat{\Re}_o\varsigma\right)\right.$$
$$\times\left(\rho G^{(o)}\left(\hat{\Re}_o\varsigma\right) - \mu G^{(i)}\left(\hat{\Re}_o\varsigma\right)\right) + 4\frac{\varsigma^4\hat{P}_i}{\varepsilon_1^2 \operatorname{Re}^2}\left(\rho H^{(i)}\left(\hat{\Re}_o\varsigma\right)H^{(o)}\left(\hat{\Re}_o\varsigma\right)G^{(o)}\left(\hat{\Re}_o\varsigma\right) - \mu^2 H^{(i)}\left(\hat{\Re}_o\varsigma\right)H^{(o)}\left(\hat{\Re}_o\varsigma\right)\right.$$
$$\left.\times G^{(i)}\left(\hat{\Re}_o\varsigma\right)\right) + \frac{\hat{\beta}^2\varsigma^2}{\varepsilon_1^3}\left(\rho H^{(i)}\left(\hat{\Re}_o\varsigma\right)\left(H^{(o)}\left(\hat{\Re}_o\varsigma\right)\right)^2 - \left(H^{(i)}\left(\hat{\Re}_o\varsigma\right)\right)^2 H^{(o)}\left(\hat{\Re}_o\varsigma\right)\right) + \frac{\varsigma^2\hat{P}_i^2}{\varepsilon_1 \operatorname{Re}^2}\left(\rho H^{(i)}\left(\hat{\Re}_o\varsigma\right)\right.$$
$$\left.\times\left(H^{(o)}\left(\hat{\Re}_o\varsigma\right)\right)^2 - \mu^2\left(H^{(i)}\left(\hat{\Re}_o\varsigma\right)\right)^2 H^{(o)}\left(\hat{\Re}_o\varsigma\right)\right) + 2\frac{\varsigma^2\hat{\beta}\hat{P}_i}{\varepsilon_1^3 \operatorname{Re}}\left(\rho H^{(i)}\left(\hat{\Re}_o\varsigma\right)\left(H^{(o)}\left(\hat{\Re}_o\varsigma\right)\right)^2 - \mu\left(H^{(i)}\left(\hat{\Re}_o\varsigma\right)\right)^2\right.$$
$$\left.\left.\times H^{(o)}\left(\hat{\Re}_o\varsigma\right)\right)\right]\hat{U}_{rel}^2 = \left[\frac{\hat{\beta}\hat{P}_i}{\operatorname{Re}}\left(\nu H^{(i)}\left(\hat{\Re}_o\varsigma\right) - H^{(o)}\left(\hat{\Re}_o\varsigma\right)\right) + 2\frac{\varsigma^2\hat{\beta}}{\varepsilon_1 \operatorname{Re}}\left(\nu G^{(i)}\left(\hat{\Re}_o\varsigma\right) - G^{(o)}\left(\hat{\Re}_o\varsigma\right)\right)\right.$$
$$\left.- \frac{\hat{\sigma}\varsigma}{\hat{\Re}_o^2}\left(1 - m^2 - \varsigma^2\hat{\Re}_o^2\right) + \left(\frac{1}{Ro}\right)^2\hat{\Re}_o\varsigma\right]\left[2\frac{\varsigma^2\mu}{\varepsilon_1 \operatorname{Re}}\left(G^{(i)}\left(\hat{\Re}_o\varsigma\right) - \frac{1}{\mu}G^{(o)}\left(\hat{\Re}_o\varsigma\right)\right) + \frac{\hat{\beta}}{\varepsilon_1}\left(G^{(i)}\left(\hat{\Re}_o\varsigma\right) - G^{(o)}\left(\hat{\Re}_o\varsigma\right)\right)\right.$$
$$\left.+ \frac{\hat{P}_i\mu}{\operatorname{Re}}\left(G^{(i)}\left(\hat{\Re}_o\varsigma\right) - \frac{1}{\mu}G^{(o)}\left(\hat{\Re}_o\varsigma\right)\right)\right]^2 \tag{15.31}$$

Here $\hat{U}_{rel}^2 = \left(\hat{V}_2 - \hat{V}_1\right)^2$

In the case of viscous medium ($\varepsilon_1 = 1, \hat{P}_i \to 0$), the relation of Awasthi et al. [15] can be recovered from equation (15.31). If swirling is absent, we can get the relation of Awasthi [7].

15.6 Results and Numerical Discussions

The parametric values for the numerical calculation are taken as

$$\mu^{(i)} = 0.00001\ poise, \quad \rho^{(i)} = 0.001\ gm / cm^3, \quad \sigma = 72.3\ dyne / cm,$$

$$\mu^{(o)} = 0.01\ poise, \quad \rho^{(o)} = 1.0\ gm / cm^3, \quad \varepsilon_1 = 0.3, \quad k_i = 0.0003\ cm^2$$

The effect of Reynolds number, viscosity ratio, Rossby number, heat transfer coefficient, porosity, etc. The arrival of instability is depicted in the following graphs.

The porous medium effect on the instability of the interface is illustrated in Figure 15.1. The marginal stability curves for the relative velocity of the interface are plotted in this figure. In this figure, we plotted asymmetric disturbances, i.e., $m = 1$. The lower portion of the curve shows a stable area, while the above portion represents an unstable area. It is noted from the figure that the stable area is more when the medium is porous and, therefore, one can say that the perturbation expands more rapidly in the pure flow than the porous flow. In the case of a porous medium, the flow area is less in comparison with a viscous medium and, therefore, perturbations do not grow as quickly as they grow in the viscous medium. As a result, the interface in a porous medium is more stable than in a viscous medium.

Now we are looking to study the swirling effect on the instability of the vapor–water interface. Figure 15.2 shows the response of the Rossby number (*Ro*) on the instability of the interface through marginal curves of relative velocity. It is detected from this figure that as *Ro* increases, stability range decreases. The Rossby number *Ro* is defined as the ratio of inertia force to the Coriolis force, i.e., *Ro* is inversely proportional to the swirling.

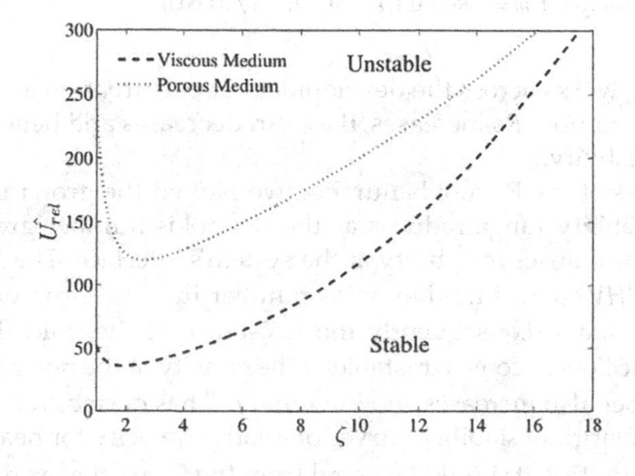

FIGURE 15.1
Marginal stability curves for viscous and porous medium ($\hat{\beta} = 1, m = 1$).

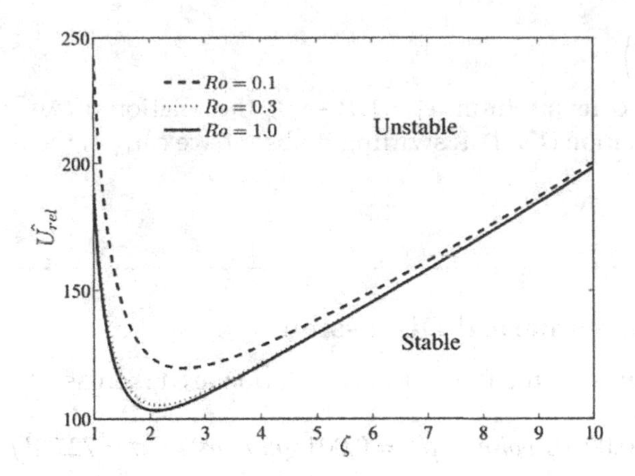

FIGURE 15.2
Effect of Rossby number when $(\hat{\beta} = 1, m = 1, \text{Re} = 100, \varepsilon_1 = 0.3, \hat{P}_i = 1/0.0003)$.

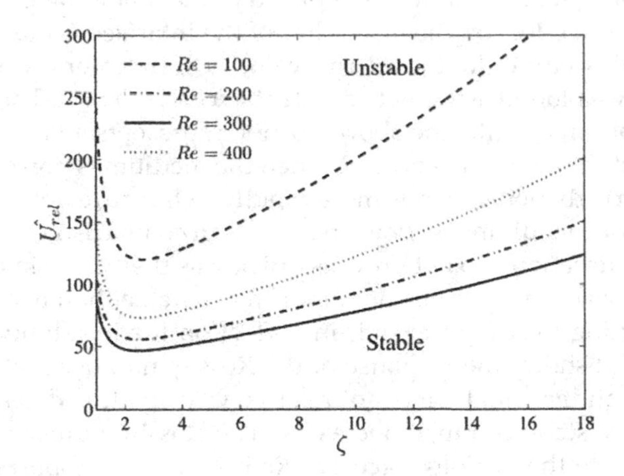

FIGURE 15.3
Effect of Reynolds number $(\hat{\beta} = 1, m = 1, Ro = 0.1, \varepsilon_1 = 0.3, \hat{P}_i = 1/0.0003)$.

Hence, the swirling works across the development of perturbation and stabilize the interface. As the Rossby number *Ro* increases, the swirl decreases and hence we concluded that the swirl induces stability.

For various values of the Reynolds number, we plotted the growth of perturbations in Figure 15.3. The stability range reduces as the Reynolds number grows, indicating that the Reynolds number causes instability at the system's interface. The Reynolds number is defined as $\text{Re} = \rho^{(o)}HV/\mu^{(o)}$. If the Reynolds number increases, the viscosity of the outer fluid $(\mu^{(o)})$ decreases, and consequently the resistance of the fluid flow at the interface decrease. Hence, the flow becomes unstable. If the density of the outer fluid $(\rho^{(o)})$ increases, the Reynolds number also increases showing that $\rho^{(o)}$ has destabilizing nature.

The variation of marginal stability curves of relative velocity for heat transport constant $\left(\hat{\beta}\right)$ is shown in Figure 15.4. It should be noted from the figure that as $\hat{\beta}$ increases, the range of instability increases, which is in the conclusion that the heat transfer at the interface

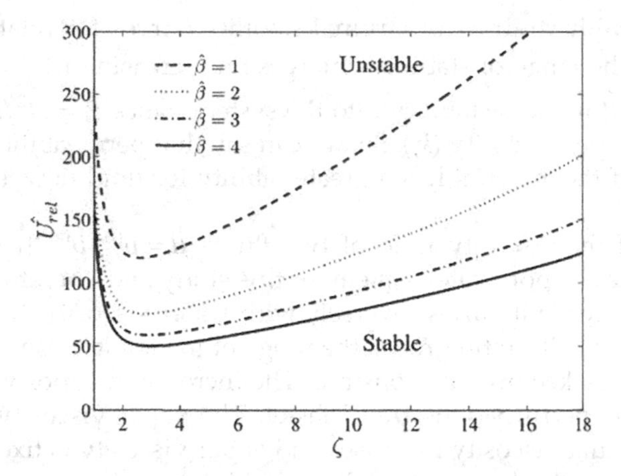

FIGURE 15.4
Effect of heat transport (Re = 100, $m = 1, Ro = 0.1, \varepsilon_1 = 0.3, \hat{P}_i = 1/0.0003$).

acts to encourage the perturbations. As the heat transport through the interface increases, more heat is conveyed to a point in a trough and less heat is passed aside from that point. Therefore, the amount of evaporation increases at the trough and it passes to the other side through the interface. This added evaporation will grow the amplitude of the perturbations and interface move toward instability. A similar result was achieved by Fu et al. [12] and Awasthi et al. [9]. One can say here that the heat transport effect does not change even in the presence of the porous medium, swirling, and/or viscosity of the fluids.

Figure 15.5 compares the marginal stability curves of relative velocity for distinct values of porosity ε_1. Since porosity is defined as the ratio of the volume of voids over the total volume and therefore, more porosity implies more void space. If a perturbation gets more void space, it rotates and, therefore, travels slower than the less void space. Hence, if the porosity of the medium increases, the void space increases, and perturbations grow slow. A similar observation has been obtained from Figure 15.5. We recognize that more void volume will form a stable arrangement.

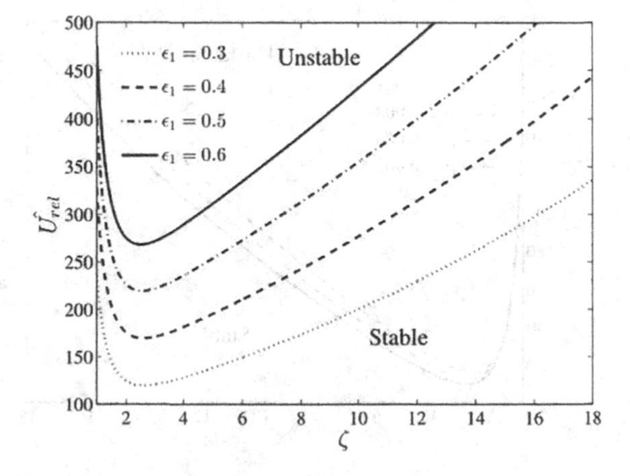

FIGURE 15.5
Effect of medium porosity (Re = 100, $m = 1, Ro = 0.1, \hat{P}_i = 1/0.0003$).

In Figure 15.6, we have drawn marginal stability curves for relative velocity for particular values $\hat{P}_i$. The range of stability enlarges on increasing the value of $\hat{P}_i$, and this is an indication that $\hat{P}_i$ induces stability into the system. Since $\hat{P}_i = H^2/k_i$, i.e., $\hat{P}_i$ is inversely proportional to the permeability (k_i). So, we can say that permeability has a destabilizing character. Hence, if the material has a greater ability for fluid flow, the interface may get easily unstable.

The response of the viscosity ratio of two fluids $\mu = \mu^{(i)}/\mu^{(o)}$ has been examined in Figure 15.7. We take a vapor–water system for this study and, therefore, μ directly depend on vapor viscosity, while it varies inversely with water viscosity. It is detected from the figure that as the viscosity ratio grows, the range of instability also grows. On increasing μ, $\mu^{(i)}$ increases while keeping $\mu^{(o)}$ constant. The increase in vapor viscosity impedes the flow and, therefore, perturbations travel faster. The vapor viscosity has a destabilizing character. If outer fluid viscosity increases and vapor viscosity is fixed, the viscosity ratio decreases, hence outer fluid viscosity induces stability.

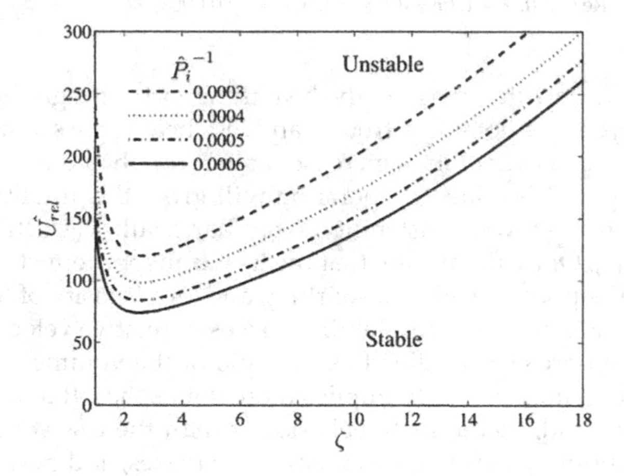

FIGURE 15.6
Effect of medium permeability ($\mathrm{Re} = 100, m = 1, Ro = 0.1, \varepsilon_1 = 0.3$).

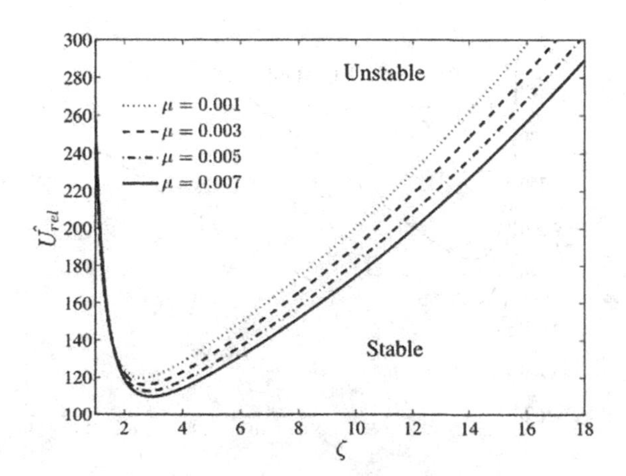

FIGURE 15.7
Effect of viscosity ratio ($\mathrm{Re} = 100, m = 1, Ro = 0.1, \varepsilon_1 = 0.3$).

15.7 Conclusions

The K-H instability at the cylindrical interface of two viscous fluids is examined. We achieve a second-order polynomial in the growth rate parameter and the imaginary part of the growth rate parameter is illustrated to investigate the effect of physical parameters like porous medium, heat/mass transfer, Reynolds number, and Rossby number. Our main result indicates that the vapor–water interface is more stable in a porous medium than the viscous medium. We found that heat/mass transfer grows the perturbation growth at the interface. The outer fluid viscosity has stabilizing nature, while inner fluid viscosity destabilizes the system. The medium porosity shows a stabilizing nature while medium permeability has destabilizing nature. The outer fluid density also has destabilizing nature.

Acknowledgment

The author S. Agarwal is thankful to CSIR, New Delhi, India (File No. 09/961(0013)/2019-EMR-I) for their financial assistantship to carry out this work.

References

1. Hsieh, D. Y., Interfacial stability with mass and heat transfer, *Phys. Fluids*, 21 (1978) 745–748.
2. Nayak, A. R. and Chakraborty, B. B., Kelvin–Helmholtz stability with mass and heat transfer, *Phys. Fluids*, 27 (1984) 1937–1941.
3. Lee, D. S., Nonlinear Kelvin–Helmholtz instability of cylindrical flow with mass and heat transfer, *Phys. Scr.*, 76 (2007) 97–103.
4. Funada, T. and Joseph, D. D., Viscous potential flow analysis of Kelvin–Helmholtz instability in a channel, *J. Fluid Mech.*, 445 (2001) 263–283.
5. Awasthi, M. K. and Agrawal, G. S., Viscous potential flow analysis of Kelvin–Helmholtz instability of cylindrical interface, *Int. J. Appl. Math. Comput.*, 3 (2011) 130–141.
6. Funada, T., Joseph, D. D. and Yamashita, S., Stability of liquid jet into incompressible gases and liquids, *Int. J. Multi. Flow*, 30 (2004) 1177–1208.
7. Awasthi, M. K., Study on Kelvin–Helmholtz instability of cylindrical flow with mass and heat transfer through porous medium, *J. Porous Media*, 17 (2014) 457–467.
8. Awasthi, M. K. and Asthana, R., Nonlinear Study of Kelvin-Helmholtz instability of cylindrical flow with mass and heat transfer, *Int. Commun. Heat Mass Transfer*, 71 (2016) 216–224.
9. Asthana, R., Awasthi, M. K. and Agrawal, G. S., Viscous potential flow analysis of Kelvin–Helmholtz instability of cylindrical flow with heat and mass transfer, *Heat Transfer Asian Res.*, 43 (2014) 489–503.
10. Awasthi, M. K. and Asthana, R., Viscous potential flow analysis of capillary instability with and mass transfer through porous media, *Int. Commun. Heat Mass Trans.*, 40 (2013) 7–11.
11. Awasthi, M. K., Effect of viscous pressure on Kelvin–Helmholtz instability through porous media, *J. Porous Media*, 19 (2016) 205–218.
12. Fu, Q. F., Jia, B. Q., Yang, L. J., Stability of a confined swirling annular liquid layer with heat and mass transfer, *Int. J. Heat Mass Trans.*, 104 (2017) 644–649.

13. Fu, Q. F., Deng, X. D., Jia, B. Q. and Yang, L. J., Temporal instability of a confined liquid film with heat and mass transfer, *AIAA J.*, 56, (2018) 2615–2622.
14. Awasthi, M. K., Rayleigh–Taylor Instability of swirling annular layer with mass transfer, *J. Fluids Eng.*, 141 (2019) 071202.
15. Awasthi, M. K., Sarychev, V. D., Nevskii, S. A., Kuznetsov, M. A., Solodsky, S. A., Chinakhov, D. A., Krampit, M. A., Kelvin-Helmholtz instability of swirling annular layer with heat and mass transfer, *J. Adv. Res. Dyn. Control Syst.*, 11 (2019) 86–96.
16. Awasthi, M. K. and Devi, M., Temporal instability of swirling annular layer with mass transfer through porous media, *Spec. Top. Rev. Porous Media Int J.*, 11 (2020) 61–70.

Index

advanced locking systems 62
agility 80
application software 3, 26
asymptotic behaviours 205
authentication and access control 22
automated lightings 62
availability 22
azimuthal velocity 204

Bassi Rebay 175
Bernaoulli's equation 234
Blockchain technology 69
Brain–Computer Interfaces 76

cloud service provider layer 19
cold booting 42
complete memory dump 47
composite velocity 195
COMSOL 203
confidentiality 22
configuration 51
conformable time-fractional 134
convective instability 165
cost and scale effectiveness 20
cost-savings 80
critical assessments 164
critical Rayleigh–Darcy number 166
cryptocurrency 67
cylindrical configuration 232

Darcy–Brinkman model 232
data dumper 44
day night sensors 62
dedicated hardware 42
DEMATEL method 86
deploy globally 80
DG discretization 175
diffusion impacts 178
digitization 9
diverse estimates 165
drift current 123

elasticity 20, 80
error estimation 140

firmware 3
Fisher–KPP 172
flow sensors 9
fuzzy DEMATEL method 85

Galerkin Scheme 171
Gaussian–Legendre quadrature 176
generalized eigenvalue condition 163
generalized heptagonal fuzzy number 89
glucose monitoring. 65
Green Chemistry 21

Hall effect sensors 9
hand hygiene monitoring. 65
HANITM 134
harsh environment sensors 10
health organizations 83
heart-rate monitoring 65
heat flux 234
heat-transfer coefficient 194
heptagonal fuzzy number 88
hibernation Files 45
hidden layer 12
highly reliable networks 8
hybrid cloud(s) 80, 82

illumination 119
industrial automation 6
influencing factors 102
infrastructure as a service 80
infrastructure scalability 20
input layer 12
integrity 22
interfacial normal stress 234
interpersonal influencing factors 106
intrapersonal factors 106
inverse conformable laplace 139

Jeffrey fluid-soaked 163
Jeffery parameter 166

kernel drivers 45
kernel memory dump 47
K-H instability 231

latent heat 233
Legendre polynomial 174
linguistic variables 86
location independence 20
low power sensors 10

magnetic sensors 9
magnet RAM capture 51
MCC privacy needs 23
mediator device 3
methodology for DumpIT 52
methodology for FTK Imager 51
miniaturization 9
mixed DG construction 173
mobile network layer 19
mobile user layer 19
monitoring the activities 62
multiclouds 80
multisensor modules 10

Newton's routine 163
NIDTM 134
normal modes technique 235

operating system 3, 26
operating system injection 45
optimal homotopy analysis 150
orthogonal properties 163
output layer 12

Péclet number 164
photoelectric sensors 9
plastic sheet extrusion 185
platform as a service 80
platform reliability 26
PMDump 44
Poisson's Equation 121
porosity 232
Prandtl number 188
presence sensors 9
privacy and data protection 26
private cloud 80
Processor Dumper Utility 44
pseudo-spectral scheme 172
public cloud 79

quasi-linearization technique 189

radiological departments 78
Rayleigh–Darcy number 165
Reynolds number 237
Rossby number 237

secondary variables 202
semiconductor 122
skin-friction 193
skin friction coefficient 189
small memory dump 48
software as a service 80
software crash dumps 45
software debuggers 45
software emulators 43
special hardware bus 42
speed tracking machines 63
stretching plate 198
symmetric component 220

tangential velocity profile 205
Telemedicine System 83
temperaments 101
time-fractional 150
toll gates 63
trapezoidal fuzzy number 87
triangular fuzzy number 87

viscoelastic cylindrical jet 221
viscoelastic fluid 215
viscoelastic pressure 220

9781032272252